DINOSAURS *WITHOUT* BONES

DINOSAUR LIVES REVEALED BY THEIR TRACE FOSSILS

ANTHONY J. MARTIN

PEGASUS BOOKS
NEW YORK LONDON

To my wife, Ruth,
whose traces of love suffuse and nourish me every day.

DINOSAURS WITHOUT BONES

Pegasus Books LLC
80 Broad Street, 5th Floor
New York, NY 10004

First Pegasus Books cloth edition 2014

Interior design by Maria Fernandez

Library of Congress Cataloging-in-Publication Data is available.

ISBN: 978-1-60598-499-5

10 9 8 7 6 5 4 3 2 1

Printed in the United States of America
Distributed by W. W. Norton & Company

Contents

CHAPTER 1	Sleuthing Dinosaurs	**1**
CHAPTER 2	These Feet Were Made for Walking, Running, Sitting, Swimming, Herding, and Hunting	**17**
CHAPTER 3	The Mystery of Lark Quarry	**56**
CHAPTER 4	Dinosaur Nests and Bringing Up Babies	**84**
CHAPTER 5	Dinosaurs Down Underground	**121**
CHAPTER 6	Broken Bones, Toothmarks, and Marks on Teeth	**161**
CHAPTER 7	Why Would a Dinosaur Eat a Rock?	**194**
CHAPTER 8	The Remains of the Day: Dinosaur Vomit, Stomach Contents, Feces, and Other Gut Feelings	**225**
CHAPTER 9	The Great Cretaceous Walk	**267**
CHAPTER 10	Tracking the Dinosaurs Among Us	**289**
CHAPTER 11	Dinosaurian Landscapes and Evolutionary Traces	**333**
	Notes	**373**
	Acknowledgments	**449**
	Index	**453**

CHAPTER 1

Sleuthing Dinosaurs

Vestiges of a Cretaceous Hour

The *Triceratops* was vexed. As the largest adult male in this part of the river valley, the challenge coming from a younger male strolling through his long-established territory was intolerable. Appropriately, then, it provoked a prompt and assertive response. From a stationary start, feet planted firmly on the damp muddy sand of the floodplain, he stared at his rival, started walking, and picked up speed. As he moved, he lowered his huge, broad head while pointing his three-horned face directly at the other *Triceratops*, communicating his intentions unambiguously. While accelerating, his rear feet registered directly on top of where his front feet had just pressed, and then exceeded them, leaving a varied pattern of tracks behind him.

From a distance, the other male seemed to stand his ground. Yet his feet shuffled, blurring their outlines in the underlying sand, as he tried to decide whether to stand his ground or turn and flee. Regardless, he was in big trouble.

In between the two *Triceratops*, a group of small feathered theropod dinosaurs with stubby forearms—similar to the Asian alvarezsaur *Mononykus*—and a nearby bunch of slightly larger ornithopod dinosaurs (*Thescelosaurus*) looked on warily. Each of these groups of dinosaurs had been striding unhurriedly across the floodplain, tolerating one another's presence, spurred on by intriguing scents wafting down the sunlit valley. Nevertheless, a charging *Triceratops* provided a good reason to temporarily abandon their long-term goals and deal with this more immediate problem.

In unison, they all looked up at the advancing *Triceratops*, its profile and rapidly increasing pace causing it to appear ever larger as it neared. Next to them, a mixed flock of toothed birds and pterosaurs all turned and aligned themselves with the wind at their backs. They began hopping while flapping their wings, and then were aloft, chattering loudly. This was all the motivation one of the more skittish theropods needed to start running, and the rest of his group followed suit. The ornithopods only hesitated a second or two before doing the same. First, though, more than a few of both species lightened the load before taking off, involuntarily voiding their bowels and leaving variably colored and sized scat, peppered with seeds, on top of their distinctive footprints. In her haste, one *Thescelosaurus* slipped on a muddy patch and fell on her side. She quickly righted herself and bolted to catch up with the others, leaving a long, smeared body impression on the sand among the tracks.

A band of *Mononykus*-like theropods became too crowded near the edge of the river and, in a moment of desperation, jumped into the water and began swimming. In the river shallows, their feet touched the bottom and claws on each toe gouged the sand, making parallel grooves with each stroke. Bodies buoyed by the water and swimming with the current, their tracks were nearly twice as far apart from one another as if they had been on land.

These three separate but related shockwaves—the aggressive movement of the big male *Triceratops*, the raucous flocks of birds and pterosaurs, and the panicked stampede of the theropods and ornithopods—triggered overt and subtle changes in the behaviors of nearly every dinosaur nearby. In this respect, it was like many other days in the Cretaceous Period, where long periods of quiet stability were occasionally interrupted by near-chaotic commotion.

On higher ground above the floodplain, a male–female pair of predatory theropods (*Dromaeosaurus*) paused from digging up small mammals from out of their burrows with their rear feet. They raised their heads, looking for whatever had provoked alarm calls from the birds and pterosaurs now passing overhead. Their footprints and excavations had disturbed a considerable amount of soil in the area. But they had much more to do before leaving. Once convinced the alarming behavior below had nothing to do with them, they went back to uncovering their furry morsels, sniffing the ground and scraping with their rear feet.

Near the *Dromaeosaurus* couple and next to the river levee, a few other theropods (*Troodon*) shifted anxiously on their sediment-rimmed ground nests. Several weeks earlier, *Troodon* mothers had laid eggs in the nests two at a time, a function of their dual oviducts. Some had deposited as many as two dozen eggs, a dinosaurian form of labor that took about a week and a half to complete. After laying, each mother *Troodon* vertically oriented the eggs in the center of the doughnut-shaped nest. Now it was the job of the male to sit above and otherwise guard the precious egg clutch for nearly fifty days. Almost nothing could motivate them to leave their nests, so they continued to squat above them, albeit nervously.

Two smaller feathered theropods, potential egg predators and only a few meters away, gave the *Troodon* fathers further incentive to stay put. These dinosaurs gnawed on a recently dead pterosaur, scraping their teeth across its limb bones to strip whatever flesh was left. One of them, though, unsettled by the ripple effects of the dinosaur duel below, succumbed to caution and scrambled up a tree, her hands and feet imparting sets of scratch marks on the smooth bark.

The spreading disturbance initiated by the *Triceratops* provoked several small ornithopod dinosaurs, distantly related to *Thescelosaurus*, to retreat into their burrows. The burrows, which they had dug previously into the banks of the levee, had entrances only slightly wider than the ornithopods' bodies, making for a tight fit as they scrambled inside. These burrows twisted to the right and then left as they descended, making S- or Z-shapes, before expanding into a main living chamber at their ends. This zigzag design effectively deterred predators while maintaining livable temperatures and humidity in each burrow, essential features for these ornithopods to safely raise their young.

Several hundred meters downstream from the two *Triceratops*—which were only seconds away from ramming each other—a group of ostrich-like theropods (*Struthiomimus*) walked along a gravel bar at the edge of the river. They stopped every now and then to swallow pea- and marble-sized quartz-rich rocks. These "stomach stones," also known as gastroliths, would lodge in a muscular gizzard just above their intestines, where they served as digestive aids to grind varied omnivorous foodstuffs like internal mortars and pestles. These theropods were aware of but mostly unperturbed by what was taking place upstream, as the "*Mononykus*" and *Thescelosaurus* groups were still far away from them. Their tracks, accented by sharp claw marks, left vague outlines on the gravelly parts of the bar, but were crisply defined on the sandier patches. The female tracks were only slightly larger than those of the males, but otherwise identical in form.

Meanwhile, located between the *Dromaeosaurus* and the *Struthiomimus*, a male–female pair of *Ankylosaurus*, joined in coitus, ignored just about everything else happening around them. These heavily armored dinosaurs, each weighing about five tons, typically would have left large, deep tracks regardless of what they were doing. But the addition of the male's weight to the rear of the female meant her hind feet sank much deeper than his into the moist sand below. Together they made a six-footed impression, the front four coming from her and two rear ones from his hind feet. The earth moved, however briefly, for both partners.

Where the "*Mononykus*" and *Thescelosaurus* once stood on the lower part of the floodplain, at least a few of their tracks and some of their scat had been squashed under the thumping feet of the older *Triceratops* as he closed the gap between himself and the younger male. The latter could no longer stay in one place and commenced walking, then trotting, toward the other. Snails, clams, and salamanders in the floodplain sediments unlucky enough to be in the same place as a *Triceratops* foot were summarily crushed.

The collision between the two male ceratopsian dinosaurs, which had a combined weight of about twenty tons, was surprisingly subdued and dull. They did not hit one another at full speed, as the additive impact would have killed both instantly. Instead, they first stopped, and then clashed their mighty heads together. Their wide head shields not only served as great billboards for attracting mates and recognizing their species, but also dissipated much of the energy when they struck. Still, the left upper horn from the older male punched through the right side of the younger one's skull, leaving a round wound about the width of the older male's horn.

Injured badly, although not fatally, the younger male turned and began limping away. His right humerus also had been fractured by the impact and started to ache. Accordingly, his trackway acquired a new asymmetry, with both legs stepping shorter on the right side and a shallower impression in the right-front footprint as he put less weight on it. The larger male *Triceratops*, satisfied for now, walked away but stopped periodically to turn and look, making sure that no further bad behavior would come from this young upstart.

This was the right moment for the *Tyrannosaurus*. She had been standing stock-still in the higher, vegetated part of the floodplain, waiting to take advantage of the chaotic situation and get an easy meal. If she could have felt disappointment, though, she would have experienced it then as she looked down from her elevated position and saw both *Triceratops* walking off the floodplain, each very much alive. She used a mixture of hunting and scavenging to feed her six-ton frame, and had dined often on *Triceratops* that eventually died from battles with one another. Whenever she found a dead one, she would take small bites of sweet meat on the face, and then grab the head shield from the back with her teeth to tear off the head, exposing its delectable neck muscles. It was tempting for her to follow the limping one to see if he would expire, but attacking too soon entailed much risk. Millions of years of natural selection had not resulted in one of the world's largest land carnivores taking on prey that could also kill it.

Flexible in her menu choices, she turned her gaze toward a nearby group of hadrosaurs (*Edmontosaurus*) of varying ages that were unknowingly sharing this grassy–shrubby part of the floodplain with her. Most of these dinosaurs were on all fours, grazing on the plants around them. But they occasionally reared up on their hind legs and looked down to their left, having been distracted by the battle below and its effect on the surrounding fauna. As the hadrosaurs pulled up low-lying plants and chewed, tiny bits of silica in the leaves and stems imparted microscopic scratches on their teeth. Unnoticed by either the hadrosaurs or the tyrannosaur, the herd was attracting the attention of thousands of dung beetles. These insects burrowed industriously into the feces and laid their eggs, the dung providing both a home and high-quality food for their progeny. Snails also grazed on the droppings, gaining plenty of nutrition from the digested plant material.

The *Tyrannosaurus* turned her massive head slowly to the left to get a better fix on a potentially easy target. There: a half-grown *Edmontosaurus* had become curious about the sounds of ceratopsian conflict and separated himself from the rest of the herd. She locked her vision onto him and shifted her weight to that side. Mud oozed between her toes and a prominent ridge formed on the outside of her left foot. Starting with her feet together, she began stalking, taking a series of steps punctuated by pauses. Right, left, stop; right, stop; left, stop; right, left, stop. The breeze flowed down the river valley in the opposite direction of her movement, masking her sounds and scent. Within minutes, she was close enough to pounce, and did.

Her tiny arms were useless for grabbing the *Edmontosaurus*, so she lunged forward with her best asset: a mouth full of stout, banana-sized serrated teeth, backed up by the most powerful bite of any land animal that had ever evolved. Unfortunately for her, but fortunately for the hadrosaur, a crunching branch on her next-to-last step revealed her close presence in an otherwise stealthy approach. The *Edmontosaurus* lurched to his left just when the tyrannosaur's jaws clamped down on the uppermost part of his tail. This action removed a chunk of the hadrosaur but left the rest of the animal running away on his back two legs, hooting loudly in pain and fright. The other hadrosaurs in the herd likewise switched to faster bipedal postures and quickly moved away, making much noise and stomping on mud, plants, dung, dung beetles, and snails with their stout feet.

The tyrannosaur trailed the retreating hadrosaurs for a few minutes, just in case the one she had bitten would falter from his wound, or some other straggler in the herd would make itself available to her. Her tracks first paralleled and then turned in harmony with the herd's trackways as she followed them. As the distance separating them became greater, though, she stopped. Going any farther would expend more resources than it was worth, so she would just have to wait for another opportunity.

She went back to the fresh, bite-sized *amuse-bouche* of hadrosaur remaining on the ground at the site of her attack, sniffed it, picked it up with her mouth, chewed, and swallowed. The next day, its remnants would come out the other end of her digestive tract with most of its useful nutrients absorbed, but with bits of etched vertebral bone and a few of the hadrosaur's muscle fibers preserved in the feces. She walked down into the lower part of the floodplain and moved along the trend of the river valley. Her huge, thickly padded, three-toed and clawed footprints obliterated some of those left only fifteen minutes before by the terrified "*Mononykus*" and *Thescelosaurus*.

Unknown to this *Tyrannosaurus* and every other dinosaur in the area that day, the river valley itself was also a remnant of a former dinosaur presence. Herds of immense long-necked sauropod dinosaurs had moved through this place several tens of million years before, but were now mostly gone from this part of the world. Trails made by these sauropods, caused by habitual movements over hundreds of generations, had breached levees and cut across river channels. These alterations provided new avenues for flooding water that eventually changed local drainage patterns, which in turn impacted regional flow.

The sauropods had grazed and trampled the local vegetation sufficiently that plant roots no longer bound sediments near the river channels. As a result, every flood rapidly eroded banks, broadened channels, and spread sand and mud farther out onto floodplains than before. A combination of flooding and wind-blown sand from this widened river built up levees, which provided banks suitable for hosting the burrowing ornithopods. Over time, this synergism between sauropods—and later, large ornithopods and ceratopsians—with sediments, and moving water irrevocably changed the neighboring habitats and became the "new normal" for all dinosaurs that lived and evolved there afterwards.

The sauropods had also made extensive nesting grounds, maintained on the upper parts of the floodplains, that covered hundreds of acres and stacked on top of one another over thousands of years. The annual visits of these huge dinosaurs and their nurseries not only inhibited plant growth, but also imbued this area with a bumpy surface, like clothes left under a blanket. In a recursive way, then, these coalescing and superimposed nests, which now formed hardened layers several meters below the *Troodon* nests, made the area amenable for laying eggs and raising young for later dinosaurs. Wasps, beetles, and other insects likewise were attracted to the well-drained soils in between the *Troodon* nests. These insects burrowed into the sand and made brooding cells, which were later occupied by their larvae and cocoons. The former nesting horizons of the sauropods also restricted the burrowing of small mammals, which brought them closer to the surface and made things a little easier for the *Dromaeosaurus* and other predators to find them.

Yet another trace left by the sauropods was an evolutionary one, evident in the plant communities. The hadrosaurs had been grazing on vegetation that was the result of intense selection pressures placed on past plants by sauropods through both stomping and eating. These changes in plant communities consequently shifted evolution in the animal communities that used them, reinventing entire ecosystems.

Thus the stage for this Cretaceous drama was constructed and its actors were unwittingly directed by the lasting marks of these vanished sauropods and other dinosaurs. In this sense, traces begat traces, and the dinosaur vestiges of the past influenced those of the present, an expansive canvas gently suggesting where the next brushstrokes should go.

Tracing Dinosaur Lives

Most of the scientists I know try hard not to write fiction. After all, our primary goal each day is to understand just a bit more about whatever we study and clearly communicate our newly realized comprehensions to others. Even so, as a practicing paleontologist, I thought that a fictional scenario would best encapsulate much of what this book is about, while also introducing its main topic in a way that engages and encourages our imaginations.

Perhaps more than any other part of paleontology, the research specialty of *ichnology*—the study of *trace fossils* (tracks, trails, burrows, feces, and other traces of behavior, including fossil examples)—is about that exciting intersection between science and flights of fancy. When applied to dinosaurs, ichnology becomes even more stimulating. In fact, I like to argue that for us to truly grasp how dinosaurs behaved, to really know how they lived as animals and interacted with one another and their environments, we absolutely must study their trace fossils, and not just their bones, in order to paint the most vivid picture imaginable of their world.

In the story above, I placed together dinosaurs that may not have been in the same time and place, although most are from near the end of the Cretaceous Period (about 70 million years ago) and in an area defined approximately by Montana and Alberta, Canada. Furthermore, even those dinosaurs overlapping in both respects still may not have encountered or affected one another. However, in this deliberate mash-up of dinosaurs and their behaviors, real dinosaur trace fossils inspired nearly every element of this story.

Even better, many of these trace fossils have been discovered or studied just recently. Because of these finds, paleontologists are reconsidering some of what we thought we knew for sure about dinosaurs, either confirming long-suspected behaviors or revealing astonishing new insights into their lives. In other words, dinosaur trace fossils very often fulfill or exceed our expectations of these most celebrated of fossil animals.

Let's start with the *Triceratops* fight as an example. It turns out a good number of *Triceratops* head shields, which are composed of paired parietal and squamosal bones, bear deformities in the squamosals. These look like former healed wounds and are consistent with injuries caused by *Triceratops* horns. Ceratopsians, a group of dinosaurs that includes *Triceratops* and related horned dinosaurs, also made tracks, which are preserved in Cretaceous rocks from about 70 million years ago in the western U.S. and Canada. Ceratopsian tracks can be identified from their size, numbers of digits—five on the front foot and four on the rear—and are preserved in rocks the same age as those with ceratopsian bones. These same tracks also show that ceratopsians walked with an upright posture. This implies that these dinosaurs could move more efficiently than previously supposed from skeletal evidence: more like a rhinoceros, and less like a lizard.

Did large ceratopsians like *Triceratops* trot or gallop? We don't know for sure yet, but their tracks would provide one of the best ways to test whether they moved faster than a walk. So even though I only imagined two ceratopsians trotting toward one another and knocking heads, it's feasible that someone could find tracks showing that such fights did indeed happen. Moreover, this possible future discovery is given hope because other trace fossils—the healed wounds—suggest that ceratopsians occasionally became cross with one another, whether over territory, mates, food, or all of the above.

Was there ever a dinosaur stampede like the one described, composed of a mix of diminutive dinosaurs and different species? Maybe, although this is now being disputed. In the Cretaceous

Period, about 95 million years ago and on a lakeshore in what is now Queensland, Australia, nearly a hundred small, two-legged dinosaurs ran in the same direction and at high speed. Paleontologists who originally studied this tracksite think that species of theropods and ornithopods were together in the same limited space. Evidently, they were then panicked by the arrival of a much larger dinosaur on the scene. These tracks also say something about different species of dinosaurs tolerating one another in the same environments, as well as reacting to the same stimuli. And just what was the identity of the large dinosaur that caused such distress? And was it really a stampede, or can this unusual tracksite be explained by other means? That's a story in itself, which, along with the science behind it, I'll gladly discuss later.

How about the scene with the swimming dinosaurs? Once again, trace fossils confirm a concept that has gone back and forth among paleontologists, but is now certain: we know that at least a few dinosaurs left land and got into the water. One grouping of swim tracks, made by seven separate theropods, is preserved in Early Cretaceous rocks (110 million years ago) of Spain. Many more swim tracks are in Early Jurassic rocks from about 190 million years ago in southwestern Utah. The latter site has shattered any doubts about dinosaurs swimming, with more than a thousand such tracks, linked to theropods and ornithopods, showing how they paddled against, with, and across currents. Until lately, swimming was thought of as an extremely rare behavior in dinosaurs. Hence, these tracks have impelled paleontologists to reexamine their presumptions, and they are now looking for more evidence that some dinosaurs were comfortable in water, or even that they may have occasionally gone fishing and taken advantage of the plethora of food waiting for them beneath the water's surface.

Dinosaur digging, whether used for making burrows in which they lived, to acquire underground prey, or to make ground nests, is yet another newly diagnosed behavior in dinosaurs, and one based mostly on their trace fossils. In 2007, I helped two other paleontologists document the first known burrowing dinosaur (*Oryctodromeus cubicularis*, a small ornithopod) from Cretaceous rocks (95 million years old) in Montana. Incredibly, this dinosaur was found in its burrow with two partly grown juveniles of the same species. Two years later, I interpreted similar burrows in older Cretaceous rocks (105 million years old) of Victoria, Australia. Why would small dinosaurs burrow? Some of the reasons were proposed in the story, such as protection of young and maintaining a controlled underground environment, although these are still subject to debate.

A different type of digging by other dinosaurs also has been inspired by unusual trace fossils found in Late Cretaceous rocks (about 75 million years ago) of Utah in 2010. These are interpreted as claw marks made by predatory theropods. The close association of these marks with underlying fossil burrows, inferred as those of mammals, adds another previously unconsidered dimension to dinosaur behavior, which was their preying on small subterranean mammals. Sediment-rimmed nests made by theropods like *Troodon* and some sauropods also imply that these dinosaurs dug up and mounded soil to make these protective structures.

Related to this, a renaissance in our understanding of dinosaur eggs, babies, and the rearing of young has revolved around their trace fossils, too. *Troodon*, a Cretaceous dinosaur from 70 to 75 million years ago and found in parts of western North America, was the first known North American example of a theropod that made rimmed ground nests. These nests also contained clutches of paired eggs, which were arranged vertically in the nests by one or both of the parents after egg-laying. All three trace fossils of *Troodon* behavior—the making of rimmed ground nests, pairing of the eggs, and their post-laying arrangement—provide insights we never would have figured out from their skeletons.

Similarly, a spectacular find of Lat[illegible] nests in Argentina from 70 to 80 million years ago and [illegible] to gigantic sauropods called titanosaurs shows that dinos[illegible] other than *Troodon* made ring-like enclosures for their eggs. T[illegible] sauropod nest structures, however, only superficially resemble those of *Troodon* and are bigger, more abundant, and stacked on top of one another, representing many episodes of sauropod breeding in the same general area. In this sense, then, did these enormous dinosaurs act like modern migratory birds, returning to the same nesting grounds for hundreds of thousands of years? Once again, this and other questions are ones that trace fossils can help to answer.

The seemingly odd depiction of the gangly theropod *Struthiomimus* consuming rocks along a riverbank and using these as gastroliths is not too far off from the truth, either. Paleontologists have long suspected that some herbivorous dinosaurs, similar to modern birds or crocodilians, swallowed rocks and used them in their digestive tracts to grind food. This especially made sense for dinosaurs with teeth poorly adapted for chewing yet somehow needing to eat difficult-to-digest plants. What has surprised paleontologists in recent years, though, is the realization that a few theropods, a group of dinosaurs once assumed to have been exclusively carnivorous, also have these "stomach stones." Paleontologists just assumed that strong stomach acids were sufficient for digesting anything consumed by a theropod. Although no one has yet found gastroliths directly associated with *Struthiomimus*, some of its relatives, collectively called ornithomimids ("ostrich mimics"), do have them. This fact has prompted paleontologists to start thinking about what these theropods might have eaten other than meat: insects, plants, or a blend of both? Or did these gastroliths have some other uses we still don't quite understand? And why would some herbivorous dinosaurs with teeth unsuited for chewing, such as most sauropods or stegosaurs, *not* have gastroliths?

Speaking of food, yet another dimension of dinosaur behavior that is much better comprehended through their trace fossils regards what they ate. Traces woven into the opening narrative, such as healed bite marks, toothmarks on bones, wear c[illegible] teeth caused by plants, and coprolites (fossil feces), tell us much more about dinosaur dietary choices than any other means of fossil evidence. For instance, we can now surmise that *Edmontosaurus* and *Triceratops* must have been quite tasty for some tyrannosaurs. This is backed by healed toothmarks caused by a large predatory theropod preserved in a few bones of

Edmontosaurus, including at least one with a smoking gun (or tooth, as it were) linking it to *Tyrannosaurus* or its close relatives.

Triceratops bones also bear toothmarks that could only have been made by tyrannosaurs, including those that mark the front of the face and others showing where they grabbed a *Triceratops* head shield to separate its head from the rest of its body. Amazingly, not one but two colossal coprolites attributed to tyrannosaurs have been documented, each with finely ground bone and one containing fossilized muscle tissue. From coprolites, we also suspect that at least some Late Cretaceous hadrosaurs ate rotten wood. (Why? Sorry, can't reveal everything just yet.) We even figured out from dinosaur coprolites that at least a few animals—namely, dung beetles—depended on dinosaur feces as "manna from heaven" to ensure their survival. Hence, these trace fossils bring us much closer to reconstructing ancient ecosystems, piecing together food webs in dinosaur-dominated ecosystems from more than 65 million years ago.

What else can dinosaur trace fossils tell us? Considering extreme ranges in dinosaur sizes, diversity, numbers, geographic dispersal—including the North and South Poles—and evolution throughout their 165-million-year history, dinosaurs clearly played key roles in the functioning of land environments. Again, this is where dinosaur trace fossils have been and will be used to augment or surpass other fossil or geological information. For example, did sauropods and other dinosaurs actually change the courses of rivers or otherwise alter landscapes through their tracks, trails, and other traces? All signs point to *yes*. Did polar dinosaurs live year-round in those icy environments, or did they migrate seasonally like modern caribou? Trace fossils, such as dinosaur tracks and burrows in sedimentary rocks from formerly polar environments, tell us that they likely stayed put during the winters. How about dinosaur evolution and extinction: What do trace fossils tell us about the timing and causes of these large-scale biological facts of life for dinosaurs? In one recent study, the earliest ancestors of dinosaurs were proposed on the basis of not-quite-dinosaur tracks in 245-million-year-old rocks in Poland from the earliest part of the Triassic Period.

Another important evolutionary step in dinosaurs we've documented quite well is that birds evolved from a lineage of small theropods. This relatedness has been certified through many lines of evidence, including fossilized feathers directly associated with the skeletons of more than thirty species of theropods. But we still have questions about this evolutionary transition that are hard to answer from just bones and feathers. For example, when did these small dinosaurs start to climb trees, or fly from the ground up, or land after flight?

The hotly debated question of whether any dinosaurs survived a post-apocalyptic landscape caused by a meteorite impact 65 million years ago is also potentially answerable by trace fossils. A single dinosaur bone in 64-million-year-old rocks is nearly always regarded suspiciously, the fossil equivalent of an online-dating ad in which a person misleadingly underreports his or her age. These bones are very likely recycled, having been eroded out of older rocks, re-deposited, and buried a million years or more after the dinosaur originally died. On the other hand, a single dinosaur track in 64-million-year-old rocks would be hard-to-refute evidence that at least one dinosaur was walking around after they supposedly all died.

So although dinosaur trace fossils certainly can answer questions that range from what an individual dinosaur was doing at a given moment during the Mesozoic to the big picture of how dinosaurs originated, evolved, and went extinct, paleontologists still hope for more. For instance, some of the trace fossils on their "wish lists" are those that flesh out some of the more dramatic encounters dinosaurs very likely had. One that comes to mind would be more trace fossil evidence supporting the oft-depicted scene of large predatory theropods stalking other dinosaurs, or pack-hunting behavior in theropods of all sizes. The first scenario is portrayed in our story toward its end as the tyrannosaur, after her botched attack on the hadrosaur, follows it and the rest of the hadrosaur herd. A similar behavior can be inferred from tracks in Early Cretaceous rocks of east Texas, in which the footprints of a large theropod paralleled and then crossed those of a sauropod, apparently shadowing it. Other compelling theropod trackways include some from the Cretaceous of

China that tell of six theropods, equally spaced and all moving in the same direction, which very much looks like evidence of pack hunting. More evidence of pack-hunting theropods is suggested by parallel trackways in Early Jurassic rocks of Utah. Oh, and I should also mention that the two-toed tracks left by the Chinese theropods show they were deinonychosaurs, sickle-clawed theropods related to *Dromaeosaurus* and *Velociraptor.* You can bet these theropods weren't digging for their food that day.

Lamentably, trace fossils clearly illustrating dinosaur mating, like those conjectured for an amorous pair of *Ankylosaurus*, have not yet been recognized. Because we don't know for sure what "dinosaur whoopie" trace fossils might look like, these will require an active (perhaps overactive) imagination to detect them, as I have tried to do above. Nonetheless, I expect such trace fossils, including those that precede mating ("wooing" traces, so to speak), will be eventually found and identified, adding to our understanding of what were very likely complex dinosaur sex lives.

In short, the story at the start of this chapter and the myriad dinosaur trace fossils that contributed to its creation demonstrate the huge advantages afforded by these sometimes-underappreciated records of dinosaurs' daily lives. The main point of the rest of this book, then, is to justify a shift in perspective and start thinking of traces. In other words, after this book, you will no longer just visualize mounted dinosaur skeletons in a museum. Instead, you will think of those skeletons covered by muscles, tendons, and skin, then moving, breathing, mating, eating, fighting, swimming, taking care of their young, and other behaviors, and of traces left by these behaviors. You will also think about the number and variety of traces dinosaurs would have left behind during their normal lifespans, from infancy to old age, and realize how these marks would far outnumber any of their bones. Once you've done all of that imagining, you're ready to look deeper at dinosaur trace fossils, a different way of thinking that's guaranteed to change what you thought you knew about these long-extinct but ever-popular animals.

CHAPTER 2

These Feet Were Made for Walking, Running, Sitting, Swimming, Herding, and Hunting

Why Dinosaur Tracks Matter

If through some miraculous disaster every dinosaur bone in the world disappeared tomorrow (or the next day, for that matter), the fossil record for dinosaurs would still be represented quite well by their tracks alone. The main reason for this is very simple: each dinosaur only had, on average, about two hundred bones per individual. Yet you could bet that those dinosaurs that made it from mere hatchling to rambunctious juvenile to surly, angst-filled teenager to a full-fledged responsible adult probably made many more than 200 tracks during their lifetimes. This supposition alone implies—although we'll never know for sure—that dinosaur tracks probably far outnumber their bones in Mesozoic rocks worldwide.

The number and variety of dinosaur tracks out there is astonishing. Thus far, dinosaur tracks have been found in eighteen states of the U.S. and on every continent except for Antarctica, with thousands of newly discovered ones each year. Dinosaur tracks range in latitude from the North Slope of Alaska to southern Argentina, and are in rocks dating from the beginning of dinosaurs, about 230 to 235 million years ago (*mya*), to their very end, 65 *mya*. Although it's tempting to think of all dinosaur tracks as potholes that would easily swallow a tricycle and its dinosaur-admiring rider, tracks also varied in size from less than the width of a thumbnail to depressions that could be used to park a Smart Car.

Other than their sheer abundance, another comforting thought about dinosaur tracks is that they very often are in places where dinosaur bones are rare or absent. Moreover, they also convey snapshots in time, reflecting a vast variety of dinosaur behaviors in the moment, telling us about a former dinosaur presence in a given place and what they were doing in whatever environment they traversed. Conversely, very few dinosaur bones were buried where a dinosaur lived; that is, most bones were likely moved some distance from their original habitats. As a result, I like to argue that dinosaur tracks constitute the "real" fossil record of dinosaurs rather than their bones, which are nice but, well, just a little too *dead*. Tracks breathe life back into dinosaurs.

Dinosaur Feet and Footprints through Time

Before jumping into a more detailed discussion of dinosaur tracks, it's probably a good idea to learn about the main groups of dinosaurs and their feet, which helps to identify dinosaur trackmakers. Paleontologists classify dinosaurs through anatomical traits, and these traits are nearly always related to dinosaurs' evolutionary history, or their shared ancestry. Ideally, then, each recognizable dinosaur bone can be correlated with about six broad groups of dinosaurs: theropods, prosauropods, sauropods, ornithopods, thyreophorans (stegosaurs, ankylosaurs, and nodosaurs), and marginocephalians (pachycephalosaurs and ceratopsians). These groupings of dinosaurs that share a common ancestor—called *clades*—are best expressed graphically through a branching diagram called a *cladogram*.

In the simplest cladogram for dinosaurs, theropods are more closely related to prosauropods and sauropods than they are to ornithopods, whereas stegosaurs, ankylosaurs, and nodosaurs are more closely related to one another than they are to marginocephalians. Also, because birds descended from theropod ancestors and thus qualify as dinosaurs, these are included on any self-respecting dinosaur cladogram. But if you want to be even more of a "lumper" with classifying dinosaurs, you could go back to their initial split into *saurischians* ("lizard-hipped" dinosaurs), which includes all theropods (birds too), prosauropods, sauropods, and *ornithischians* ("bird-hipped" dinosaurs), which are all ornithopods, thyreophorans, and marginocephalians.

Dinosaur feet can be roughly correlated with the evolutionary history of dinosaurs, based on the appearance and disappearance of these clades in the fossil record. For example, the hypothetical "first dinosaur," which would have evolved about 235 *mya* and was the common ancestor to both saurischians and ornithischians, probably walked on its rear two legs (bipedal) and its feet would have had three prominent toes (digits) pointing forward, one toe off to the side, and all toes tipped with claws. Its front feet, had it also used these for walking (making it quadrupedal), would have had five fingers (also digits), again all pointing forward and with claws. All subsequent dinosaur feet were modified from this basic body plan, whether certain dinosaur lineages stayed bipedal, went to quadrupedal, or used some mixture of the two. As a result, dinosaur tracks made by feet that had more than four digits in the rear are rare—happening only with sauropod tracks—and more than five digits in the front would be really weird. If anything, most dinosaur feet reduced or lost digits throughout their evolutionary histories. Unneeded toes or fingers, which can be evolutionarily expensive, were weeded out by survival and propagation of species, in which it was advantageous to get rid of these over time.

Just as modern trackers might classify mammal tracks by the number of toes, the same can be done with dinosaurs. For the major evolutionary groups of dinosaurs, the following modes of movement and digit numbers, with only a few exceptions, can be applied to help with identifying their tracks:

- Theropods—rear feet only (bipedal), three digits pointing forward, and sometimes another small one off to one side.
- Prosauropods—rear and front feet (bipedal or quadrupedal), four digits on the rear, four on the front.
- Sauropods—rear and front feet (quadrupedal), five digits on the rear and five on the front (although digits are almost never visible in sauropod front-foot tracks).
- Ornithopods—rear feet only or all four feet (bipedal or quadrupedal); on the rear feet, three digits pointing forward, sometimes another digit off to the side, whereas front feet had five or less.
- Stegosaurs—rear and front feet (quadrupedal), rear feet with three digits and front with five.
- Ankylosaurids and nodosaurids—rear and front feet (quadrupedal), four digits in the front and four or three in the rear; most ankylosaurids (perhaps all) had three.
- Ceratopsians—rear and front feet (quadrupedal), four digits in the rear and five in the front.

Granted, paleontologists, just like any other scientists, delight in pointing out exceptions. So just to get those distracting thoughts out of the way, here are a few dinosaur-foot oddities to keep in mind:

Therizinosaurs—These were unusual theropods, so unusual that no self-respecting dinosaur paleontologist discussing them can complete a sentence with these as the subject without also saying "strange," "odd," "weird," "bizarre," and other synonyms denoting their differences from other theropods. As an example of their unusualness, therizinosaurs, unlike all other theropods, had four toes pointing forward on their rear feet. As of this writing, a few therizinosaur tracks are known, with the most astonishing recently found in Cretaceous rocks of Alaska. These four-toed feet may represent some ancestral condition in theropods that was somehow retained in therizinosaurs well into their evolutionary history.

Dromaeosaurids—These were bipedal theropods with three toes retained on their rear feet, but only two of those digits contacted the ground. One of the three digits was raised and sported a nasty-looking claw, which evidently was used for gripping and holding down prey, climbing, or slashing. Consequently, dinosaur trackers sometimes do a double-take when they spot dromaeosaurid tracks, which look as if their theropod trackmakers somehow lost a digit. This notion is quickly discarded, though, once those paleontologists realize both feet are missing exactly the same toe, which surely is not a coincidence. Once considered rare, dromaeosaurid tracks are now being discovered in some of the same places and geologic formations as dromaeosaurid bones, helping paleontologists to fill out more complete pictures of their life habits.

Pachycephalosaurs—these dinosaurs, often nicknamed "bone-headed dinosaurs" because of their thick, bony skulls, share a common ancestor with those other big-headed dinosaurs, ceratopsians. Yet we know nothing about their tracks, because we know nothing about their feet. So far, no pachycephalosaur foot parts have been discovered, meaning that we can only speculate about what their tracks look like. Based on what we do know from their skeletal remains, we think they were mostly bipedal, so their tracks mostly just show alternating right–left rear foot impressions. Such tracks possibly might resemble those of their ceratopsian cousins, which have four toes on the rear foot. If so, that difference would be helpful when distinguishing their tracks from those of ornithopods or theropods.

So you might get the impression that, armed with digit numbers and knowledge of the basic groups of dinosaurs, identifying their tracks will be oh-so-easy, a leisurely stroll in the park while also attended by servants providing tea, backrubs, and answering your e-mail for you. Once you come back from your saunter down a fantasyland version of how dinosaur ichnology is done (hey, I've been there, and visit it often) and are ready to do a little more thinking, here is what else you'll need to know.

For instance, you've probably already noticed that most theropod and ornithopod tracks are three-toed. Then how do we tell the difference between them, especially if they're preserved in rocks of the same age and place? This can get tricky, especially if the tracks are poorly preserved. But the easiest way to tell the difference between a theropod track and an ornithopod track is to apply three criteria: (1) look at the length of the foot versus its width; (2) check out the width of its toes; and (3) see whether it has claws that end in points or if they're a little more blunt. Theropods usually left tracks that are longer than they are wide, with thin toes and sharp claws, whereas ornithopod tracks are typically wider than they are long, with thicker toes and blunt tips. Again, there are exceptions to this generality, and even dinosaur-track experts have doubts about the identity of some three-toed dinosaur tracks, especially if a rival dinosaur-track expert identified them. But with application of these three criteria, their distinction is a little more assured. The only other complicating factor is where theropod and three-toed bird tracks overlap in size and shape, but that's another story and one far too long to tell here and now.

Prosauropod tracks are also challenging to identify, partly because they weren't made for very long, geologically speaking. Prosauropods only lived from the Late Triassic through the Early Jurassic periods, from about 230 to 190 *mya*. These relatives of sauropods, the largest land animals of all time, developed into the world's largest herbivores during their geologically brief 40-million-year time span, and were among the first dinosaurs to get around on all fours. However, many prosauropod tracks also show them walking on their rear feet only.

Sauropod tracks are not so difficult to spot, but are tough to recognize for what they are. In the early days of dinosaur tracking, probably more than one paleontologist or geologist walked by their footprints without a second glance, thinking they were some sort of large erosion-caused features. Once these footprints were correlated with the sizes and shapes of sauropod feet, though, this oversight was quickly rectified, and sauropod tracks magically appeared in the search images of paleontologists worldwide. So where we originally had none, we now frolic in the land of plenty, as sauropod tracks have been found on all continents except for Antarctica, and in rocks ranging from the Late Triassic (230 *mya*) through the Late Cretaceous periods (65 *mya*). Other than the extraordinary size of the largest tracks, sauropod footprints are oblong (rear) to crescent-shaped (front), and often form two-by-two diagonal patterns. The best-preserved rear-foot tracks show claws at the end of each digit, too. Despite artistic recreations and skeletal mounts depicting sauropods rearing up on their hind legs, no tracks have yet demonstrated that sauropods did anything more than walk on all fours.

Stegosaur, ankylosaur, and ceratopsian tracks, like sauropod tracks, were similarly considered rare, but were not identified until just in the past few decades. For example, the first undoubted stegosaur tracks were not found until 1994, in Middle Jurassic (about 170 *mya*) rocks of England. Now stegosaur tracks are becoming more readily recognized, also having been found in Spain, Portugal, Morocco, and the exotic far-off land of Utah. These tracks have also filled gaps in the known geologic history of stegosaurs, handily augmenting or surpassing their skeletal record in some places. Some of these tracks even include skin impressions, the first known glimpse at the scaly feet of stegosaurs. For ankylosaurs, their tracks weren't noticed until the 1990s in western Canada. Now their tracks are documented from places as widespread as Bolivia, British Columbia (Canada), Colorado (USA), and elsewhere; oddly, all found thus far are in Early Cretaceous rocks. Ceratopsian tracks, also unknown until the 1990s, are still apparently uncommon, but now that people know what to look for, more of these trace fossils are being discovered each year, too.

So now you have an overview of the major dinosaur groups, what their tracks look like, and the current state of their record. However, simply answering the question "Who?" should not halt all further inquiry. Here is a small sample of the questions that could be asked of any given dinosaur track:

- How old were these dinosaur trackmakers: hatchlings, juveniles, sub-adults, or adults?

- Can we tell dinosaur genders from their tracks?
- How did dinosaurs move: did they ever do anything more than just walk, such as lope, trot, or gallop?
- Did dinosaurs ever stop to take a break from their daily activities and sit down?
- What did dinosaurs do when encountering a body of water: did they walk around it, or swim across it?
- How about their social lives: were some dinosaurs "rugged individualists" who shunned the company of others, or did they seek out and travel with their own kind, whether in small or large groups?
- What about one-on-one encounters, such as those between predatory dinosaurs and their prey?
- Can we even discern a given dinosaur's medical history, that is, did it have some injury or other affliction that modified its behavior enough that we can notice its effects?
- On a much grander scale, what do dinosaur tracks tell us about the timing of their origins or demise?

As you can see from these questions, just identifying what dinosaur made a track is actually a very small part of understanding how dinosaurs behaved. So with this humbling thought in mind, let's go on to those most exciting facets of divining dinosaurs' lives from their tracks, starting with their evolutionary origins.

First Steps of the Dinosaurs: Origination

When were the first dinosaur tracks pressed fresh into the ground? Naturally, the answer to this question also depends on when the first dinosaur existed, a difficult problem to address. It's like trying to answer the question "When did we first become human?" But the dinosaur-track one has the decided advantage of lacking all of the anthropocentric baggage accompanying the latter inquiry.

The current claim for "oldest dinosaur from the fossil record," a label guaranteed to cause a fight among dinosaur enthusiasts, lies with *Eodromeus* ("dawn runner"). This dinosaur, which was discovered in Late Triassic (230 *mya*) rocks of Argentina, was a small bipedal theropod that weighed about the same as a big turkey. Despite its antiquity, *Eodromeus* is nicely preserved, with about 90% of its skeleton known, including its hind limbs.

From these bones, we know it had four toes on its rear feet, and three of those toes would have likely left impressions as it walked. Thus we can use its feet as a predictor for what its tracks looked like, or those of its close kin. Other dinosaur fossils from rocks of nearly the same age in Argentina include one other theropod, *Herrerasaurus*, a basal sauropodomorph, *Eoraptor*, and a primitive prosauropod, *Panphagia*. Although *Eoraptor* was chicken-sized and *Herrerasaurus* was more like a two-legged German shepherd, they both had three prominent toes on their rear feet, with four total. Unfortunately, the skeleton of *Panphagia* did not include its foot bones, so we don't know for sure the forms of its tracks or those of its relatives at the time.

Using this combination of skeletal data and knowing that evolution often preceded the oldest preserved fossils, paleontologists figure that dinosaurs actually originated at around 235 *mya*, toward the end of the Middle Triassic Period. Sure enough, three- and four-toed tracks similar to those predicted for primitive dinosaurs are fairly common in some Middle Triassic rocks. Because these tracks were so similar to those of known dinosaur tracks, their discoverers excitedly pronounced them as "dinosaur-like" and hinted that these footprints extended dinosaur lineages to well before their skeletal record. Unfortunately, such claims were thoroughly trounced, flogged, ridiculed, and otherwise treated as unworthy of any encouragement whatsoever. The bulk of this disdain, of course, came from paleontologists who studied dinosaur bones, not tracks. As a result, advocates of trace fossil evidence for dinosaur ancestry were stymied, as they also somehow had to connect tracks to feet foretold—but not yet found—for animals that heralded the arrival of true dinosaurs in the Late Triassic Period.

Some of this disrespect for all things ichnological was allayed in 2010 when a team of paleontologists, led by Stephen Brussatte, published a paper in which they proposed the oldest "dinosauromorph" tracks from the fossil record. *Dinosauromorph* refers to the clade Dinosauromorpha, which includes all animals more closely related to dinosaurs than other non-dinosaurs, such as pterosaurs and crocodilians. This means that ancestral dinosauromorph tracks almost look like dinosaur tracks, but not quite: sort of how a primitive human's tracks would differ from those of a same-sized modern human. Some of these tracks, which were from Early Triassic (about 245 *mya*) rocks in Poland, precede *Eoraptor* by more than 15 million years. These paleontologists also pointed out, somewhat indignantly, that ". . . footprints are often ignored or largely dismissed by workers focusing on body fossils, and are rarely marshaled as evidence in macroevolutionary studies of the dinosaur radiation." Yes, indeed. Knowing they would face resistance from their body-fossil-focused brethren, they carefully linked the tracks to likely anatomical features of dinosauromorphs. They also pointed out that the oldest probable dinosaur tracks were from the end of the Middle Triassic Period, about 235 *mya*, which were geologically younger than their tracks, but still before the earliest dinosaur body fossils by about 5 to 7 million years. Dinosaurs later became much more abundant and diverse by the end of the Triassic Period, at about 200 *mya*. Accordingly, their tracks were common and varied by then, too, and became much larger. Within only about 50 million years, dinosaurs began making the largest footprints of any animals that ever lived.

These Triassic dinosauromorph tracks point toward what paleontologists call, intriguingly enough, a *ghost lineage*. A ghost lineage is one for which we have evidence that ancestral members of a clade and their descendants lived at a certain time in the geologic past, but we so far lack corporeal evidence for its existence. Given these dinosauromorph tracks, somehow this phrase seems even more appropriate as we consider these ethereal, disembodied traces as evidence of the first proto-dinosaurs.

Four Legs Good, Two Legs Better, or Does It Matter? Trackway Patterns and Dinosaur Gaits

Thus armed with all of this knowledge about dinosaur tracks and how they record when dinosaurs first evolved, it's tempting to think that you can just identify a dinosaur by looking at one track, state confidently "theropod," "ornithopod," "sauropod," or whatever other dinosaur clade you think it belongs to, identify the geologic age of the rocks hosting this track, take a photograph, and be done with it. But that would be a sad state of affairs, utterly lacking a sense of adventure and curiosity, living in a bland world filled with beige tones and unseasoned instant grits. In other words, you don't want to do that.

Instead, you want to know more about how this theropod, ornithopod, sauropod, prosauropod, stegosaur, ankylosaur, or ceratopsian behaved. What was it doing in the place where it left its tracks? When did it arrive on the scene relative to other dinosaurs, insects, or worms living in the same area? Where did it go after it made the tracks? Could its body be nearby, or did it travel a long way before dying? Was it with any others of its species, or looking for love in all the wrong places? How long were these tracks there before they were buried and preserved for us to see them millions of years later? You want to know more. Much, much more.

To understand dinosaur behavior from their tracks, one must absolutely study sequences of tracks, or *trackways*. Knowing that most dinosaurs either got around quadrupedally or bipedally, trackways therefore can be expected to show right–left rear foot impressions or a combination of all four feet. However, a few dinosaurs mixed it up, switching from bipedal to quadrupedal and back again, just like how someone can go from walking upright while filled with pride to crawling on hands and knees begging for forgiveness to walking tall again. In a dinosaurian sense, though, a change from a four-legged to a two-legged gait meant that a dinosaur was *facultatively bipedal* (became bipedal when it wanted) and a normally two-legged dinosaur going on all fours was—you guessed it—*facultatively quadrupedal*. These changes in which limbs touched the ground were likely related to dinosaurs altering their speed, foraging, or other such behavioral shifts necessitated by daily life.

Every four-limbed animal has a baseline gait, or how it normally moves around on those limbs. In quadrupedal animals, such as canines, felines, bovines, or other domestic mammals, a few examples of gaits include: slow walking, normal (average) walking, fast walking, trot, lope, or gallop. For example, cats normally walk and dogs normally trot. When teaching these patterns to my students, I emphasize how gaits translate into distinctive track patterns, much like letters put together to form words. In these instances, trackway patterns read as "slow walk," "fast walk," "trot," and so on. Once these students apply this knowledge to different animals' baseline gaits, they then can more readily glance at and discern a trackway pattern, rather than stopping to measure track sizes and count toes, and much later saying "raccoon," "coyote," "deer," or "grizzly bear." (In my experience, the last of these is a very handy one to identify quickly, especially in a remote field area.) We also can get a better understanding of gaits by measuring distances between alternating feet (*pace*), between the same foot (*stride*), and the width of the trackway (*straddle*). Many other measurements can be taken from a trackway, but these three are essential and constitute a good start in their study.

Can these same principles be applied to dinosaur trackways, in which you can just glance at a dinosaur trackway and excitedly shout "Theropod!" "Ornithopod!" or "Barney!" (whatever the heck he is)? The answer is, mostly, yes. Part of this identification is aided by the obviousness of some tracks, which are then confirmed by trackway patterns. For example, if you see bathtub-sized depressions that express themselves in an alternating diagonal pattern, you will probably not shout "Baby theropod!" Furthermore, sauropod tracks normally show a slow walking or "understep" pattern in which the rear foot did not quite fall in the same place as the front foot; appropriately, its stride is short, too. In a few instances, their rear tracks registered directly on top of (and hence wiped out) their front tracks, indicating a slightly less sluggish pace. So far, I have only seen one sauropod trackway in which the rear feet were placed ahead of the front feet on the same side, approaching what we might call a "trot." This trackway was from a relatively small sauropod, which might have been a juvenile that didn't know it wasn't supposed to run.

Speaking of sauropod trackways, their patterns can be further placed into two categories based on their widths: *narrow gauge* and *wide gauge*. These terms are borrowed from railroads, in which the rails are either narrowly spaced (light rail) or widely spaced (freight trains). For bipedal dinosaurs like theropods and most ornithopods, their normal trackway patterns show they were walking, although some have been interpreted as slow walking, fast walking, or running. Their trackways have an alternating right–left–right pattern, and most are probably narrower than the body width of the dinosaur that made them, especially for theropods. They did this by rotating their legs inward with each step forward, as if they were fashion models sashaying down a runway.

To figure out the approximate size of a dinosaur from its footprint, you'll need to use a formula. No worries, it's an easy one. Take the length of a theropod or ornithopod track and then multiply it by four. The resulting number gives the approximate hip height of the dinosaur:

$$H = 4\,l$$

where H = hip height and l = footprint length.

For example, let's say you find a definite theropod track—longer than it is wide, relatively thin toes, with sharp claw marks—and it is about 35 centimeters (14 inches) long. Let's see: 35 cm × 4 = 1.4 m, which translates to about 55 inches, or four and a half feet off the ground. That's an intimidating height, especially if you think about it in human-meets-dinosaur terms, as a horizontally oriented predatory theropod would be staring directly in the faces of most people, and with the rest of its body behind it.

Okay, now for a more complicated formula:

$$V = 0.25g^{-0.5}s^{1.67}h^{-1.17}$$

In this, *V* is velocity, *g* is the acceleration of a free-falling object, *s* is stride length, and *h* is hip height, which you've already tackled. Originally devised in 1976 by a paleontologically enthused physicist, R. M. Alexander, this formula took into account the size (mass) of an animal as part of its forward momentum (the "*g*" in the equation relates to gravity), while also expressing the common-sense principle that, all other things being equal, short strides between tracks means an animal was moving slower, and longer strides means it was moving faster.

But not all things are equal in this relationship, either. For instance, if a chicken were forced to race against an elephant, it has a decided disadvantage of its leg length being much shorter than that of a typical elephant. Chicken leg lengths are more or less proportional to their foot lengths, which can be readily seen in their growth from a small chick to a full-sized roaster. In other words, relatively long strides measured between tracks made by a small-footed and short-legged animal implies it was moving faster. Alexander's formula was also based on modern animals, in which he used measured speeds, body masses, and stride lengths of many two- and four-legged animals to establish a baseline for comparing these to dinosaur trackways.

Fortunately, there is a simpler way to express this equation and get a quick-and-dirty sense of whether a dinosaur was walking slowly, trotting, or running. This is to look at stride length versus hip height as a ratio, called *relative stride length*. As an example, let's take our previously mentioned theropod with the 35-cm long footprint and 1.4 m hip height. Let's say its stride was measured as 2.8 m (about 9 feet). So its relative stride length is 2.8 m/1.4 m, which = 2.0. Basically, Alexander proposed that relative stride lengths of 2.0 or less reflect walking, 2.0–2.9 trotting, and >2.9 running. This means our hypothetical theropod was likely walking. Using the full formula, this corresponds to a calculated speed of about 3 m/s, or 10.6 kph (6.6 mph).

How fast is that in practical everyday terms? Olympic racewalkers regularly exceed 15 kph, which they can keep up for 20 km (12.4 mi). However amusing it might be to visualize, a good racewalker would cross the finish line of a 20-km race a half hour before our imaginary dinosaur. Even more entertaining, human racewalkers would have even outpaced huge dinosaurs—such as sauropods—with leg lengths more than double the heights of those people, as their trackways also indicate slow speeds for these dinosaurs, too.

Nonetheless, let's go back to that theropod trackway and follow it for a while. Along the way, you might notice its stride increased to a maximum of, say, 4.0 m. Consequently, its relative stride increases to 2.9. Now we're talking "run," and the calculated speed would be 5.3 m/s, which is about 19 kph (12 mph). Our Olympic racewalkers would badly lose a race of any distance to this theropod, and many well-conditioned runners would too.

Since Alexander first came up with this formula, it's been prodded, probed, tweaked, and otherwise tested to see how well it works, or not. Not surprisingly, then, other paleontologists have come up with their own formulas for estimating dinosaur speeds. A few have even tried to say that some dinosaurs were not capable of running at all, a supposition based on analyses of dinosaur skeletons, probable ranges of motion, and muscle masses that would be required to propel a multi-ton animal forward at high speed. Nonetheless, Alexander's original formula still endures and is normally the first that paleontologists reach for when they find a dinosaur trackway and want to know how quickly that dinosaur was moving.

This is probably where the gentle reader, who has seen supposedly sophisticated mathematical models undergo complete failures, might wonder if other features in a dinosaur trackway tell whether a given dinosaur was speeding up, slowing down, or otherwise varying its pace. For one, the width of the trackway should get narrower as a dinosaur increased its speed, and wider if it slowed down. Think about how a full stop by a *Triceratops* would show all four feet planted at the width of the shoulders and hips; in contrast, a full gallop would have registered the feet more toward the centerline of its body. A general rule for both bipedal and quadrupedal trackways: the narrower the trackway, the more likely the trackmaker was moving quickly.

Don't believe me? Fine with me, as science thrives on disbelief and testing. If you have a dog, have it walk, trot, and then run down the beach, and you will see its trackway width become visibly narrower along the way. The reverse will happen as your dog slows down to a stop, feet planted at shoulder and hip widths, tail wagging happily.

For another independent way to test dinosaur speed, you can look at how each footprint is a record of how much ground beneath a dinosaur was disturbed by its movement. This is an experiment that can be readily performed on the same sandy beach you used for your dog experiment, and you can use your dog again, or even yourself. Tracks made by walking show little disturbance of the sand around and inside of the tracks, but jogging results in ridges, mounds, and plates of sand caused by each foot as it pressed against the sand. Sprinting imparts more prominent structures, and maybe sand kicked completely out of the track. In other words, tracks and these structures caused by the applied and released pressure—which some trackers call *pressure-release structures*, *pressure releases*, or *indirect features*—can be applied as another type of speedometer and used as independent checks of relative stride lengths.

These structures, however, depend on many factors for their formation and preservation, such as the type of sediment (mud, sand, pebbles?), moisture content (dry, slightly moist, saturated?), packing (loose sand, hard-packed sand?), slope angle (flat surface, heading uphill, going downhill?) and not just the speed of the trackmaker. Imagine if a theropod ran across moist and slippery mud, and then dry sand; then visualize how different its tracks would look in each type of sediment. Add in little behavioral nuances such as head position (up, level, down, right, left?), turning to one side or another while moving, stopping to scratch its back, or bending down to nab a small mammal with its mouth. All of these factors culminate in what paleontologists consider as "track tectonics," miniature landscapes wrought by a dinosaur's foot pushing or twisting against a muddy or sandy medium.

Another important factor in the formation of such structures around a track is the size and anatomy of the trackmaker's feet. For instance, think of how a sauropod's elephant-like foot made far different structures compared to the foot of a small, thin-toed theropod. This all means that tracks can be quite complicated; and to better understand them, they are the subject of much experimental work with real animals (more on that later) and three-dimensional computer simulations. In contrast, drawing a simple cartoon outline of a track would be like summarizing a Salvador Dali painting as "art," a terrible misdeed that omits all of its colors, hues, gradations of tones, and themes. What a pity to miss all of those metaphorically melted watches.

Before moving on to excitedly interpreting dinosaur behavior from their tracks, though, I should point out one minor disadvantage caused by how most tracks were preserved. Dinosaur tracks, just like modern ones, were probably weathered soon after they were made, placed under assault by the erosive effects of rain, wind, gravity, and other animals stepping on them. This means that dinosaur tracks may not reflect the original surface impacted by a dinosaur. Consequently, a common way for dinosaur tracks to have made it into the fossil record was as *undertracks*. That is, many of the tracks we see were actually transmitted below the surface where a dinosaur walked, trotted, or ran.

This phenomenon is similar to how, when exerting pressure with a pen or pencil while writing or drawing on a sheet of paper, an image of the writing or drawing is also impressed on underlying pages. The decided preservational advantage of this phenomenon is that such tracks were already buried, protecting them from destruction. Hence, all paleontologists who study dinosaur tracks first assume they are looking at undertracks, and only modify these realistic expectations if confronted by the delightful details of skin.

In essence, if a dinosaur track shows pad and scale impressions from the foot of its maker, only air separated the flesh of that dinosaur from the sediment preserving it.

Any dinosaur track we find in the geologic record, though, is at the end of its history. So we also have to throw in an additional dollop of skepticism when divining any given track found at the same earth's surface we now occupy. Undertracks made by dinosaurs may have been spared weathering for as much as 200 million years, but once exposed they can quickly become blurred or erased completely by modern weathering. This is where paleontologists depend on the sharp eyes and good graces of the general public, who far outnumber paleontologists and perhaps are outside more often. Many a dinosaur track or trackway has been found by hikers, bikers, or other recreationalists who recognized their patterns and then did the right thing by reporting their locations. Maybe it will be your turn some day.

Dinosaurs on the Run

You might be thinking by now, enough about the hypothetical and anachronistic races with humans and talking about dinosaur tracks as theoretical objects. What was the fastest dinosaur ever interpreted from a dinosaur trackway? How fast were they, really, especially when compared to the speediest of humans or modern mammalian predators such as lions or cheetahs?

Most animals spend much of their time moving at a normal pace. Among humans, even marathon or ultramarathon runners normally do not spend more than 15% of any given day running and, let's face it, they're not sprinting throughout even that time. Hence, a reasonable reflexive prediction about dinosaurs is that whenever they were making tracks, they were walking. Running would have been quite rare and reserved only for emergencies, such as chasing down prey, not becoming prey, or avoiding other dire threats such as forest fires, floods, or storms. Sure enough, these expectations about dinosaurs mostly taking leisurely strolls are probably right. Whenever the Alexander formula or variations of it are applied to dinosaur trackways, most dinosaurs were simply walking.

Still, finding exceptions to a norm is one of the joys of science. Thus when paleontologists encounter tracks that indicate sprinting dinosaurs, they justifiably get really excited. So far, almost all trackways interpreted as those made by running dinosaurs were made by bipedal ones, theropods and ornithopods. In contrast, only a few trackways of trotting quadrupedal dinosaurs are known. One of these, preserved in Late Cretaceous rocks of Bolivia, was made by an ankylosaur—a big, armored dinosaur—which must have looked like a living tank as it ambled along. Sadly, not one trackway yet shows that sauropods, ceratopsians, or other quadrupedal dinosaurs loped, galloped, pranced, or minced. However, this does not necessarily mean these animals only walked. Just remember that running is rare in all land animals, which means the likelihood of finding fossilized tracks of running quadrupedal dinosaurs is also quite small.

The first discovered running-dinosaur trackways were from a site in Texas, where at least three theropods moved at high speed. These dinosaurs had footprints ranging from 29 to 37 cm (11.4–14.5 in) long, which are not much longer than many people's shoe sizes. Yet they were taking strides that measured 5.4 to 6.6 m (17–22 ft)! Once these numbers were crunched through Alexander's formula, the tracks spoke of speeds of 8.3 to 11.9 m/s (27–39 ft/s), or 30 to 43 kph (19–27 mph). To put it into a bipedal-human perspective, the top speed recorded by Usain Bolt over 200 m (656 ft) during the 2012 Olympics was also 27 mph, meaning that these dinosaurs and he would have had a very good race, perhaps with an outcome only affected by who was pursuing whom. But we also have no idea if these running dinosaurs were actually reaching their top speeds or not on whatever day their tracks happened to get preserved for us to see millions of years later. Theropods and humans alike, though, would be humbled by the top speed recorded for a cheetah (*Acinonyx jubatus*), which from a standing start and over 100 m (328 ft) has been clocked at 98 kph (61 mph).

Nonetheless, a dinosaur tracksite in Queensland, Australia outdoes the Texas tracksite for sheer numbers of running dinosaurs.

This tracksite is worth its own chapter, because it wasn't just one dinosaur trackway but nearly a hundred, suggesting dinosaurs all with relatively small footprints skedaddling, scampering, booking, bolting, or otherwise traveling quickly (for them), which was at about 12 to 16 kph (7.5–10 mph). Even more significant, it was a mixed group of small ornithopods and theropods all moving in the same direction. However, this interpretation has undergone some recent scrutiny and reinterpretations, which we'll take up in detail later.

To better understand what running-dinosaur tracks look like, we often turn to modern examples that can serve as search images for finding these, and not looking to cheetahs. Instead, we observe large bipedal flightless birds. For example, the fastest bipedal land animals today are ostriches (*Struthio camelus*), which have maximum speeds of 45 mph (72 kph). Less speedy—but still much quicker than humans—are emus of Australia (*Dromaius novaehollandiae*), which can move as fast as 40 mph (64 kph), and rheas of South America (*Rhea americana* and *R. pennata*) at about 35 mph (56 kph). And because all of these flightless birds are actually modern theropods, they act as marvelous proxies for how some Mesozoic theropods could have reached similar speeds when running.

Of course, modern animals have their limits when applied to the geologic past, especially once we start looking at gigantic dinosaurs, for which we have no modern analog. Hence, one of the more imaginative scientific questions applied to running dinosaurs has to be one posed of nearly everyone's favorite dinosaur, *Tyrannosaurus rex*. The question was this: If *T. rex* could run at about 45 mph—as depicted so memorably in the first *Jurassic Park* movie—what would have happened to it if it tripped? One paleontologist, Jim Farlow, and two other colleagues figured out that given the average mass of an adult *T. rex* (about 6 tons) and a speed of 45 mph, its forward momentum, halted suddenly by a fall, would have instantly killed it. Given this, the notion of a flat-out running *T. rex*, however entertaining (or frightening), is not very likely if it or any other large predatory theropod died every time it tripped while running at full speed—which, let's face it, is a poor adaptive strategy for passing on genes. However, we should also keep in mind that animals run fast in terrains and substrates conducive to such behavior, but do not in, say, gooey mud or ice-covered lakes.

Other scientists figured out how much musculature a *T. rex* would have needed to move such a massive body at high speeds and came up with numbers for the mass of muscles for its legs and around its tail. What they found was that a 45-mph-running *T. rex* would have required about 85% of its entire body mass concentrated in its legs and muscles around its tail, giving this dinosaur more than just a little "junk in the trunk." Based on their calculations of more realistic proportions of muscle mass in those areas that correlated to speed, they instead proposed that *T. rex* moved at speeds of about 10 to 25 mph.

So as a result of this thought experiment and that of the running-stumbling-dying *T. rex* or a huge-booty *T. rex*, paleontologists are not expecting to find tracks of a running *T. rex*, *Spinosaurus*, *Gigantosaurus*, or other massive theropod anytime soon. But of course we would be delighted to be proved wrong and would be the first to applaud anyone who discovered such a trackway.

Sitting Dinosaurs

Watch nearly any documentary film that uses CGI (computer-generated imagery) to recreate dinosaurs in their natural Mesozoic habitats and you will almost never see a dinosaur sitting, lying down, sleeping, or otherwise taking it easy. This is understandable on the part of the director and animators, because the attention span of viewers would decrease in inverse proportion to the length of such a segment and they would quickly switch the channel to watch their favorite reality-TV stars. (Coincidentally, these "stars" will be mostly sitting, lying down, sleeping, or otherwise taking it easy.) Yet dinosaurs must have slept, rested, or paused, however briefly, in their daily activities.

How can we know for sure dinosaurs took a breather in their lives? Surprisingly, the skeletal evidence is scant, although two examples are exquisite. Both of these skeletons belong to the same species of dinosaur, *Mei long* ("soundly sleeping dragon"). Each was found with its long tail wrapped around its body, looking very much like a sleeping duck or goose. Even more amazing, the two specimens are mirror images of each other, one with its head turned back between its left arm and torso, and the other with its head to the right side.

A few other dinosaur skeletons, such as those of the theropod *Citipati*, have been found preserved in sitting positions, which also might be construed as "resting." But these were positioned over nests with eggs, hence these dinosaurs were probably staying put to protect their eggs and died trying. As any expectant parent can tell you, though, taking care of your potential offspring should never be considered as "resting." For instance, some modern flightless birds such as emus stay seated over their eggs for 50 to 60 days.

I suppose, then, that we must once again resort to dinosaur trace fossils to learn more about how they rested. Sure enough, we do know of some so-called "resting" or "crouching" traces—made when a dinosaur sat down—although these are quite rare. As of 2013, only ten had been found in the world, with three in Massachusetts, three in Utah, two in China, one in Poland, and one in Italy. For some unknown reason, these trace fossils are time-restricted, as they are only preserved in Early Jurassic rocks. Of the three from Massachusetts, two are from medium-sized theropods and one from a small ornithopod, whereas all the rest are from medium-sized theropods. So far, we do not know of any resting traces made by quadrupedal dinosaurs such as sauropods, ankylosaurs, stegosaurs, or ceratopsians.

Identifying a dinosaur resting trace is relatively straightforward. First of all, look for a side-by-side pairing of the rear feet. Most bipedal dinosaur trackways show a diagonal-walking pattern, meaning that any pattern deviating from that catches our attention and is examined more carefully. Hence, where dinosaurs stopped, they would have pulled up their trailing leg next to their lead leg, meaning their footprints will be parallel and adjacent, or slightly offset. Upon sitting down, the dinosaur will have lowered the long parts of its legs just above its feet, which consist of its metatarsals (equivalent to our heels). This pressing of its metatarsals onto the ground would have given its tracks elongated extensions. Once seated, the dinosaur didn't stop moving and may have shifted its position as it settled in, causing multiple prints in a small area. Additional parts of the body might have made contact and left their marks too, such as the rear part of its anatomy—which is properly called an *ischial callosity*, and not the more appealing term "dinosaur butt"—as well as its tail and front-foot (hand) impressions. Of these, the ischial callosity is most likely to be preserved, although tail marks and hand imprints have been recorded in a few, too. Incidentally, dinosaur tail impressions are quite rare, with fewer than forty reported from the entire geologic record, and many of these are associated with resting traces.

The most recently discovered dinosaur-resting trace, and probably the best, is a spectacular one. Reported in 2009 in southwestern Utah, this Early Jurassic trace fossil not only shows where a theropod approached a sitting spot and sat down, but also got up and walked a ways afterwards. Just like how we would adjust our sitting position to a more comfortable one, this dinosaur shuffled forward twice after lowering itself to the ground, evidenced by repeated prints of the same feet. It also includes impressions of its metatarsals, ischial callosity, and two thin slices left by its tail.

An even more remarkable aspect of this sitting trace, though, is that the theropod put its hands down in front of it and left impressions of these. The traces showed the positions of the theropod's hands with its "palms" turned inward toward the center of the body, almost as if it were measuring the width of the trackway. For too many years, paleontologists have cringed at reconstructions of theropods walking around limp-wristed, palms down: a posture sometimes derisively labeled as "bunny hands." In fact, skeletal evidence indicates this was anatomically impossible, and that the hands must have been held with the palms turned inward, not downward. Thus these two handprints vindicated critics' previous assertions of theropod hand positions. This combined resting trace and trackway, along with hundreds of other dinosaur tracks, warranted enough importance to have a building constructed around them for protection (the St. George Dinosaur Discovery Center), ably providing public education about the tracks in St. George, Utah.

Nevertheless, as wonderful as this trace fossil might be, my favorite dinosaur-resting trace is one made by an Early Jurassic theropod in what we now call Massachusetts. On display at the Beneski Museum of Natural History at Amherst College, this specimen, designated specimen AC 1/7 by paleontologist Edward Hitchcock in the 1850s, is a near-perfect record of where a human-sized theropod sat down on a muddy lakeshore just a little less than 200 million years ago. Unlike the St. George example, this trace fossil is quite limited in its area, preserved in an isolated slab of rock about the size of a coffee table. At some point after its discovery in the 1850s, it was framed like a work of art (which it is). It has two slightly offset pairs of feet and rear "heel" (metatarsal) impressions, and between those, an oval, apple-sized impression from a svelte part of its rear end. The detail associated with these traces is incredible, accompanied by wrinkle structures formed as the theropod shifted its weight from one side to the other when sitting down and getting up.

In 2004, I studied this specimen intensively with a colleague, Emma Rainforth, in which we tried to figure out the sequence of movements made by the theropod that would have produced such a trace fossil. We were also testing an audacious idea that some of the wrinkle marks near the edge of the leg impressions were actually from feathers. This was an extraordinary claim at the time because feathered theropods, although then-recently discovered in Early Cretaceous rocks of China, were completely unknown from the Early Jurassic anywhere in the world. Yet other paleontologists who had examined the trace fossil just a few years before us concluded that the odd wrinkle marks were "feathers." The surface preserving the trace fossil also had little pockmarks, which had been interpreted as "raindrop impressions" imparted by a Jurassic shower.

We wanted these structures to be feathers, too; but in science, reality does not always live up to our wishes. Once we looked at the trace fossil more carefully, we realized that a thin algal film covered the original muddy (but firm) surface. This film acted like plastic wrap covering a dish: any pressure exerted laterally against it caused the film to deform and wrinkle. Debunking further, we also concluded that supposed "raindrop impressions" were more likely gas-bubble escape structures. These were made when the theropod stepped onto the surface and pressed its full body weight on the mud when it sat down; this in turn caused trapped gas in the mud from underneath to bubble up to the surface. In short, this specimen records a full sequence of movement by the theropod and how it altered the ground beneath it, recorded in exquisite detail because it was preserved under the right conditions.

As fascinating as these dinosaur-sitting traces might be, though, we are still puzzling over why they only seem to be in Early Jurassic rocks. Did dinosaurs just stay upright through the rest of the Jurassic and Cretaceous periods, too busy to sit down or otherwise take it easy? This scenario seems absurd, although it also poses a good question as to how some of the largest of dinosaurs, especially those with small arms, would have managed to both lie down and get up (I'm looking at you, *T. rex*). As many of us experience each morning, getting up is the hardest part following our resting. Still, the Middle and Late Jurassic, as well as the Cretaceous, abounded with small dinosaurs, too, which would have had no problem stopping and becoming supine. So perhaps it's only a matter of time before paleontologists start recognizing more such trace fossils that record when a dinosaur took the pause that refreshed.

Swimming Dinosaurs

Dinosaurs and water have had an odd back-and-forth relationship in our imaginations. At some point in the initial studies of sauropod and "duckbill" dinosaurs (hadrosaurs), paleontologists started wondering how such large animals kept themselves upright on land without also placing incredible stress on their muscles, bones, and joints. So all paleontologists needed was a little bit of suggestive evidence to nudge these big animals into the water, where their weights would have been supported through buoyancy.

For hadrosaurs, this evidence was scanty but persuasive for those who wanted these dinosaurs to be aquatic. For example, one hadrosaur trace fossil specimen had skin impressions around its hand that stretched between its fingers. This led paleontologists to conclude that this skin was webbing that aided it in paddling around in bodies of water. Only later did paleontologists realize this "webbing" was actually a result of skin drying around its bones after the dinosaur had died. Another hadrosaur, *Paralophosaurus*, also had a tall hollow crest on its skull, which was explained as a "snorkel" that allowed the dinosaur to breathe while most of its body was hidden underwater from predators. A major flaw in this seemingly marvelous adaptation was that the hollow tube in the center of the crest, once studied in more detail later, actually makes a U-turn which would have constituted a perfectly inept snorkel. (If you don't believe me, try making one like this and let me know how that worked out for you.) Yet another anatomical trait was an elongated snout that led to the nickname of "duck-billed dinosaurs" for hadrosaurs, which imagines them as favoring soft aquatic plants as food. Again, a reexamination of their teeth and jaws as well as their trace fossils (coprolites and microwear on their teeth, explained in a later chapter) revealed that hadrosaurs could eat all sorts of land plants. In short, just calling a hadrosaur "duck-billed" doesn't make it a duck.

This explanation of body fossil evidence favoring aquatic lifestyles for dinosaurs was even extended to dinosaur tracks. In 1938, paleontologist Roland Bird of the American Museum of Natural History learned that the area around Glen Rose, Texas, had lots of dinosaur tracks. Once he investigated, he confirmed the presence of exquisitely preserved three-toed theropod tracks, but also made an astonishing discovery: the first known sauropod dinosaur tracks from the geologic record. These huge tracks faithfully matched the size and anatomy of sauropod feet: five toes in the rear, and a rounded pad in the front. However, among these sauropod trackways were ones in which only the front feet registered. Why would the weightiest part of a sauropod—its rear end, with long tail—not connect with the sediment surface? Bird surmised that this was a result of a sauropod floating along, only touching the bottom with its front feet.

Later, a closer look at these tracks showed that the missing tracks in the sequence of steps could be attributed to differences in track preservation. If these sauropods had applied more pressure in the front while walking on land, these would have been more likely to be preserved as undertracks than the rear feet. Hence, Bird had not been looking at tracks from the original surface where sauropods placed their feet (or not), but more at the ghostly prints below. Once this alternative explanation caught hold, people realized that Bird was likely wrong about "swimming sauropods" at the Texas site.

Ironically, Bird's recognition of sauropod tracks in the first place led from an initial view of sauropods as aquatic dinosaurs that, with more such discoveries, shifted them onto the land. Once paleontologists had the right search images for sauropod trackways, they started finding them outside of Texas. In the U.S., sauropod tracks are also in Colorado, New Mexico, and Utah, as well as in Argentina, Australia, China, France, Korea, Mexico, Morocco, Switzerland, the United Kingdom, and Zimbabwe, among other places. These tracks are also in rocks from near the start of sauropods in the fossil record (Late Triassic) to their very end (Late Cretaceous). Something noteworthy about these sauropod trackways found thus far, though, is that nearly all show these massive animals walked on emergent surfaces, such as along coastlines, lakeshores, or river floodplains.

Still, paleontologists wondered: What if dinosaurs other than hadrosaurs or sauropods went for a swim? How would we know from looking at their bones? For example, even the most skilled anatomists would be hard pressed to demonstrate from an elephant's skeleton that these multi-ton animals are capable of swimming long distances. Yet Indian elephants (*Elephas maximus*) can swim as far as 25 miles (40 km), a feat far better than most humans are capable of. In fact, elephant swimming abilities show one of the probable ways mammoths dispersed to islands during the Pleistocene Epoch, where some isolated populations lasted until only about 4,000 years ago. (These elephants also became much smaller after generations of living on these islands, leading to the oxymoronic condition of becoming "dwarf mammoths." But that's another story.)

Just in case you were wondering whether trace fossils might come to the rescue again to solve this dinosaurian mystery, you would be right (again). First, as early as 1980, a paleontologist interpreted swim tracks from Early Jurassic rocks of Connecticut as made by theropods, and provided a fine argument as to how such dinosaurs would have made these tracks while partially buoyed by water. More than twenty years later, in 2001, paleontologists working in separate studies and places (Wyoming and the U.K.) interpreted Middle Jurassic tracks as possible dinosaur swim tracks. Soon after that (2006), hundreds of much better examples were discovered and documented by Andrew Milner in Early Jurassic rocks of southwestern Utah at and near the St. George site that also has the dinosaur-sitting traces mentioned earlier. The next year (2007), dinosaur swim tracks were again interpreted from long linear marks on an expansive surface of Early Cretaceous rock in Spain. In 2013, yet more dinosaur swim tracks were reported from another Early Cretaceous site in Queensland, Australia. Suddenly, dinosaurs seemed to be swimming everywhere.

How would you know whether a dinosaur was swimming by looking at its tracks? Well, for one thing, you wouldn't know it at all unless its feet touched the bottom of the water body it crossed. If the water were too deep, buoyancy would have kept dinosaur bodies—along with their feet—above any sediment surface that would have recorded their tracks. But through a combination of legs long enough to reach the bottom and water shallow enough to allow this, they would have made tracks.

Why should a dinosaur swim at all? Or as an actor might ask, what was their motivation? Getting from one place to another is a likely reason, instead of walking around a shallow lake or stream, or the old "to get to the other side" answer. Yet another argument relates to their attraction to aquatic environments as great sources of food. For theropods, this might have been fish, but other aquatic animals also might have served as tasty treats. For hadrosaurs and sauropods, though, which were (as far as we know) all herbivores, this is not such a good explanation. Not surprisingly, recreational purposes have never been suggested for swimming dinosaurs, but who knows whether an occasional dip might have also relieved any dinosaurs suffering from skin parasites or a hot day in the Mesozoic.

The Not-So-Secret Social Lives of Dinosaurs

Tracks also tell us about dinosaur social lives, and thanks to these trace fossils we are confident that many dinosaur species moved together as herds, packs, flocks, congresses, murders, or whatever group name seems appropriate. Assemblages of dinosaur bones composed of many individuals but only representing one species also support this idea, and we now take for granted that the stereotype of the "lone dinosaur" is not so likely in many species. Because of this combination of trackway and skeletal evidence, we now nonchalantly discuss the ecological effects of vast herds of sauropods, hadrosaurs, and ceratopsians, or the pack-hunting behavior of theropods, on Mesozoic ecosystems. "Strength in numbers" is a strategy used by many animals today, from schooling in fish to herding in caribou to pack hunting in wolves.

So let's say that paleontologists find a dinosaur tracksite with hundreds of tracks preserved in it. They can then test whether these dinosaur trackmakers moved together as a large group or not. This is based on whether the following questions receive an answer of "yes" or "no": (1) Do the tracks look alike? (2) If the tracks look alike, do they also show variations in sizes? (3) Are the tracks all heading in (more or less) the same direction? (4) Do the tracks show any other harmony of movement, such as staying parallel to or following one another? (5) Do the tracks seem to have been made at about the same time?

Here's what those questions are testing: The first—do the tracks look alike—is examining whether they belong to the same dinosaur species or not. If they do have the same basic form but also show a range of sizes (question 2) from small to large, then they also could represent growth stages of the same species, from babies to full-grown adults, and perhaps gender differences as well. The third question is key, then, as this sorts out whether the dinosaurs were truly moving together or not, and not just randomly milling about. The fourth question further clarifies the third, as it asks about more nuanced behavior such as whether dinosaurs in the group maintained a consistent "personal space" from one another or whether they were following leaders. Finally, the fifth question addresses whether these tracks were all made by a sizeable group of dinosaurs moving through the area, as opposed to, say, a dinosaur family consisting of just two adults and two juveniles that neurotically walked through the same spot every day for several weeks.

Given this idealized dinosaur tracksite in mind, do any fulfill all of the criteria? I think you suspect the answer to that, but let's look at a few examples anyway. The first recognized example of herding behavior shown by dinosaur tracks, and still one of the best, stems from a series of Early Cretaceous sauropod tracks at Davenport Ranch in Texas, found by Roland Bird in 1941. This site has trackways of more than twenty sauropods walking in the same direction and apparently made at about the same time. A Late Jurassic sauropod tracksite near La Junta, Colorado, also shows the tracks of five sauropods moving in the same direction, spaced at regular intervals, and turning in harmony along the length of their trackways. One time, when visiting this site with students, I asked them to walk alongside the tracks. By observing them this way, the lesson became much more visceral for these students as they could experience the subtle changes in movement made by the sauropods, rather than just gazing at them from afar or listening to me babble on about them.

Sauropod tracks, in fact, provide the best evidence that these dinosaurs herded. So far, sauropod trackways indicating group behaviors are documented from Jurassic–Cretaceous rocks of the

U.S. (Arkansas, Colorado, Texas, Utah), Brazil, Bolivia, China, Portugal, and the U.K. The tracksite in the U.K., preserved in Middle Jurassic (about 165 *mya*) rocks, was likely made by dozens of sauropods moving together, a truly awesome spectacle to imagine. Even better, some of the same sauropod tracksites show smaller tracks along with larger ones, suggesting that sauropods of different ages were traveling with one another, perhaps with multiple families.

For ornithopods, similar sorts of trackways also point toward their herding. These include several sites from the Cretaceous of Korea, one of which has tracks of about twenty large ornithopods heading in the same direction, and another from the Cretaceous of Canada, which had 10 to 12 ornithopods. A few ankylosaur tracksites have parallel trackways, suggesting that at least two ankylosaurs were traveling together at the same time. So far, stegosaur tracks are so rare we don't know if these animals were social or not. Ceratopsian tracks are also uncommon enough to withhold judgment on that aspect of their lives too, although rocks bearing hundreds of bones of the same ceratopsian species tell us these dinosaurs were likely group-oriented also.

Knowing that many mammalian herbivores today travel in sizeable groups, the preceding insights on the social lives of sauropods and ornithopods are probably not big surprises. But what about theropods? Do their tracks ever show that they hung out with one another, shared time in the same place, and hunted together like wolves or other social predators? Oh yes, and in some instances these trackways paint nightmarish scenarios for their intended prey.

At one site in Middle Jurassic (about 165 *mya*) rocks of Zimbabwe, trackways of at least five large theropods, with calculated hip heights of more than 2 m (6.6 ft), indicate they were traveling together. At this site, some of the theropods stepped on each others' tracks, further suggesting they had some sort of pack arrangement with one or more theropods taking the lead in a loose formation. An Early Cretaceous (about 125 *mya*) site in China also shows six theropod trackways, equally spaced and pointing in the same direction, all of their tracks about the same size, 24 to 28 cm (9–11 in) long, and with only two toes on each foot. Yes, that's right, these are dromaeosaurid tracks. The track sizes further indicate hip heights of about 1 to 1.1 m (about 3.5 ft), or slightly smaller than the fictional "*Velociraptors*" of the *Jurassic Park* movies. Yet these tracks reflect a chilling reality, one in which small or large prey would have been stalked and terrorized by sickle-clawed predators. Running dromaeosaur trackways—which I've seen at a spectacular tracksite in Utah—amplify this empathy, whether for the hunter or hunted.

Dinosaurs Who Stalk and Feed

Based on what you just learned about dinosaur pack-hunting behavior from tracks, along with other evidence, paleontologists have no doubt that some dinosaurs hunted other dinosaurs, birds, mammals, lizards, and additional animals. For example, gut contents in a few rare specimens tell us directly about a dinosaur's last meal (which will be detailed in a later chapter). But what happened just before that meal was acquired? How did predatory dinosaurs hunt their prey? And how do tracks provide some insights on dinosaur hunting behaviors? All of these are good questions, and I'll do my best to answer them through what dinosaur tracks tell us.

Probably the most famous and longest-known example of a possible "stalking theropod" trackway comes from near Glen Rose, Texas. This trackway, discovered by paleontologist Roland Bird in 1938, was in a limestone bed cropping out in the Paluxy River. He noticed the trackway in direct association with a sauropod trackway, and was thrilled to see how the theropod tracks at first paralleled and then intersected those of the sauropod; at this point, the theropod tracks ceased. Bird surmised that this was where the theropod leaped onto the left side of the sauropod to bring it down, just like a lion would with its prey.

"Wow, that sounds incredible!" you think. Yes, it does, which also means the story may not be so simple. For one thing, the sauropod tracks continue on past where the theropod tracks stopped, and show no alteration of gait or depth of tracks on the left side. One would think that the addition of a multi-ton predator on one side would cause some reaction, or at least a little bit of imbalance. So now the more reasonable explanation is that, yes, the theropod might have been stalking the sauropod but did not jump onto it there. Instead, the tracks just weren't preserved after the point where they disappear. So we don't really know whether the theropod was going after this sauropod or not. (By the way, if you want to see this trackway, don't go to Texas, unless you like looking at rectangular holes in riverbeds there. In 1940, Bird and many laborers extracted the trackway and took it to New York City, where it is now displayed in the American Museum of Natural History.)

Within a mile of this trackway, though, are elongated dinosaur tracks that may reflect evidence of stalking behavior, connecting to speculations mentioned previously that some theropods went fishing. Glen Kuban, a paleontologist who has studied and mapped the Paluxy River dinosaur tracks for more than twenty years, wondered whether a few of the theropod tracks there were a direct result of their stalking fish in shallow water, like modern grizzly bears seeking salmon. His reasoning was based on how a few trackways show where their makers went from normal digit impressions to those including metatarsal prints. This meant the dinosaurs were sinking more deeply into the mud, causing their footprints to become elongated.

Wait, metatarsal impressions? Where have we heard about those before? That's right, these are associated with trace fossils where theropods voluntarily squatted. So these elongated tracks could have been made either through the dinosaurs walking from spots with firm mud to spots with squishier mud (involuntary), or they could have occurred from these dinosaurs lowering their bodies while walking (voluntary). When fishing without a pole or net, the best way to catch fish is to get your tools closer to the water surface, and for a theropod that would have meant lowering its arms and mouth. Although this is a difficult hypothesis to test further, it is nice to have a more dashing alternative to one that simply states, "They waded into squishy mud and kept going."

Have paleontologists ever discovered theropod tracks connected to a kill site? Not yet, but some Late Jurassic theropod tracks and other circumstantial evidence of feeding are associated directly with sauropod bones. I was lucky enough to see these trace fossils during the summer of 2005 while passing through Thermopolis, Wyoming. My wife Ruth and I had stopped to see the Wyoming Dinosaur Center there, but also took a guided tour of a nearby dinosaur dig site. The site not only had an impressive array of sauropod bones exposed but also included (more significant) dinosaur tracks.

There, clustered around the skeleton of a juvenile *Camarasaurus*, were three-toed tracks. Researchers who worked on this site had also found sauropod tracks, showing they also lived in the area, but found ominous toothmarks on the bones along with shed teeth of juvenile and adult *Allosaurus*. The tracks were also about the right size and shape for this formidable predatory dinosaur. Additional trace fossils in the sauropod skeleton included large rounded stones. These were interpreted as gastroliths originally within the body cavity of the sauropod, exposed by the theropods as they chowed down on the *Camarasaurus*. As a result, the researchers concluded that this was a feeding ground, perhaps after a kill or—perhaps less exciting to most people—following some other cause of death for the sauropods, which brought in allosaurs to scavenge. More compelling, it conjures a scene of an adult *Allosaurus* and its offspring dining together on the remains of a young and then-recently departed sauropod.

Other Cool Stuff You Probably Didn't Know about Dinosaur Tracks

Because there are so many dinosaur tracks recorded in Late Triassic through Late Cretaceous rocks worldwide, the chances are good that a few unusual or otherwise stimulating insights might be conveyed by some of them. Sure enough, some dinosaur tracks tell amazing stories. The following is a short selection of what I think are provocative perspectives on dinosaurs bestowed by their tracks, and that we might not have ever figured out from bones alone.

Imagine you are a small theropod dinosaur, out on a walkabout one fine Jurassic day, on the prowl for food or mates, and not necessarily in that order. As you stride over some slightly bumpy terrain, you smell something interesting in a meter-wide mud-filled pit. As soon as you jump into the pit to investigate, your feet start sinking into the saturated mud and you suddenly realize you're stuck. Even worse, as you struggle to free yourself, you've further liquefied the mud and caused it to envelop you even more. At some point you just stop flailing because you can't move. You've just been buried alive.

During the next few days, several others of your species make the same mistake. A few make it out, but most others, including additional species of dinosaurs, likewise become stuck and die. Meanwhile, the sauropods that unwittingly made these traps are already miles away, having no idea that they contributed to the asphyxiation and fossilization of these small theropods. Their deep tracks, pushing down into the moist substrate, had created new hazards for smaller animals, which fell in and became mired in the squishy track interiors.

Oddly enough, the preceding scenario is exactly what three paleontologists proposed to explain how some exquisitely preserved small theropods in Middle Jurassic rocks of western China were unwillingly ushered into the fossil record. In a 2009 journal article with the beguiling title of "Dinosaur Death Pits from the Jurassic of China," the researchers—David Eberth, Xu Xing, and Julia Clark—proposed that specimens of the small theropod *Guanlong wucaii* had fallen into 1 to 2 m (3.3–6.6 ft) deep mud-filled pits made by large sauropods similar to the Late Jurassic *Mamenchisaurus*. *Guanlong* wasn't the only animal to suffer such a fate at the feet of these sauropods, either, as the theropod *Limusaurus* and about twenty other species were entombed in these tracks. What all of these dinosaurs had in common was that they were small and bipedal, meaning they lacked sufficient limb strength to yank themselves out of their morasses. More than 160 million years later, those dinosaurs' misfortunes became paleontologists' treasures, as this unusual mode of death and fossilization resulted in beautifully complete skeletons of scientifically significant species. All thanks to the sauropod tracks.

But let's say your timing was even worse, and you just happened to be in the same place where a dinosaur foot—with the rest of a living dinosaur attached to it—happened to land. If the dinosaur was *Microraptor* or some other similarly puny feathery theropod, then no big deal, unless you were a small insect. But if the foot belonged to a tyrannosaur, ceratopsian, ankylosaur, stegosaur, or sauropod, then that was a problem for any animal smaller than that foot. Sure enough, a few fossils, such as snails and clams, found in dinosaur tracks were crushed underfoot, although we can't tell for sure whether they were alive or already dead when stomped.

Dinosaur tracks can also sometimes tell us if a given dinosaur had any health problems. We know a fair number of dinosaur bones show evidence of having been broken and later healed, but it's difficult to say whether these had an effect on an animal getting around, either while healing or afterwards. Fortunately, dinosaur trackways are abundant enough that some with asymmetrical gait patterns pop out. In these trackways, one side of the body stepped less distance than the other, resulting in noticeably unequal strides. The easiest explanation for these imbalanced patterns is that these dinosaurs were compensating for a weak leg, such as one injured from an accident or combat.

One theropod trackway in a Late Jurassic stratum in Utah leaves no doubt that its maker had a tough time walking, showing alternating long and short steps that demonstrate how it was hobbling along with just one good leg. Because this trackway is relatively easy to access and on public land, I've taken university students to it, given them a tape measure, and said, "Measure this, then tell me what you think this dinosaur was doing," without saying anything more about it. One time there, within about fifteen minutes one of the students, after writing down a few measurements, looked up from the trackway with a huge grin on her face. "It was limping!" she shouted triumphantly. (It was one of those proud moments that, as a teacher, I'll never forget, and hopefully the students have not either.)

Once the students had all of the information they needed to further confirm and test this hypothesis, I then asked them, "Okay, which leg was injured?" It took a few more minutes for them to figure this out, but they did. The students reasoned that the strong (non-injured) leg of the theropod enabled it to take a longer step, whereas the hurt leg could not as easily bear the theropod's weight, so it would have taken a shorter step when pushing off with that leg. Using this logic, they figured the left leg was hurt, whereas the right leg was normal.

Other maladies show up in dinosaur tracks, such as missing or deformed digits. Paleontologists have to be careful when making such judgments, of course, because many, many dinosaur tracks did not faithfully preserve a record of every digit on the original feet, especially those made as undertracks. Nonetheless, once in a while we notice that a trackway consistently shows all expected toes on one foot, but one less on the other foot. One such trackway is from the Early Jurassic of Massachusetts, in which three tracks in sequence—right, left, right—have a perfectly fine three-toed theropod track on the left, but a two-toed one on the right, missing its innermost digit.

Expanding from a single toe to entire ecosystems, dinosaur tracks are also incredible tools for showing us exactly where dinosaurs lived, which seems to have been nearly every lowland environment and with occasional forays into water bodies. Their tracks are interpreted from geologic deposits formed in deserts, river and delta floodplains, swamps, lakeshores, and tropical shorelines, and from formerly equatorial regions to near the North and South Poles. Because of the preservation problems of mountainous areas—their rates of erosion are too high to preserve tracks very well—we don't know whether dinosaurs lived in alpine environments or not. Still, the nearly ubiquitous occurrence of their tracks elsewhere implies that they would not have been restricted from occupying higher-altitude places, either. Dinosaurs, according to their tracks, covered the landscape.

The Last Steps of the Dinosaurs

So what about at the other end of dinosaur history, when they all died out? Do their tracks help to answer the seemingly unanswerable? Well, sort of. As nearly every child can tell you, non-avian dinosaurs all went out with a cosmically delivered bang from a meteorite impact at the end of the Cretaceous Period and the start of the Paleogene Period, around 65 *mya*. This impact left an ash layer enriched with a rare element—iridium—that is only found abundantly in meteors. It also left lots of other evidence for its impact, not least of which is a huge crater of the right size and age in the Gulf of Mexico next to the Yucatan peninsula of Mexico.

To make a long story short, the effects of this impact likely caused what ecologists call a *trophic cascade*, in which the demise of each plant or animal that played a key role in an ecosystem led to further extinctions. Based on the impact scenario, paleoecologists figured that dust caused by both the impact and widespread fires blocked out most sunlight for several years, creating a Mesozoic version of a "nuclear winter." This effectively shut down photosynthesis in most land plants, thus depriving herbivores of their basic foodstuff. Nearly anything big that required lots of food to sustain its biomass, such as *Triceratops* or *Tyrannosaurus*, quickly went extinct. Yet smaller dinosaurs went with the bigger ones, too. For some reason, and we're still not sure why, birds were the only dinosaurs to survive this global disaster.

Seemingly every few years, though, paleontologists who intensively search the boundary between Cretaceous and Paleogene rocks end up uncovering a few bits and pieces of dinosaurs from above the boundary, suggesting dinosaurs were still living 63 to 64 *mya*. If these paleontologists publish a peer-reviewed article reporting such a find, the peer review does not end then and there. Post-Cretaceous dinosaur bones are always contentious, and in nearly every instance are explained as reworked older material that was deposited again, into a younger formation. In short, there's just no way to avoid an argument about the veracity of post-Cretaceous dinosaurs if your claim is based on bones.

If only we had some foolproof way of knowing whether any dinosaurs lived into the Paleogene Period. What evidence would convince even the most hardcore, pessimistic paleo-curmudgeon? You guessed it: dinosaur tracks. Tracks would show not only dinosaurs actively behaving, but also would be in the same place where dinosaurs walked. In contrast, a transported, out-of-place and hence out-of-time track is very easy to identify. Tracks found in situ might even be linked with a general clade of dinosaurs, such as non-avian theropods, hadrosaurs, or ceratopsians, whose bones are in strata just below.

So with this goal in mind, several places have what are among the last dinosaur tracks, coming from just below the Cretaceous–Paleogene boundary. One such site in Utah has hundreds of deeply impressed tracks preserved in strata at multiple levels, one of which is only a few meters below the boundary. Although the authors couldn't identify the specific dinosaurs that made these tracks, their huge sizes and numbers pointed toward herding herbivores, such as ceratopsians or hadrosaurs. A single—but huge—theropod track from New Mexico, attributed to *Tyrannosaurus rex*, also came from a layer just below the boundary. In Spain, abundant dinosaur tracks, some ascribed to hadrosaurs and sauropods, are preserved only a few meters below strata that preserved fossils of Paleogene fish and mammals. At one site, more than forty horizons contained hadrosaur and sauropod tracks below the boundary, and paleontologists estimated that some tracks were made only 300,000 years before the end of the Cretaceous. Alas, there were no more dinosaur tracks in the Paleogene rocks.

So these few examples show that dinosaurs were indeed alive just before the meteorite impact and mass extinction, which helps us to better clarify whether an overall dinosaur extinction was already happening before the impact or not. If someone does find a dinosaur track above the Cretaceous–Paleogene boundary, it would be a fantastic discovery, worthy of praises and raises. As of now, though, it just doesn't seem likely that any dinosaur walked for very long after the earth had its very bad day, the last of the Cretaceous. But if you think that is heartbreaking, imagine what it must have been like to be that last dinosaur: walking alone and seeking its own kind but never finding them, its tracks the only ones dotting a desolate and mostly silent landscape.

CHAPTER 3

The Mystery of Lark Quarry

The Case of the Mistaken Dinosaur

Like all stories in paleontology, that of Lark Quarry—a world-famous dinosaur tracksite in Queensland, Australia—has a vivid geologic history, beginning during the Cretaceous Period at about 98 million years ago. The cast of this tale consisted of more than a hundred dinosaurs, played by:

- One species of a large herbivorous dinosaur (an ornithopod), a mere bit actor, with a cameo appearance before the main act.
- Two species of small dinosaurs, one a theropod and the other an ornithopod, who composed most of the cast.
- The star of the show, a tyrannosaur-sized theropod who makes a grand entrance toward its climactic finish.

The story is filled with dramatic flourishes and tells of a tranquil scene shattered by brutal carnivory in which a quiet Cretaceous lakeshore became a killing ground. The former lakeshore, though, left no bodies, only tracks. Consequently, we must reconstruct what happened there by using ichnology, a little knowledge of modern animals and their behavior, and our imaginations. And in the spirit of imagination, I will tell the story as a stage play, titled *The Mystery of Lark Quarry*, and attended by an audience of enthusiastic Australians.

The curtain opens, revealing a subtropical lakeshore, and a large ornithopod, similar to the Australian dinosaur *Muttaburrasaurus langdoni*, emerges from a fog and lumbers onto the dimly lit stage. His arms are long and nearly touch the ground, but all of his walking is done with his rear legs. He stops briefly, bellows a low-frequency, didgeridoo-like moan, turns to the left, and exits, enveloped by the mist. Minutes slip by as the stage lights swell in imitation of a Cretaceous dawn. Insect-like buzzes and chirps intrude on the silence, accompanied by bird songs of the Mesozoic.

Soon thereafter, small two-legged theropods and ornithopods, cavorting and gamboling splendidly, enter from stage right in twos and threes. Each pays no heed to the presence of the other species, as neither poses a threat. They wander about and occasionally halt to sip water from the shoreline, but whether stopped or moving, they frequently pop their heads up, down, and around, ever alert to danger. The little theropods behave much like chickens in a . . . well, a chicken yard. The ornithopods, most about the size of modern emus of various ages, also remind the audience of these modern animals.

At some point the insects, which had reached a cacophonous peak through their blended mating calls, suddenly cease their racket and are quickly joined in silence by the birds. This hush puts the theropods and ornithopods on high alert, the Cretaceous equivalent of a fire alarm. They all look up and to the left simultaneously, where the quiet first descended and then spread like a wave. They begin to sniff the air, but delay expressing full-fledged panic until they actually feel the danger in their feet. There it is, detected by the audience as a Sensurround-like special effect, the *basso profundo* vibrations imparted by each footfall of an approaching multi-ton animal, coming from the left and offstage. Boom. Boom. Boom.

His imposing head appears first, with a mouth full of bladed teeth and nostrils snorting steam into the cool morning air. The rest of his massive body reveals itself with each halting step, his long tail held straight behind him, tiny arms dangling in front. The smell of rotten meat wafts across the stage and into the rest of the theater, a sample of this carnivore's bacterially fueled bad breath. After stopping once, he crosses his right foot over his left, stops again, crosses his left foot over his right, and then quickens his pace once a target is acquired. He is now stalking his prey.

The small dinosaurs react appropriately to such a sudden and rude interruption of their morning. Acting as one panicked mind, they run to the left as quickly as their short legs can go, straight toward, around, and then behind the marauder. Some of them step into and out of his tracks, leaving their own comparatively tiny footprints inside his. In the slippery mud, though, one ornithopod loses its balance and lands on its side, hapless against the assault, bleating like a sheep about to be sheared.

The theropod roars, pins its struggling prey underneath one 65-cm (26-in) long clawed foot and lowers its head to deliver a fatal chomp. The stage darkens, stirring theme music rises, the curtain closes, and the audience erupts into applause. Peaceful reverie shattered by gruesome death, in which the fleet of foot and the strong survive and the weak perish: what's not to like?

Then just as the applause begins to subside, the curtain opens again to reveal the star of the show, the villain/hero theropod. The audience claps madly, thinking this is the start of the curtain call. Only, their enthusiasm is abruptly curtailed when they realize the theropod is not bowing, but instead is looking directly at them with a mischievous smirk. This is *not* a tyrannosaur!

Their shock is doubled when the "theropod" pulls off an elastic mask, revealing that it was, all along, a bigger version of the ornithopod seen at the opening of the play. Within seconds, rotten vegetables and fruit, brought in by the otherwise finely attired and upscale crowd, shower the stage. The curtain closes, and a full-fledged riot engulfs the theater.

A year goes by, and although this displeasing conclusion to the play has not been entirely forgotten, it has been forgiven. Hence, the former audience is glad to see advertisements concerning a remake of the original play. So once again they fill the theater for the grand premiere, an excited buzz filling the air. Maybe it will be better this time, they all say with hopeful expectation.

The stage lights dim and the curtain parts. At first, the audience is reduced to squinting into the darkness, confused by strange reflections coming from the stage. The familiar insect and bird sounds from the previous production are present, but something is amiss. As the lights gradually brighten, again in imitation of a dawning day, realization strikes the audience. A large see-through glass container filled with water occupies the stage. The buzz turns to murmuring, followed by shocked silence as the *Muttaburrasaurus* emerges, splashes into the water, and swims from left to right. It is quickly followed by a hundred small ornithopods and theropods which, from the right side, dive one after another in a perfect domino-like effect, then paddle in unison to a blaring waltz-like number, their toes delicately touching the bottom of the pool. The play has been transformed into an Esther Williams-style aquamusical.

"Bloody hell! It's a pool!" someone shouts. "They're swimming!" someone else yells. "The music and choreography are dreadful!" screams a third. "Where are the sharks?" demands a fourth. Pandemonium descends, the play comes to an abrupt end, and police—who are waiting in the wings, expecting trouble—politely escort everyone out before the unrest escalates. Needless to say, the reviews are awful.

So with these dramatizations in mind, imagine the first part—large theropod causes a stampede among more than a hundred smaller dinosaurs—being retold thousands of times. Think of this story generating museum dioramas, artwork, a rollicking song, computer-animated recreations, dramatic scenes in mainstream movies, and a specially designed building with national-heritage status in Australia to help preserve the dinosaur tracksite that inspired it. Except . . . what happens if the last plot twist is inserted—large theropod was actually a peaceful herbivore—after having been told with a different ending for decades? Even worse, what to do with the remake, with its shocking aquatic take on the tale? Furthermore, what happens when the new stories are scrutinized and found lacking in details, further confusing matters for anyone who wants to know what really happened at Lark Quarry? Not that anything will ever be straightforward about a 98-million-year-old "crime scene," and especially one with no bodies left behind.

The History of the Mystery of Lark Quarry

Maybe a visit to Lark Quarry will help us to better understand the conflicting stories gleaned from the dinosaur tracks there. Lark Quarry Conservation Park—the full official name for the tracksite—is in central Queensland, about 110 km (68 mi) southwest of Winton, the nearest town of any size in that part of Queensland. The tracksite is protected from the elements of the Australian Outback by a beautiful, spacious, and environmentally friendly building constructed over and around the tracks in 2002. During my first visit in 2007, I wondered how comfortable I might be spending a night there alone with the dinosaur tracks. This unusual thought came to me upon realizing that I had been accidentally locked in, just after the last tour of the day had departed.

Before talking about that embarrassing episode, though, let's go back to justifying why these dinosaur tracks deserve such renown and reverence, including their own building in the proverbial "middle of nowhere" and National Heritage status in Australia. I remember first learning about Lark Quarry as a geology graduate student, soon after beginning my studies in ichnology while at the University of Georgia in the mid-1980s. Imagine how thrilling it was for a nascent ichnologist in Georgia to read about a site that held more than three thousand Cretaceous dinosaur tracks in faraway Queensland, Australia. I likewise dreamed of the ichnologically induced ecstasy that surely would result from a pilgrimage to see it in person. This aspiration became real for the first time in 2007 (that's when I was locked in) and was fulfilled twice more, once in 2010 with my wife Ruth (who successfully kept us from getting trapped) and in 2011 when I brought a group of university students there as part of a study-abroad program I taught in Queensland that year.

How was such a remarkable dinosaur tracksite discovered? As is still typical for fossil finds in many parts of the world, it was spotted by a sharp-eyed amateur, cattle station manager Glen Seymour, who lived and worked in the area. When Mr. Seymour first saw the tracks in 1962, only a few were visible on a rock surface sticking out of a hillside. I imagine the resemblance of these three-toed tracks to those of emu or brush-turkey tracks caught his eye, his perceptions honed by much time outdoors and looking at the ground. Indeed, he later said he considered them to be fossil bird tracks and was flabbergasted when Peter Knowles, a local photographer and naturalist, identified them as dinosaur tracks. I also can't help but think, though, that the indigenous people of this area and other parts of Australia, many of whom were expert trackers, probably also noticed these marks of long-gone animal life in the Cretaceous rocks, perhaps incorporating these trace fossils into Dreamtime stories.

Although this might seem odd to people outside of paleontology, nine years passed from the discovery of this world-class tracksite until paleontologists arrived to evaluate them. Nonetheless, this lag time reflected the reality of how long it took news to travel from rural Queensland to the "ivory towers" of academia, especially in the days before such academics could be contacted more easily by e-mail. The only hint of the trace fossils' presence was conveyed a few years after their discovery. In this instance, Ron McKenzie, a station hand of Mr. Seymour's, took natural casts of the tracks to the Queensland Museum and had them confirmed as dinosaur tracks.

Otherwise, nothing much in a scientific sort of way happened until 1971, when Mr. McKenzie and Mr. Knowles took paleontologists to the site: Richard T. Tedford of the American Museum of Natural History, Alan Bartholomai of the Queensland Museum, Patricia (Pat) Vickers-Rich of Monash University, and her husband Thomas (Tom) Rich of the Museum of Victoria. Nonetheless, despite the presence of so many prominent paleontologists in that part of Queensland, they weren't there to look at dinosaur tracks. Instead, they were prospecting for body fossils of Cretaceous mammals, whose remains are much smaller and rarer than dinosaur tracks. Nonetheless, the tracks, mentioned by Mr. Knowles in conversation with the paleontologists, made for an interesting non-mammalian fossil diversion, and they readily agreed to look at the site where the tracks had been discovered.

Once these paleontologists visited this place with Mr. Knowles, Dr. Tedford, through application of his well-practiced geological reasoning, traced the small outcrop of the track-bearing layer into an adjoining hillside. He figured that the stratum continued underneath the surface of the hill, which they tested by walking to the other side of the hill. Sure enough, the same track bed was there, just where Dr. Tedford predicted it would occur. Thus the entire track-bearing bed was likely more extensive than originally surmised, and would need to be excavated before it could be investigated in detail.

This is all well and good, but still does not tell how the tracksite gained its current nickname of "Lark Quarry." When paleontologists from the University of Queensland and Queensland Museum—Drs. Tony Thulborn and Mary Wade—decided to study the site in the mid-1970s, one of their most dedicated volunteers was Malcolm Lark, a local resident. Accordingly, they named the site in his honor. The "quarry" part of its moniker came from what these paleontologists and volunteers did during the field seasons of 1976 and 1977. In order to expose and see more of the track-bearing surface, they had to strip off a 30-cm (12-in) thick sandstone bed overlying the tracks. By the time they finished, these industrious folks estimated that they'd moved more than 50 tons of rock, which revealed an area of 210 m^2 (2,250 ft^2) holding more than 3,300 dinosaur tracks. This was a hard-earned mother lode of dinosaur trace fossils.

You might think that such a concentration of dinosaur tracks—on average about 15 tracks per square meter—would be enough in itself to guarantee everlasting paleontological fame. But it got even better. Once the surface was cleared of its overburden and its tracks could be mapped and described in detail, a remarkable story came to life. And storytelling is what ichnology is all about, and why it is probably among the oldest sciences known to humankind. Moreover, science ideally should surprise us, and these tracks certainly astonished the original investigators, along with succeeding generations of paleontologists and the public afterwards.

Based on Thulborn and Wade's analysis of the site, the area was a lakeshore that had been submerged regularly by a nearby stream emptying into it. Fluctuations between submergence and emergence of the shoreline could be discerned from alternating sand and mud layers in the rock, as well as long parallel grooves caused by logs that dragged across the lake bottom during flooding of the stream and lake. Once emergent, this shoreline would have served as a water source for local animals, and its moist, silty, and sandy sediments may have registered tracks over the course of several days or weeks. The tracks would have dried slightly underneath the warmth of the Cretaceous sun, but not so long as to bake them and cause them to crack around their edges. A minor flood from the lake or stream later covered the tracks with a protective layer of sand, which, along with the track-bearing layer, hardened and became part of the geologic record.

At the time of their excavation, the forms and sizes of the tracks pointed toward four species of bipedal dinosaurs as the trackmakers: a medium-sized ornithopod, about the same mass as a hefty heifer; a smaller ornithopod, close to the size of a modern emu; a much smaller theropod, only slightly bigger than a chicken; and the source of dramatic tension in the story, a large *Allosaurus*-scale carnivorous theropod. Most of the tracks on the surface were from the small theropods and ornithopods, and they were not lolling about in idyllic harmony at the lakeshore. Instead, these dinosaurs were running flat-out, clearly motivated by something that caused all of them to move in the same direction at high speed.

Drs. Thulborn and Wade came up with this scenario once they measured the directions of the individual trackways and distances between tracks within each trackway. Using the formula devised by R. McNeil Alexander for calculating dinosaur speed based on track length, hip height, and stride, they figured these dinosaurs were moving at about 12 to 16 kph (7.5–10 mph), which is quite fast for short-legged dinosaurs. Furthermore, the tracks mostly showed toe-tips touching the ground instead of entire digits. Lastly, some tracks were greatly elongated, indicating that slipping and sliding happened during their journeys. Very simply, when small animals take big strides, get up on their toes, and lose their footing on moist ground, they are running. To find out just how many dinosaurs were involved in this collective action, they walked a line perpendicular to the direction of movement and counted a minimum of 130 animals; about 55% of these were theropods and 45% ornithopods.

So instead of my recounting more of their evidence, I will let words from Drs. Thulborn and Wade's first research paper about the site, published in 1979, speak for the feelings these tracks evoked. In reading this, note how they broke an unspoken rule in paleontology, in that they expressed an emotional empathy with animals from nearly a hundred million years ago:

> *Persuasive circumstantial evidence leads us to conclude that they represent a stampede—that is, a wild, unreasoning and panic-stricken rush to escape the threat of danger. What could have caused such presumed panic?*

It was a great question, and in their attempt to answer it Thulborn and Wade pointed to the large three-toed tracks that entered the scene from the left. These tracks originated from the same direction taken by more than a hundred small dinosaurs, some of which ran around and onto the tracks of the big dinosaur, interpreted by the paleontologists as those of a big theropod. How big? As mentioned before, dinosaur sizes can be estimated by their tracks: just multiply the length of the footprint by 4.0, and you have the approximate hip height of the dinosaur. In this instance, the best-preserved track was 64 centimeters (25 inches) long, so its hip was about 2.5 meters (8.3 feet) off the ground. Just to put this in perspective, this is higher than the tallest basketball player in the NBA, and would have been big enough to make you and me run, too. So imagine the fright felt by a much smaller ornithopod or theropod from the approach of such an imposing predator.

Compounding this effect would have been the contagious fear spreading instantly through a sizeable group of small dinosaurs, instigating a chain reaction of similar behaviors. Think of the arrival of a fox in a chicken yard, or even a human walking up to a group of shorebirds, and how the jittery reaction of one bird is enough to spook the others, thus causing all to share the terror.

Bolstering this idea of a panic-inducing theropod was its trackway pattern, which was unusual. Based on their measurements, Thulborn and Wade figured it was moving slowly, at only 8 kph (5 mph), and because it crossed its right foot over its line of travel, then its left foot, it may have even paused at some points before picking up speed and turning sharply to the right. Frustratingly, this is also the point where the trackway reaches its end, forever eroded. This odd trackway pattern, expressed by only eleven footprints, led Thulborn and Wade to propose that the theropod might have been stalking, looking for prey in a target-rich environment.

Thulborn and Wade's map of the tracksite and other analyses of the tracksite constituted a masterpiece of meticulousness, supplying a wealth of data to support their interpretations. From their work they published two peer-reviewed papers, the first with the provocative title of "Dinosaur Stampede in the Cretaceous of Queensland," published in 1979. The second, published in 1984, was a much longer and more detailed report modestly titled "Dinosaur Trackways in the Winton Formation (Mid-Cretaceous) of Queensland."

I still point to the 1979 paper as one of the most compelling I have ever read in ichnology and often re-read it to remind myself of what constitutes a "gold standard" in the traditional study of dinosaur tracks. I also keep in mind that Thulborn and Wade completed their study of 3,300 tracks without the benefit of tools we now take for granted: no high-resolution digital cameras, digital calipers, global-positioning systems (GPS), geographic information systems (GIS), image analysis, laser scanners, 3-D modeling, Internet, or other technological shortcuts that would have made such a study far easier. Just heaps of hard physical labor, lots of surveying and other measurements, and scientific reasoning, served with a healthy dollop of intuition on top. And the end result was a tale that still astonishes.

This story of a dinosaur stampede in Queensland was so intriguing that it is rumored to have inspired one of the more spectacular scenes in the movie *Jurassic Park* (the first one, that is, not its awful sequels). In this scene, a flock of the Late Cretaceous theropod dinosaur *Gallimimus* ("ostrich mimic") stampedes in fear at the nearby presence of *Tyrannosaurus rex*, the most redemptive character of the movie, and who was also from the Late Cretaceous. (The casting of both dinosaurs, along with *Triceratops*, *Paralophosaurus*, and *Velociraptor*, also illustrates why *Jurassic Park* should have been titled *Cretaceous Park* instead.) Unfortunately for the prey but fortunately for the predator, one member of the flock is separated from its compatriots and singled out for sacrifice, while the others continue running for safety.

Although perhaps more than a hundred million people have watched this scene in the movie, I'll bet a six-pack of my favorite adult beverage that less than 1% of these people know how it mirrors the initial scientific interpretation of the Lark Quarry tracksite. Of course, locals in Winton know about this piece of fiction imitating fact, and I have heard them joke about how nice it would have been for the Lark Quarry Conservation Centre to receive a tiny royalty from the film profits as a sort of honorary tithe. Regardless, I have been pleased to contribute to the local

Winton economy in a non-Spielbergian way by visiting the tracksite, bringing others there with me, staying in Winton, purchasing pub meals, singing *Waltzing Matilda*, and of course quenching my thirst with Queensland-produced beers, all ably demonstrating some of the touristic and economic impact of dinosaur tracks, or what I like to call "ichnotourism."

What Made the Three-Toed Dinosaur Track?

In paleontology, just like any other science, a scientific hypothesis might rule for several decades or more, enter the public realm, and become part of popular lore. Nonetheless, because the science behind such stories is always changing, this means that what we took for granted as a "true story," even one we like very much, can be upended in a way that surprises everyone, including the paleontologists revising its narrative. Such is the situation with the "giant stalking theropod dinosaur causing a small dinosaur stampede" story of the Lark Quarry dinosaur tracksite in Queensland.

This tale, which has reigned for more than thirty years and is known worldwide by paleontologists and laypeople alike, faced a radical makeover in the light of new evidence presented in a paper published by paleontologists Anthony Romilio and Steven Salisbury in 2011. The fresh hypothesis states that the "dinosaur stampede" was not triggered by the arrival of a predator, and no stalking of other dinosaurs by a voracious predator happened either. Yet this new evidence and its accompanying explanation has not been completely accepted, and in fact is facing a fierce challenge from one of the original researchers, Dr. Thulborn. In other words, don't grab an eraser just yet, as the controversy may not be resolved.

The story of Lark Quarry and its dinosaur tracks illustrates very well a constant feature of science, which is that it is always evolving. This basic principle certainly has been realized in a grand way with evolutionary theory, gravitational theory, plate tectonic theory, and quantum theory. Lots of testing, scrutinizing, and arguing took place before the majority of the worldwide scientific community accepted any of these broad-sweeping, unified explanations for much of how our world works. But these theories also underwent considerable changes—an ongoing process.

The same standard applies to more focused hypotheses of all shapes and sizes, including those concerning dinosaur tracks and other trace fossils, which is that new evidence or perspectives can modify these, making for better fits. Along those lines, in science we do not prove, we disprove, meaning that old stories are sometimes revised in the face of novel finds. Furthermore, answers invariably generate more questions, meaning that although we may get ever so much closer to the truth with each investigation, we also keep in mind that we may never quite reach it absolutely, especially when we deal with events of the distant geologic past.

The overturning of long-held paleontological hypotheses by trace fossils is not uncommon, either. As mentioned previously, fossil tracks reported from Poland dating to the Early and Middle Triassic Period (about 245 *mya*) indicated that the immediate ancestors to dinosaurs were extant earlier than previously thought. In my own research in 2009, I proposed the oldest known dinosaur burrows in Cretaceous rocks of Victoria, Australia, one of which was eerily similar to the first known dinosaur burrow, reported from Cretaceous rocks of Montana (USA) in 2007 (explained more in the next chapter). In the non-dinosaur realm, the oldest tracks attributed to four-legged animals from 395 *mya* precede their body fossil record by about ten million years. In short, trace fossils matter in paleontology, whether for supporting new hypotheses or knocking down old ones.

So just what constitutes the evidence and scientific basis for a dramatic re-write in the story of the Lark Quarry dinosaur tracksite? Also, how could this new interpretation be wrong, exemplifying the fair warning that in science, just because something is newer doesn't mean it's also better? The thought of an ornithopod impersonating a large theropod—the dinosaurian equivalent of it wearing clown shoes (or, more aptly, "Bigfoot" shoes)—is not only a shocker for anyone who has grown up hearing the tale of Lark Quarry, but it borders on heresy. In science, though, we rather enjoy slaughtering sacred cows, making burgers out of them, adding a couple of slices of bacon, and putting a fried egg on top. After all, hypotheses are only accepted conditionally, and then are subject to further testing so we can find out whether or not they still hold up to scrutiny.

Sometimes these hypotheses continue to stand (so far, so good) but more than a few get modified or knocked down completely. And if they get knocked down, it's often because someone found data that better supports an alternative hypothesis, or "another story." Granted, "recycling" also happens sometimes, especially with scientists who get a little too attached to a pet hypothesis, perhaps long after it's been pronounced dead by everyone else (hey, egos happen). But for the most part, consensus is based on the evidence—not the people, their degree of self-promotion, or the volume of their message.

So just what was the evidence at Lark Quarry supporting the previous hypothesis (which can be summarized as "big theropod maybe caused a panic because it was preparing to kill and eat an ornithopod"), and how does this contrast with the evidence supporting the new hypothesis ("big ornithopod maybe caused a panic, but for different reasons than eating another dinosaur")? It all comes down to a common dilemma in dinosaur ichnology, and one that has been around for more than two hundred years, which is how to distinguish three-toed dinosaur tracks from one another and interpret who made the tracks.

So let's look at how the continuing dilemma of three-toed dinosaur tracks figures in all of this. The 2011 study of the Lark Quarry dinosaur tracksite by Romilio and Salisbury involved: a close look at the dinosaur tracks as they are preserved today at Lark Quarry; studying casts that were made of the dinosaur tracks soon after they were excavated in the 1970s; and lots of statistics, which I will do my best to explain to any non-scientific (yet admirably geeky) readers who might need it. Still, the heart of Romilio and Salisbury's 2011 interpretation was a focused reexamination of the large dinosaur tracks. Although these large tracks dwarf the others at Lark Quarry, as mentioned before, they are relatively few in number, with only eleven such footprints recorded on a surface that contains more than 3,300 tracks. That's right: The key plot element of the original story of Lark Quarry hinges on a sequence of only eleven tracks, and identifying what made those tracks.

Why identifying the maker of these big tracks is so difficult is mostly attributable to the tracks only having three toes, a trait also known as *tridactyl*. These tracks are also more or less mirror images on either side of the middle toe, a condition called *mesaxonic*. Furthermore, because bipedal dinosaurs made such tracks, their trackways normally show an alternating right–left–right diagonal walking. (Incidentally, now that you know these nifty terms, be sure to incorporate them in your daily conversations, such as "Wow, your chicken leaves some of the best tridactyl mesaxonic tracks in a bipedal trackway I've ever seen!")

This little checklist helps to narrow down the possible dinosaur trackmakers, in that we know it is definitely not a stegosaur, ankylosaur, sauropod, or ceratopsian, all of which walked on four legs (quadrupedally) and had feet with more than three toes. Well, except for stegosaurs which had tridactyl rear feet, but as far as we know stegosaurs never walked bipedally, so they're eliminated as suspects too. This means you should be thinking "ornithopod" or "theropod" for nearly all three-toed fossil tracks in rocks from dinosaur times. But also keep in mind how some ornithopods also left small front-foot impressions, a result of *sometimes* walking quadrupedally. Theropods, in contrast, almost always moved on just two legs, although, as you learned earlier, a few rare instances of hand impressions show up in trace fossils made where they stopped briefly to sit.

So now let's say you found some three-toed dinosaur tracks and you want to figure out whether these are from a theropod or an ornithopod. The most basic way to tell the difference is to measure the track length and width, then compare the two. On average, theropod tracks are longer than they are wide, whereas ornithopod tracks are wider than they are long. This means a length:width ratio for a theropod track will be >1.0, and <1.0 for an ornithopod. The tracks you've examined have ratios of 0.8 to 0.9, so these *must* be from ornithopods! Right?

Oh, if only ichnology were so simple, where we could all be so satisfied with our results, teeming and preening with confidence. Okay, time for an exercise in humility. Let's go through a few questions and see how you do in answering them:

- How would you describe the toes relative to the overall length of the track: thin, medium, or fat?
- Did those toes end with sharp clawmarks or blunt ones?
- What did the "heel" (posterior of the track) look like?
- Did you take into account how the substrate preserving a track, which might have been wet mud, dry sand, or moist muddy sand, might have affected the overall outline of a track?
- Did you think about how the dinosaur stopping suddenly, turning to the right, or moving its head might have distorted the track outline?
- Did the dinosaur, when extracting its foot from the mud or sand, cause sediment to collapse into the toe impressions and thus change the character of the track?
- Did you also think of how the substrate drying out might have changed a track outline before it was fossilized?
- Are these tracks undertracks, and if so, how far below the original surface were they formed?
- Do you feel clueless yet? (Welcome to my world.)

Along these lines, three Spanish paleontologists—Josè Moratalla, Josè Sanz, and Santiago Jimenez—tried to take some of this guesswork out of distinguishing theropod and ornithopod tracks. In an article published in 1988, they used a sample of 66 Early Cretaceous tridactyl dinosaur tracks from Spain, all of which had been identified confidently as either ornithopod or theropod tracks on the basis of their qualities. With these tracks, they measured nearly every parameter they could imagine: digit lengths, digit widths, angles between digits, widths of the foot between digits, and more. Moratalla and his coauthors then compared ratios of these parameters—such as digit length:digit width—to see which ones were significantly different from one another (statistically speaking).

From these analyses, they figured out "threshold" values and probabilities for some of the ratios and calculated probabilities of a ratio belonging to an ornithopod or theropod. For instance, if the length:width ratio of a tridactyl track is above 1.25, or 25% longer than it is wide, then there was an 80% probability that the track belonged to a theropod. Fortunately, they didn't just stop with the length:width ratio. They also checked all other ratios to see whether these consistently show a high probability of a theropod trackmaker or not, just to retest their initial identification.

With the publication of this study in 1988, dinosaur ichnologists had a quantitative checklist they could apply to three-toed dinosaur tracks. Of course, whether all dinosaur ichnologists actually read this paper, applied its methods, or tested their applicability to dinosaur tracks other than the ones they studied is another matter. Regardless, numbers produced by such a study also could be combined with non-numerical observations to test whether or not a fearsome carnivore or a peaceful herbivore had made a given series of three-toed dinosaur tracks. One example of such an observation is whether a track has sharp clawmarks or not. This feature is present in theropod tracks, whereas ornithopods tend to have more rounded or blunt ends to their toes.

Just to make a long number-laden story a little shorter, Romilio and Salisbury calculated threshold values based on every parameter they could measure from the eleven big "theropod" tracks. Once these numbers were compared to those that Moratalla and his colleagues figured for their Spanish dinosaur tracks, the numbers fell into the "ornithopod" range rather than "theropod." Accordingly, Romilio and Salisbury then concluded that the original trackmaker had been wrongly identified as a theropod and proposed that it must have been a big ornithopod.

To further back up their claim, they pointed toward a possible perpetrator, *Muttaburrasaurus langdoni*. Skeletal remains of this ornithopod were discovered in 1963 near the small town of Muttaburra in central Queensland, northeast of Lark Quarry, and like many dinosaur finds in this part of Australia a rancher, Doug Langdon, spotted its bones. Still, it wasn't named officially until 1981, when the paleontologists took care to honor Mr. Langdon's discovery by naming the species after him. Based on its body parts, *Muttaburrasaurus* was a big dinosaur. Its femur alone was about a meter (3.3 ft) long, which meant its hip height would have been close to the calculated value of 2.5 meters for the Lark Quarry trackmaker. Hence, it seemed a perfectly reasonable dinosaur to pick as a possible large three-toed trackmaker for Early to mid-Cretaceous rocks in this part of Australia.

In contrast, bones of only one large predatory theropod have been found in the same area and rocks of similar age: *Australovenator wintonensis*, lovingly nicknamed "Banjo" after Australian poet and writer of "Waltzing Matilda," Andrew Barton "Banjo" Paterson. Despite the unabashed wickedness of this dinosaur—gorgeous big-clawed hands attached to long grasping arms, a mouth full of nasty pointed teeth, and a lightweight frame which suggested that it was a speedy predator—this theropod seemed a bit too small to match the trackmaker at Lark Quarry: more *Allosaurus* and less *Tyrannosaurus*. Otherwise, body parts of tyrannosaur-sized theropods are unknown in Queensland, but also in much of Australia with the exception of a few isolated bones in Victoria. Nearly all that is known about big Cretaceous theropods of this entire continent comes from their tracks, which Thulborn and others described from Western Australia in the 1990s, and a few that I and others found in Victoria in 2007–2008.

When Australian paleontologists Tony Thulborn and Mary Wade proposed the "dinosaur stampede provoked by a large stalking theropod" idea in 1979, this was nine years before the publication of Moratalla and his colleagues' study and thirty years before the world learned about *Australovenator*. So Thulborn and Wade can't be faulted for what they didn't yet know, and they were using the best science known then to interpret the tracks. Also, like I said before (and it bears repeating), they did fantastic work. I consider their 1979 paper a classic in dinosaur ichnology, which they also followed up in 1984 with a considerably more detailed 105-page report.

Nonetheless, it's certainly possible they made a mistake in identifying the maker of the large three-toed tracks. At the time, they tentatively identified these on the basis of their resemblance to large theropod tracks known elsewhere, which had been allied with tyrannosaur-like tracemakers. Interestingly, objections to this identification, voiced formally by Romilio and Salisbury, were not new, as a few dinosaur paleontologists questioned it soon after Thulborn and Wade's second article came out in 1984. Romilio and Salisbury were just the first to rigorously test the original hypothesis using statistical methods as opposed to just the scientific equivalent of name-calling.

Backtracking the Walking Dead

So the hypothesis is dead; long live the new hypothesis! This means the popular story of Lark Quarry needs to be revised, and the tour guides there must alter their spiels or at least present "both sides" of the argument. Furthermore, the *Gallimimus*-panicked-by-*Tyrannosaurus* scene in the movie *Jurassic Park* needs to be redone with an alternate version, showing the *Gallimimus* running for some other, unspecified reason. And, without a doubt, a stage play is absolutely necessary: to do anything less would be irresponsible. (I say the last facetiously, but a small troupe of creative folks actually did write and perform a short musical number retelling the original story of the dinosaur stampede, shown at the 2012 Museum Australia National Association meeting in Adelaide. Could a Broadway production be far behind, but now with the anticlimactic ending inserted?)

Not so fast. There's one little problem with such simple and definitive declarations. What if Romilio and Salisbury were also wrong? What if they had neglected a few clues in their study, or carried out the research in a less-than-careful way? For instance, we've seemingly forgotten all about one of the cast of characters in this drama, which was the medium-sized ornithopod that first left tracks at the site, before the arrival of the terrorized flock and the big . . . well, whatever. Were its tracks studied in the same way, to test whether it was also an ornithopod or not? Also, how about revisiting *Australovenator* as a possible trackmaker?

First of all, the tracks made by the medium-sized ornithopod at Lark Quarry present what scientists might call a "control" at this site. In other words, these could be compared to the larger three-toed tracks to see whether or not they match in form. If they did match, this would have strengthened Romilio and Salisbury's assertion that the statistical technique they used could be applied to three-toed tracks of differing sizes, and that they could distinguish more than just the one trackmaker at the tracksite. Alas, they did not do this, or perhaps they did but it was not mentioned in their published paper.

How about *Australovenator* as the new antagonist (or protagonist, depending on perspective) for the Lark Quarry tracks? This isn't a completely crazy question to ask, because "Banjo" is currently the largest known theropod in that part of the world, and its bones come from rocks of about the same age as the tracksite. However, its feet, which are known from well-preserved foot bones, seem too small to fit the 60+ cm tracks, a sort of Cretaceous Cinderella story in which the slipper (footprint) does not fit.

Unfortunately, a documentary film trumpeted the conclusion that *Australovenator*'s feet did fit the tracks, and before yet another scientific investigation had been completed. Imaginatively titled *Dinosaur Stampede*, the ABC (Australian Broadcasting Corporation) broadcast this made-for-TV film in 2011, adding further confusion and controversy to public perception of a fossil site that didn't need more of either. This film, which used a combination of slick computer-generated graphics, dramatic music, lots of footage of paleontologists reexamining the tracksite with new technologies (lasers and computers and 3-D models—oh my!), and a fully fleshed-out recreation of an *Australovenator* leg and foot, culminated in one main conclusion: Yes, "Banjo" could have indeed been the culprit for the large tracks. Now there is a third possible solution to the mystery of Lark Quarry, and one that honors the original story but with a slightly modernized and updated perspective.

But here's the problem. Before allowing preliminary results out into the public realm, we scientists are supposed to do peer review. This means not just doing the initial research, but also scrutinizing the results, summarizing these in a formal article, submitting the article to a journal, having that article scrutinized by experts in the topic of your article, and in the end possibly being told, um, no, you got this all wrong, do it over. Is peer review a perfect process? No, and if for no other reason, because the scientists who review journal articles are doing it as unpaid volunteers, finding time to perform this important duty in between all of their other tasks such as teaching, grading, research, walking the dog, or (most heinous of all) sleeping. Still, more often than not, peer review provides an outside perspective that ultimately results in scientists honing their argument or even rejecting it outright. In the latter case, the research may never make it into print. All of that work for naught, so to speak.

Because of this possibility of being wrong, paleontologists and other professional scientists are generally discouraged from announcing their preliminary results to the public, and especially through popular media. An exception would be if these scientists deliver a talk or present a poster at a professional meeting in front of their peers, where an accelerated and more direct in-your-face form of peer review can take place. But to have a documentary film made about your research before the actual paper is published? This cheeky behavior often results in raised eyebrows and much harrumphing, mostly prompted by honest skepticism but also perhaps a wee bit of jealousy.

So as far as most paleontologists are concerned, a documentary, no matter how entertaining and informative, is never on equal footing with a peer-reviewed paper. Now, what if the research results were reported in a reality-TV series in which all of the competing paleontologists are forced to live together in the same house and argue over their respective results? I'd watch that for sure, but it still wouldn't count as real science either. In the end, we have to write it down, have it reviewed, and go through the formal process.

However, once an article is accepted by a journal and published, this triumph does not imply that its methods and conclusions are inviolable. All published articles are subject to further peer review long afterwards, or at least they should be. Indeed, this is exactly what happened to Romilio and Salisbury. Just like a boomerang, a major challenge to the revised hypothesis came from an unexpected source, which was Dr. Thulborn himself. Sadly, Thulborn's colleague on the original two milestone studies, Dr. Mary Wade, passed away in 2005, meaning her original impressions and insights on the tracksite died with her. Thus it was up to Thulborn alone to respond to the critique of their work.

Respond he did, and with ferocious intensity. Most devastating of all, his reaction to the restudy of the tracksite was in print and published as a peer-reviewed journal article and not aired as a scientific grievance through the more instant media of news stories or blogs. In this manuscript, which was posted online in the same journal (*Cretaceous Research*) that published the re-study, Thulborn pointed out several significant flaws in the rebooted version, effectively sowing doubt on its new conclusions. A minor tussle resulted, magnified by online discussions, thus provoking much head-shaking and sighing amongst other paleontologists and dinosaur-track enthusiasts who wondered how this would end.

One potential source of error related by Thulborn was that the 1988 study on dinosaur tracks done by the Spanish paleontologists was performed on a mixture of only sixty-six ornithopod and theropod tracks, and only from one formation in Spain. In other words, it was likely never intended as a universally applicable, one-size-fits-all method that could be used on dinosaur tracks of varying geologic ages anywhere else in the world. A second possible flaw is that the tracksite, once uncovered in the 1970s, had been outdoors for several decades before having the protection of a building around it. (A roof covered the tracksite for a while, but this shady spot in the middle of the desert attracted kangaroos, which found it a most excellent spot to sleep and defecate. Trust me, this is not good for dinosaur tracks.) Their exposure meant that the tracks were weathered and hence altered since the original measurements were taken on them, decreasing the accuracy of any new numbers derived from them.

A third problem was that some of the "new" measurements must have been taken from Thulborn's original line drawings rather than directly from the tracks. This sort of extrapolation again results in diminished accuracy. A fourth problem—and in my mind, this is the biggest one—is that the foot structure of *Muttaburrasaurus*, the alternative trackmaker, is based on just five bones: three metatarsals (heel bones) and two phalanges (toe bones). Yes, that's right: We don't know the actual size of its foot or other details of its anatomy. It would be like saying a suspect in a crime was identified by his tracks, but the suspect only has two toes on one foot and lost the other foot in an unfortunate accident. A fifth (but not final) problem was an oversight, which was that a set of perfectly fine but smaller ornithopod tracks was only a few meters away at the same tracksite (remember the first actor from the stage play?). Yet these were neither studied nor compared to the large tracks to see whether they matched in their dimensions. In short, the re-study had been re-studied, but this iteration put everything back to where we started, with the original story of Lark Quarry mostly intact.

One problem, though, was that Thulborn's critique was acerbically worded and pointedly critical of the new results, including the inflammatory word "fabricated," implying that some of the data may have been invented. Consequently, Romilio and Salisbury complained vehemently to the journal, and as a result an editor or employee of the publishing company that owned the journal decided the article was too hot to keep in print. Hence, it did what very few academic journals do, which was pull the article offline.

A few people (including me) were lucky enough to have downloaded it before it vanished from the Web, which felt vaguely like owning a bootleg Bob Dylan album. (Although, based on the tone of the article, Iron Maiden might be a better analogy.) Suddenly, the former existence of the article or people possessing a copy of it did not matter, as no one can cite it in their own research. Its abrupt absence from the journal was thus the academic equivalent of "disappearing" Thulborn's work and all it portended.

Nevertheless, this was not the end of this scientific squabble. Thulborn submitted a revised version of his article to the Australian paleontology journal *Alcheringa*. It went through peer review again and was published in early 2013. (Full disclosure: I was one of the reviewers of this paper.) So once this critique was back in the public realm and available for the paleontological community to assess, the controversy over Lark Quarry and its mysterious dinosaur tracks resumed again, and probably not for the last time.

Then, also early in 2013, the aquamusical version was unveiled. In another article by Romilio and Salisbury, joined by paleontologist Ryan Tucker, they proposed that the "dinosaur stampede" was actually made by small dinosaurs—either theropods or ornithopods—swimming downcurrent in a river, not a lakeshore, and over days, not in a moment of panic. In this paper, they explained that long shallow digit impressions left by the feet and the long distances between each impression were a result of buoyant (but not necessarily ebullient) dinosaurs. This was the most radical of all reinterpretations for Lark Quarry, serving up a one–two punch to the public consciousness of this world-famous tracksite: no giant carnivore, and no stampede.

What do I personally think happened? Based on the previous evidence, I used to have little doubt that a dinosaur stampede took place, and that it happened after the arrival of the big three-toed dinosaur. But now that I've also seen hundreds of dinosaur swim tracks from the Early Jurassic of Utah, I also wonder about this as a possibility too. As to whether the large dinosaur was a large ornithopod or theropod, I'll continue to reserve judgment on that.

A "submerged" scenario complicates any detailed analysis of the tracks, because tracks made underwater often become vaguer or distorted compared to those made out of the water.

I'll further state, somewhat impudently, that in the "stampede" hypothesis, the identity of the big trackmaker doesn't matter. After all, a lumbering ornithopod also could have provoked panic in smaller dinosaurs, which may have been just sensibly avoiding being in the same spot as an ornithopod foot. Or a stampede could have been prompted from some other cause that happened a day after a large theropod or ornithopod walked through the area. The direction of movement also might have been an artifact of landscape, in which a nearby lake shoreline kept all of the dinosaurs moving in one of two directions. Thus I am left with more questions than answers, a common problem in science but one particularly exemplified by the dinosaur tracks of Lark Quarry.

Regardless, this whole affair also serves as a reminder that almost no one gets the final word in such debates, and that some problems surrounding dinosaur trace fossils defy solving while still tantalizingly beckoning us to solve them. This is surely very good for paleontology, for long after I have written these words we will have learned much more about the dinosaur tracks of Lark Quarry and other known and still-undiscovered dinosaur tracksites.

Sleeping with the Dinosaur Tracks

Just how did I get trapped in the building housing these fabled and controversial dinosaur tracks? In July 2007, I was in Queensland teaching a study-abroad program through my university that was hosted by James Cook University in Townsville. One day, after looking at a map and realizing that Lark Quarry was a mere seven-hour drive from my location, I decided to make a pilgrimage there. So once I was done with teaching my part of the program, I rented a car and drove west from Townsville to Winton. Once in Winton, I took a left, and drove another hour to Lark Quarry, most of it on a dirt road, braking and swerving madly whenever kangaroos bounded across the road. (Please don't tell the car-rental agency.)

Having successfully made it to the last afternoon tour of the day, many other visitors and I were met by a tour guide in the main lobby of the building. After a brief introduction, she unlocked a door and led us inside a vestibule. With one turn around a corner, we were in a very large room and on an elevated walkway, looking down on thousands of dinosaur tracks. Once there, our guide told the familiar, traditional yarn of Lark Quarry to a large and eager crowd of Australians ranging from children to grandparents. It was warmly reassuring for me to see so many people there, all admiring these trace fossils and marveling at the gripping narrative they inspired.

During this time, I kept quiet, a stealth paleontologist not wanting to affect the tour or otherwise call attention to myself. Once the tour ended, I purposefully dawdled, waiting for other people in the tour to leave so I could soak in the beauty of these traces alone and in silence. I took many photographs and otherwise gazed longingly at the tracks, beheld at long last after reading about them for more than twenty years. It was a lovely moment, one that needed savoring.

Nonetheless, while in this ichnologically induced nirvanic state of mind, I neglected to hear the clicking of the door—locked by the tour guide who had not bothered to check whether anyone was still inside with the tracks. The immoveable door invoked a panic akin to what might have been experienced by the small dinosaurs there about 98 million years ago. After all, I was about 110 km (60 miles) from the nearest town of Winton, without food, water, blanket, toilet, or other amenities, and would not be rescued until mid-morning the next day, and in front of an audience of about a hundred people. On the bright side of things, I was not locked in overnight with a voracious predator, although I was concerned about a fitful sleep filled with bad dreams taking me back in time to that Cretaceous lakeshore. It looked to be a long night.

Fortunately for everyone, I recalled how the tour guide had used a portable PA (public address) system during the tour. I quickly commandeered this unit, cranked it up to its highest volume, and commenced hooting and hollering into the microphone. This commotion managed to attract the attention of the tour guide, who was about to leave for Winton but luckily was still talking with a visitor outside the building. When she opened the door with a bemused (but amused) expression, I thanked her, introduced myself to her and the other visitor who had stayed behind, and we all had a good laugh about how this crazy Yank paleontologist got himself in such a predicament. I had been saved. Hallelujah! So however appealing it might have been in a spiritually enriching way to spend a night with these much-adored tracks, I was also happy to say "good-bye for now" and hope that I would see them another day.

As mentioned before, I did see them again, twice more in fact. The next time was with my wife Ruth in 2010, when we passed through while doing a grand tour of the paleontological and cultural riches of the Queensland outback, and the most recent time in 2011 was with students from that year's study-abroad program. It was a proud educational moment to be there with my students at the tracksite and to discuss its then-refreshed notoriety with them. Adding to their educational experience, a graduate student working with Romilio and Salisbury was there, and he graciously volunteered to talk with my students about other research he had been doing in the area. Even better, my students later witnessed him engage in a disagreeable discussion (one might term it an argument) with our tour guide about the "old" versus "new" stories of Lark Quarry and its tracks. This squabble served as a living example for my students of how one step forward in scientific investigations can cause two steps back in public relations.

Perhaps the most important perspective gained from this last visit, though, came when one of my students later related an epiphany he had while we stood there on the walkway, looking at the tracks. At some point he realized, "Wait a minute. There were dinosaurs *right here!*" With that stunning insight, he felt the primordial and compelling power of dinosaur tracks, behavioral remnants showing us exactly where dinosaurs had lived, breathed, walked, run, or swum, pulling us back into those freeze-frame moments of time yet still leaving us with unanswered questions. Was there a stampede? Who left the big tracks? Did they all go for a swim? What else can we learn from Lark Quarry and its tracks as a place where dinosaurs were right there and then? One thing is for certain, though: without these footprints there would be no such questions, and our lives would be considerably duller.

CHAPTER 4

Dinosaur Nests and Bringing Up Babies

Parenthood, Dinosaur-style

Dinosaurs were good parents. Although such a statement is imbued with anthropomorphism and hints at unrealistic expectations of "family values," we now know that at least a few species of dinosaurs protected and nurtured their young. Even better, we know that those same species of dinosaurs had evolved the instinct to change their surroundings in a way that protected their potential offspring before a single egg left a dinosaur mother's body. Both trace and body fossils for such sophisticated behaviors show that foresight and planning for the next generation was naturally selected in these dinosaurs.

Yet dinosaurs did not always have the winning reputation of the Cleaver family. Instead, their parenting skills were thought of as more along the lines of *The Addams Family* or *The Simpsons*. Previously viewed as "lay 'em and leave 'em" parents, they were assumed to have been like lizards, snakes, or turtles, depositing their eggs in a nest, then waving good-bye to them, never to find out whether they hatched or not, or watch them grow up. We imagined dinosaur mothers thinking "You're on your own, developing embryos!" as they walked away from a clutch of eggs, possibly laid out in the open for ravenous insects, mammals, and other dinosaurs to enjoy as meals. Continuing such a dire state of affairs, dinosaur fathers were regarded as ne'er-do-wells, long gone after performing their (very) brief service to dinosaur reproduction.

Fortunately for dinosaurs, though, these unseemly character flaws have undergone a massive makeover during the past thirty years or so. The pivotal shift for this still-new perspective on dinosaurs came about from a combination of evidence: from dinosaur eggs and embryos that have been connected to adult dinosaurs (their parents); to dinosaur babies; and, yes, trace fossils made by juveniles and parents.

However, probably the most important trace fossils that prompted paleontologists and the rest of the world to change their minds about dinosaur parental behavior are dinosaur nests. Although dinosaur eggs had been known since the mid-19th century, actual nests—structures deliberately built by dinosaurs that held and protected their eggs, and were also used to raise dinosaur hatchlings into responsible members of Mesozoic ecosystems—were not described until 1979, and definitive examples remained unrecognized until the 1990s. Still relatively rare compared to dinosaur tracks and some other trace fossils, these nests and what they contained nonetheless propelled and supported the revolutionary idea that dinosaurs cared for their young, behaving less like lizards and more like modern crocodilians and birds. Also, now that paleontologists know what to look for, I have every confidence that more dinosaur nests will be found in upcoming years, expanding our understanding of this formerly enigmatic dimension on how dinosaur lives began.

Mom, Dad: Where Did Baby Dinosaurs Come From?

Before talking about dinosaur eggs, babies, and nests, a discussion of the potentially weighty subject of dinosaur sex is warranted, if for no other reason than that these behaviors were a necessary prelude to fertilization, but mostly because such banter is also great fun. For example, whenever dinosaur sex comes up in conversation, its mysteries are summarized succinctly by the question "How did they do it?" The incredulity behind such a question might reflect the wonder inspired from a trip to a zoo or on a safari, where if one was timely enough to witness elephants, giraffes, and rhinoceroses in an amorous moment, it also provoked ideas on how other big animals, like dinosaurs, mated. Alas, these couplings can never quite satisfy, no matter how much we might like to watch. After all, the largest bipedal dinosaurs (theropods) may have been close to 8 tons, and the largest quadrupedal ones (sauropods) were 4 to 5 times that weight, dwarfing the largest of mammals and birds we have today. There simply are no modern equivalents for dinosaur mating.

So how did two-legged male and female dinosaurs manage to keep their respective balances while mating? How did four-legged males pick their front feet up from the ground? How did male dinosaurs get past the thick tails of some females to reach their goals? More intriguing, how did male dinosaurs get past the spiky or clubbed tails of some females? For the larger dinosaurs, how did a female dinosaur support the weight of a male without him breaking her back? How did the male get off, or rather, disengage from the female? Sure, we have a few ideas that approach these questions from watching what we consider today as "charismatic megafauna." But elephants, giraffes, and rhinoceroses are inadequately sized and inappropriately equipped when compared to the largest of dinosaurs. It would be like comparing the mating habits of kangaroo rats to kangaroos, which is not just about size but also anatomy and physiology.

So knowing that modern sex surrogates and zoo-based voyeurism just aren't up to the task of giving us proper analogs, we must look at the fossil record. The first and most obvious question would be: Have we ever found a male and female dinosaur skeleton of the same species together in what might be construed as a compromising position, depicting the bawdy aphorism "the beast with two backs"? Frustratingly, as far as we know no body fossils have been found of a closely joined male and female dinosaur of the same species, together forever. This disappointment is not surprising, as such a pair of fossils would have required an extraordinary set of circumstances to have preserved them: mating, then both suddenly dying while mating, an example of *la petite mort* extended a bit too literally. For ideal fossilization conditions, this sudden death of both partners would have been followed by a quick burial, one that managed to keep them in close contact.

Still, paleontologists hold out hope that such fossils will be found some day and are encouraged by an incredible find, which was of the "fighting dinosaurs" found in Late Cretaceous rocks of Mongolia in 1971. This discovery was of a small ceratopsian dinosaur, *Protoceratops*, on top of a dromaeosaur, *Velociraptor*, with both locked in an embrace but a pointedly deadly one. The right forearm of the *Velociraptor* was in the mouth of the *Protoceratops*, which had apparently clamped down on it, fracturing the arm. Meanwhile, the formidable raised claw on the *Velociraptor*'s left foot was positioned just below the *Protoceratops*' head shield, close to its neck; its left hand was grasping the head shield. Rest assured, this was not like a kitten and puppy wrestling with one another, nor was it like any other hint of interspecies play, but more emblematic of Tennyson's "nature, red in tooth and claw." Paleontologists who studied these fossils agree that both dinosaurs must have been buried instantaneously, which probably happened from the collapse of a nearby sand dune, a geological referee abruptly and permanently stopping the fight.

So one would think by now that trace fossils would again come to the rescue with evidence of dinosaur mating, arising and proudly pointing to dinosaurian congress. Surely sediments somewhere, sometime, during the more than 160-million-year history of dinosaurs recorded visible signs of a male and female dinosaur conducting a Mesozoic version of "business time." Alas, the answer thus far eludes us, a postponing of evidence that tries the already-stretched patience of intellectually lusty paleontologists who desperately seek the warm afterglow of scientific gratification that comes from a sudden, ecstatic release of suspense.

Fortunately, ichnology, like all sciences, has predictive power, promising deeply satisfying scientific insights that happen multiple times and in rapid succession. Through the use of modern traces and animal behavior, we can paint pictures of what dinosaur mating traces should look like, and use those conjectured images as guides for detecting mating trace fossils preserved in the geologic record. Of course, when looking for such double-entendre traces, paleontologists will have to adjust their search images according to dinosaur anatomies, sizes, and behaviors that may have left their marks while making the horizontal bedrock.

For theropods, such traces would have varied with the size of the mating dinosaurs involved. For example, the mating traces of a pair of *Microraptor*—each weighing less than a kilogram (2.2 lbs.)—should have left much less of an impression compared to traces made by a pair of *Spinosaurus*, which together would have made for 8 to 15 tons of convivial society. Still, mating traces of these bipedal dinosaurs would show a basic pattern of one or two footprints adjacent to one another (male tracemaker) just behind but slightly offset to the right or left of two other paired footprints (female tracemaker). All four tracks should be nearly identical in size and form, but perhaps with slight differences accounting for gender differences (*sexual dimorphism*). Bipedal ornithopods would have had similar patterns, although distinguishing these from those of theropods feeds into the old "what made the three-toed track" dilemma discussed in the previous chapter. Furthermore, many of us sincerely hope that no mistaken attempts at interspecies mating took place and became registered in the fossil record, wherein a theropod found a same-sized ornithopod winsome, or vice-versa.

For quadrupedal dinosaurs, such as sauropods, ceratopsians, ankylosaurs, some ornithopods, and other dinosaurs that got around on all fours, mating traces would differ appropriately from those of bipedal dinosaurs by adding a few more tracks to the mix: two, specifically. This situation would have been a result of the male, having approached his potential partner from behind and not being rebuffed—sexual selection or rejection writ large—then likely rearing up and putting his front feet onto the back of the female. This positioning also means that her rear legs would have supported at least part of his weight, which should have deepened her hind tracks relative to her front tracks. Hence the full sequence of trace fossils might be:

- Normal quadrupedal diagonal walking pattern by both dinosaurs, either subparallel to each other or perhaps showing a shortening of pace as the male approached from behind.
- Two sets of four (front–rear) tracks, one behind the other and perhaps offset to the right or left—accounting for bypassing the female's tail on one side—denoting pre-mating positions. These tracks also might be accompanied by a central mark behind the male's tracks, caused by his tail pressing against the ground as he adopted a tripod-like stance.
- Noticeably deeper rear tracks in both sets of tracks (mounting in progress), but maybe with one of the male's legs draped over the female's tail, and some evidence of shuffling forward if she needed adjusting to his weight or otherwise accommodating his anatomical attributes. Such shuffling also might have left an elongated drag mark or more intermittent bounce marks (from his tail, that is).
- The resumption of normal diagonal walking patterns, with hers showing a more rapid pace to get away from her partner, and his showing deeper front tracks as his weight was properly redistributed upon dismounting.
- A resting trace connected to his tracks, post-coitus.

Of course, for mating traces to get preserved in the first place, such activities should have taken place in environments with the right sediments to mold every little (or not so little) parry, thrust, twitch, shudder, or shrug. Among the best environments for registering these traces would have been river floodplains or shorelines with firm mud or sand. On the other hand, the worst places to have made a lasting mark for future generations of humans to see would have been in trees, water bodies—especially if doing it while floating above a lake or river bottom—or burrows.

So with these search images in mind and an active lust for such trace fossils, paleontologists should have all of the right tools for ensuring their explorations all have happy endings. Here's to getting lucky.

Dinosaur Eggs, Over Easy

Dinosaur mating surely happened many times during their 165-million-year history, and more than a few of those couplings successfully resulted in fertilized eggs getting laid by dinosaur mothers. So are dinosaur eggs—which are normally preserved as fossil eggshells and filled with sediment—considered trace fossils or body fossils? One would think these and eggshell material in general would qualify as trace fossils, because they seem like indirect evidence that doesn't involve any obvious body parts, except for rare instances when they also hold embryonic dinosaur bones.

Nonetheless, eggs and eggshells are actually body fossils. The reason why is because eggshells served as extra body parts for developing embryos (similar to how skin functioned in an adult dinosaur), protecting softer parts inside while also allowing an embryo to breathe through pores in the eggshell. Thinking more ecumenically, though, I like to consider dinosaur eggs as traces of the mothers that developed them. By extension, then, every dinosaur body is a trace fossil of its mother, and ultimately of both parents mating. But such thoughts might be a little too metaphysical, so let's just go back to thinking about eggs as body fossils.

Despite their lowly status as body fossils, dinosaur eggs provide valuable clues about dinosaur reproduction, brooding, and their surrounding environments, which all supplement information provided by the dinosaur fossils that really matter—namely, their trace fossils. In fact, the mere pattern shown by a number of eggs laid by a dinosaur mother in one egg-laying episode (known as a clutch) constitutes a trace fossil in itself, regardless of whether a nest structure is associated with the clutch or not. Or if a dinosaur parent actively pressed an egg against the side of a nest and thus made an external mold of that egg, that would count as a trace fossil of the parent, too.

Vertebrates that lay eggs today, such as reptiles, birds, and monotremes (egg-laying mammals, such as the platypus and echidnas), are all classified as *amniotes*. This means they share a common ancestor that evolved an *amnion* (a sac containing a developing embryo), which was enclosed by an egg. This novel trait likely developed in response to widespread arid conditions, which naturally selected amphibian ancestors capable of putting watery conditions inside an egg, regardless of whatever climate might have been outside of them. This new reproductive strategy evolved in vertebrates by about 320 to 330 *mya*, or about 100 million years before dinosaurs existed.

Later in their evolutionary history, amniotes evolved myriad reproductive strategies, including live birth which cut out the "middle egg" (so to speak) by developing an embryo inside the mother until it's ready to emerge, instead of trusting this to all take place in an egg. We currently have no reasonable evidence that dinosaurs gave birth to live young, but this trait (*viviparity*) showed up in ichthyosaurs—marine reptiles unrelated to dinosaurs—by the Early Jurassic, around 190 million years ago. Live birth in ichthyosaurs made sense evolutionarily, because it kept ichthyosaur mothers in the seas instead of having to lumber up on land to dig nests and lay eggs. Although that strategy has worked very well for sea turtles for the past 100 million years, it apparently was selected out of ichthyosaurs early on in their evolutionary history. Viviparity also occurs in some modern species of lizards and snakes, including sea snakes; the latter shows how other marine reptiles came up with the same reproductive mode.

The oldest known dinosaur eggs come from Early Jurassic rocks, or about the same time ichthyosaurs were using live birth. Dinosaurs, however, were very likely laying eggs in the Late Triassic Period, too; but these may have been soft, leathery eggs that were not as easily preserved as the Jurassic ones. The Early Jurassic fossil eggs, which were preserved because of calcite in their shells, give us a minimum time for when dinosaurs had developed mineralized eggs. Basically, this made eggshells that went from "squish" to "crunch" when another dinosaur stepped on them. Hence, the eggs we are so familiar with in our everyday lives, thanks to those most famous of modern dinosaurs (chickens), got their origins relatively early in the evolutionary history of dinosaurs.

Despite the fact that dinosaur eggs are technically considered body fossils, paleontologists were still motivated enough to devise a classification scheme that helped distinguish different types of eggs. Paleontologists, using a system parallel to a biological classification, came up with *ootaxonomy*, in which the "oo" prefix refers to eggs. These categories, based on shell microstructure, arrangements of the pores, and overall forms, were given linguistically daunting names such as Spheroolithidae, Ovaloolithidae, Megaloolithidae, Dendroolithidae, Faveoloothidae, and at least a half dozen others. Despite such names causing computer spell-checkers to run and hide, they are essential for the small and dedicated group of dinosaur-egg paleontologists to better communicate with one another.

Unfortunately, these egg categories do not always exactly match clades of dinosaurs. Although we can be reasonably sure that theropods, sauropods, ornithopods, and ceratopsians produced certain types of eggs or egg clutches, initial identifications are occasionally tested by the discovery of embryonic bones in an egg. Moreover, as of this writing, no one has yet identified an undoubted stegosaur, ankylosaur, nodosaur, or pachycephalosaur egg, which make for surprisingly big holes in our knowledge of the life cycles for these dinosaurs.

Given that eggs are body fossils, is there any way to elevate their importance further by somehow making trace fossils out of them, too? The answer is yes. At least two species of dinosaurs, *Citipati osmolskae* and *Troodon formosus*, provide evidence of post-laying movement of eggs in a nest, an action that must have been done by one or both dinosaur parents. This sort of arrangement constitutes a trace fossil of that parent's behavior like how a flower arrangement reflects the handiwork of a florist.

In the case of *Troodon* nests, distribution patterns and orientations of the eggs provided marvelous clues about female *Troodon* reproductive anatomy, as well as what she did with the eggs after they exited her body. Paleontologists who studied *Troodon* egg clutches were surprised to notice that not only were eggs paired but also aligned vertically. Based on their statistically significant pairing, paleontologists surmised that this pattern must have been caused by the mother laying eggs two at a time. Secondly, the eggs were longer than they were wide (elliptical), which meant they naturally should have rolled onto their sides once deposited onto the ground surface. As a result, their vertical orientation means they were righted after laying.

To best accomplish this feat, the mother or father *Troodon* likely would have used their hands to turn the eggs upright, then buried the lower end of each egg so that they wouldn't just roll onto their sides again. No one knows whether *Troodon* mothers did this with each pair of eggs—two at a time—or whether they waited until the entire clutch was laid before turning them all sunny-side up. In an evolutionary sense, the turning-two-at-a-time scenario seems more likely than the let's-wait-until-they're-all-out one. But sometimes evolution works in mysterious ways, going counter to the expectations of us geologically short-lived primates.

Did dinosaurs ever engage in cuckoo-like behavior, laying their eggs in with the eggs of other species in a nest, and letting some other species' parents do all of the work of raising the young? Thus far, we only know of one instance in which an egg assemblage contained embryonic bones of more than one species of dinosaur, which was found in Late Cretaceous rocks of Mongolia. The bones showed that one dinosaur was probably *Oviraptor* (closely related to *Citipati*) and the other was *Byronosaurus*, a dinosaur more closely related to *Troodon*. One would think a discerning mother *Oviraptor* would have been able to tell whether she was brooding and feeding a *Byronosaurus* child. But such is the insidious nature of nest parasitism in modern birds: most bird parents don't know they have a changeling in their midst until it is too late. However, this one instance of possible parasitism is hard to test further, as eggs also could have easily been transported and deposited together in the same place by currents. One way to test this hypothesis, though, would be through trace fossils, such as two distinctive sets of hatchling tracks in a nest. Have hatchling tracks ever been found in a nest? Not exactly, but they have been found very close to nest sites, as will be explained later.

Another type of interpretable hatchling trace fossil would be an exit hole ("hatching window") in an eggshell, made by hatchlings as they emerged from their temporary confinement. Sea turtles, for example, have a temporary extension of their beaks when born, which is applied like a can opener from the inside of the egg to open it. Sea-turtle researchers can thus pick up an empty sea turtle egg from a previous year and instantly tell whether its former occupant successfully hatched, died in the egg, or an egg predator got to it before hatching. Similar parts have been described from Late Cretaceous sauropod embryos in Argentina, so these baby dinosaurs could have made such traces in eggs. Have hatchling trace fossils been identified in dinosaur eggs? Yes, which is amazing when one considers how much more fragile an egg would have become once abandoned. Indeed, telling the difference between fractures caused by a hatchling coming out of its egg versus those inflicted afterwards would be very challenging. Nonetheless, trace fossils of "hatching windows" were interpreted in 2002, evident in Cretaceous eggs from Mongolia and China. More such traces could be reasonably proposed for fossil eggs with localized and consistently sized hatchling-appropriate holes in them, especially if more than one is seen in the same clutch of eggs within a nest structure.

Other trace fossils that might be preserved in eggs would be those from the aforementioned egg predation, where another animal—whether a dinosaur, crocodilian, lizard, snake, mammal, or insect (depending on when in geologic time this might have happened)—nibbled, gnawed, bit into, or chomped an egg. Yet these trace fossils have not been discovered yet either, despite many decades of our demonizing mammals for eating too many dinosaur eggs toward the end of the Mesozoic Era.

We also currently do not know whether dinosaurs laid unfertilized eggs—like chickens or some other birds—but we presume that if their eggs are found in a nest structure, these were probably fertilized. The extra effort required to build a nest represents an investment of time, resources, and energy and would have exposed parent dinosaurs to predators by keeping them in the same place for a while, thus making them predictable. Instincts that would have impelled parent dinosaurs to make nests with every laying of an unfertilized egg clutch would have been quickly selected out of those lineages.

An adage I often tell my students when discussing extinction is: "If you want to make a species go extinct, stop it from reproducing." The huge success of dinosaurs throughout nearly every land environment within much of the Mesozoic Era attests to how they reproduced just fine, probably aided through the vast majority of them making nests for their egg clutches at the right times and in the right places.

Starting a Dinosaur Family: Building Nests

Given that at least some, if not most, dinosaur mating successfully resulted in fertilization, and a mother dinosaur started internally producing eggs soon afterwards, the next step for either her or both parents should have been nest building. Although this supposition might be taken for granted, what is surprising to many people is that actual trace fossils of dinosaur nests are only known for a few dinosaurs, and the details of how these were made are still a bit murky. The known nest builders are:

- The ornithopod *Maiasaura* from the Late Cretaceous in Montana.

- The theropod *Troodon*, also from the Late Cretaceous in Montana.
- Unidentified therizinosaurs from the Late Cretaceous of Mongolia, which may have even formed nesting colonies.
- Titanosaur sauropods (not yet identified precisely) from the Late Cretaceous in Argentina and Spain.
- The prosauropod *Massospondylus* from the Early Jurassic in South Africa.
- The ceratopsians *Psittacosaurus* from the Early Cretaceous in China, and *Protoceratops* from the Late Cretaceous in Mongolia, although their nests are interpreted more on the basis of many same-sized hatchlings, tightly packed in a small space.

Other dinosaurs that we are sure made nests, even though we do not have direct evidence of a nest structure, are the Late Cretaceous theropods *Oviraptor* and *Citipati* from Mongolia. In a famous case of dinosaur-parenting misattribution, paleontologist Henry Fairfield Osborn, of the American Museum of Natural History, found an *Oviraptor* skeleton on top of an egg clutch during an expedition to Mongolia in 1923. Nonetheless, he figured the eggs were those of the small ceratopsian *Protoceratops*, whose bones were abundant in the area, too. In Osborn's scenario, *Oviraptor* died and was buried while raiding the nest of another dinosaur, not while taking care of its own eggs. This idea was disproved much later, and through a closely related dinosaur, *Citipati*. In one of many spectacular finds from American Museum of Natural History expeditions to Mongolia in the 1990s, a nearly complete specimen of *Citipati osmolskae*—missing only its head—was found in a sitting position above a clutch of long oval eggs. Even better, the clutch was oriented radially. This suggests that the mother or father *Citipati* arranged the eggs after laying, with such a pattern constituting a trace fossil in itself. When viewed from above, the skeleton and eggs paint a striking portrait. The dinosaur's arms are held wide in a semi-circle around the eggs, evoking images of a protective mother or father that died in a last vain effort to keep their unborn children from harm. Paleontologists have also documented direct associations of dinosaur parents and eggs—some of which have embryonic bones linking them with their parents—with another specimen of *Citipati* and several examples of *Oviraptor*.

Take away a parent dinosaur's skeleton, though, and discerning a dinosaur nest becomes a little more challenging. For instance, a cluster of identical dinosaur eggs is sometimes considered as indirect evidence of nests. However, unless these eggs are inside a bowl-like structure, show some sort of post-laying arrangement, or have remains of a parent dinosaur on top of the eggs, a collection of eggs may not be a nest after all.

This doubt is cast because the eggs may have been moved after laying. As anyone who has done an Easter-egg roll can attest, or had an egg fall from a kitchen counter to the floor while trying to make an omelet, well-rounded eggs can move easily from one place to another under the influence of gravity. In nature, eggs can be rolled or floated by currents, then deposited in low-lying areas with other debris that, to an untrained eye, might look like someone's idea of a nest. Indeed, the discovery of a dinosaur egg—containing embryonic bones, no less—in Late Cretaceous shallow-marine deposits in Alabama shows that at least one dinosaur egg was capable of floating a long way from land. As we learned previously, dinosaurs may have occasionally swum, but like modern birds or reptiles, were not inclined to make underwater nests.

In order to better understand what constitutes a dinosaur nest as a trace fossil, a good starting place is to look at the traces associated with modern vertebrates that lay eggs and make nests, such as those of reptiles and birds. If talking just about reptile nests, these can be summarized into two broad categories: ground nests and hole nests, with ground nests made on the ground surface and hole nests below the surface. From there, each type can vary. For example, ground nests can range from simple hastily scraped depressions made by some lizards, to large mounds of vegetation amassed by alligators.

Hole nests are also diverse, ranging from the blunt, shallow holes of freshwater turtles and tortoises, to the more elaborate excavations of sea turtles and crocodiles.

On the other hand, bird nests are as crazily diverse as birds themselves, almost defying facile categorization. Bird nests range from simple scrapes in the ground, to better defined excavations, to tunnels in the ground or in vertical bluffs, to elaborately woven and architecturally complex structures made of a wide variety of natural and man-made materials. Some birds resemble reptiles in their nesting behavior by constructing nests on the ground or as underground burrows, but flight has also made it easier for birds to build nests well above ground surfaces, such as in trees, cliff faces, or buildings. Furthermore, by "in trees," this is sometimes literal for woodpeckers that actually bore into tree trunks with their beaks, hollowing out areas to lay their eggs and raise hatchlings inside trees.

How did dinosaurs fit in "reptile vs. bird" models for nesting behavior? It turns out they were somewhere in between, although definitely leaning toward behaviors we observe today in some reptiles and ground-nesting birds. For one thing, every dinosaur nest structure recognized thus far is a ground nest. Hole nests like those of sea turtles or crocodilians, or more fancy structures such as those of some tree-dwelling birds, aren't yet known for dinosaurs. Of course, this current lack of evidence does not necessarily mean that no dinosaurs made either underground or arboreal nests. For instance, one good candidate for underground nesting would have been the small Cretaceous ornithopod *Oryctodromeus cubicularis*, which made dens for raising its young (discussed more in the next chapter). For arboreal nests, a few small feathered tree-climbing, gliding, and flying non-avian dinosaurs are known from Early Cretaceous rocks of China. So these dinosaurs feasibly could have made their nests in trees, just like modern birds. However, until we have some evidence for these, we can only speculate.

For those people who might wonder how to tell a hole in the ground from, well, other things, how would someone recognize a dinosaur nest in the fossil record? Fortunately, paleontologists who have interpreted the few indisputable dinosaur nests in the geologic record made a nice little checklist for the rest of us to follow, which helps considerably. Here is a summary of that list, but posed as questions to ask when encountering a depression in a Mesozoic rock that might be a dinosaur nest:

1. Does the depression cut through any layered sedimentary rock below it? This is an especially good question to ask when investigating whether the rock that preserves a potential nest has different colors and grain sizes than the sediment above it. For example, the sediment below the depression might be a green rock mostly made of mud (mudstone), whereas the sediment filling it might be a red rock composed of sand (sandstone).
2. Does the depression hold lots of entire (or nearly entire) eggs, along with skeletons of baby or otherwise young dinosaurs, and all of the same species? Be careful with this one, though. As mentioned before, eggs can be transported far away from where a mother dinosaur originally laid her eggs, meaning such a depression might simply be where these eggs and skeletal remains accumulated after being washed about by currents.
3. Does the depression also have a raised rim around it, accentuating its basin-like appearance? A related question to ask is, is this rim also made of sediment differing from what is underneath or outside of it? These rims would have served an important primary purpose, such as preventing rounded eggs from rolling away as a mother dinosaur laid them.
4. Does the rock filling in the depression and covering the rim have its own distinct traits, like different colors and textures (mentioned before), bedding, or insect burrows and cocoons? Such features might hint of a fossil soil (*paleosol*), which would tell paleontologists

> that the depression was exposed at the ground surface before being covered, as opposed to, say, at the bottom of a river.

However facile this checklist might seem, it represents a lot of previous work on dinosaur nests. Knowledge of some of the history behind it may provide a perspective on how what we know now about dinosaur nests also reminds paleontologists how much we still need to learn about these trace fossils.

The First Recognized Dinosaur Nests

Anyone who has not studied the history of dinosaur studies might be astonished to know that the first genuine dinosaur nests were not interpreted until 1979. At the time of this discovery, more than a hundred years had elapsed between paleontologists first linking fossil eggs to dinosaurs, which was in 1869. Furthermore, only a few of the previously mentioned criteria were applied to these nests, showing how far paleontologists have come since then in defining them.

In what became a revolutionary discovery, inspiring nearly everyone to reconsider what they thought they knew about dinosaur parenting, John ("Jack") Horner and his friend Robert ("Bob") Makela, while prospecting Late Cretaceous rocks in Montana, found depressions filled with eggs and partly grown juveniles of the large ornithopod dinosaur *Maiasaura peeblesorum*. Horner and Makela inferred the presence of nests on the basis of many nearly entire and identical eggs, some containing embryonic skeletons of *Maiasaura*, which were in the bottom of what looked like indentations to them. These former hollows were visible as differently colored sediment above and below the eggs; unfortunately, though, no sedimentary rims were present. So either these dinosaurs did not make such rims, the rims had been eroded and not preserved, or these researchers missed them because of too-subtle differences in the sediment.

Still, on the basis of localized assemblages of eggs and baby dinosaur bones in the same small areas, they hypothesized that this part of Montana was a *Maiasaura* nesting ground about 75 to 80 *mya*. The close spacing of the nests also hinted, for the first time, that dinosaurs nested communally. Subsequent field work done in the same area yielded more and different eggs, which at first were linked to the small ornithopod dinosaur *Orodromeus makelai* (yes, this species was named in honor of Makela), although later they were connected to *Troodon formosus*. Hence, for the first time paleontologists began thinking of dinosaurs and their nesting behaviors as more akin to those of ground-nesting birds and not reptiles. Even more exciting, the find of one nest with fifteen juvenile *Maiasaura* in it, all the same size and apparently in the same age range, was strong evidence favoring extended parental care in this species of dinosaur. Otherwise, why stay in a nest unless your parents are feeding you? (All of you parents with children who have graduated from college yet are still living at home, you may now wearily nod your heads in agreement.)

Despite this seminal work by Horner, Makela, and others, the full checklist for interpreting dinosaur nests as trace fossils would not come about until more than twenty-five years later, stemming from studies of nests of two dinosaurs very different from *Maiasaura* and from one another: the small Late Cretaceous theropod *Troodon formosus* of Montana, and massive Late Cretaceous titanosaurs (sauropods) of Argentina. Amazingly, their nests are not so different in their overall sizes, despite titanosaurs weighing about a thousand times greater than *Troodon* (50,000 vs. 50 kg, or 110,000 vs. 110 lbs). However, each of their nests have their own distinctive forms, leading to better understanding of how each type of nest was made.

First, let's talk about *Troodon* nests. In a geological coincidence that was not a coincidence, paleontologists discovered these nests in the same location as the *Maiasaura* nests—a place that was nicknamed "Egg Mountain"—and in the same sedimentary rocks of the Late Cretaceous Two Medicine Formation. Recognized by David ("Dave") Varricchio, who was a Ph.D. student of Horner's in the early 1990s, these structures were the first dinosaur nests defined solely as trace fossils. Interestingly, until Varricchio and his colleagues worked on these nests, Horner and Makela attributed *Troodon* eggs to the small ornithopod *Orodromeus*, a case of mistaken identity that was fairly common with dinosaur eggs until just recently.

Full disclosure: Dave and I overlapped academically when we both attended the same graduate school (University of Georgia, Athens) in the late 1980s. In between our respective research projects, as well as going to local music clubs to watch melodic toe-tapping performances by bands such as Agent Orange, Black Flag, and other colorfully named troubadours, we both took a class from Robert ("Bob") Frey, simply titled *Trace Fossils*. At the time, Frey was one of the foremost experts in ichnology, so we felt privileged to take a class from him. But it also resulted in Dave and me sharing an ichnological worldview that stuck with both of us. Thus I was not surprised to later read about his recognition of the *Troodon* nests, which he and his colleagues described quite properly as dinosaur trace fossils. Dave also dedicated the first paper about the *Troodon* nests to Frey, an acknowledgment of how good mentorship can later contribute to discovery.

The first *Troodon* nest structure noted at Egg Mountain surrounded a clutch of 24 asymmetrical eggs, which, like chicken eggs, were wider at one end than the other, but more elongated and voluminous, having held about 0.3 liters (1.2 cups) each. Eggs in the clutch were oriented almost vertically, narrow ends down but also pointing toward the center. They were unhatched and clustered in an oval space smaller than most serving trays, measuring only 45 × 56 cm (18 × 22 in). The eggs defined a ring-like pattern, with what Dave and the other paleontologists called an "egg-free space" in the middle. This clutch was surrounded by a semi-circular structure that was about 1.6 to 1.7 m (5.3–5.5 ft.) wide, about the diameter of a small kiddy pool. The "egg-free space," egg clutch, and rim shared a north-south orientation, making concentric oval rings; this orientation was also apparent in another *Troodon* egg clutch at the same site.

The rim of this structure sloped abruptly, and then more gradually inward to a flattish area about a meter (3.3 ft) wide, where the egg clutch was centrally located. The exterior of the nest was defined by a rim that was 10 cm (4 in) tall and about twice as wide. The rim was composed of micrite, a hard, resistant, fine-grained limestone; it and the egg clutch were buried under a mudstone. Unhappily, in recovering the egg clutch, the field crew had to destroy part of the nest to extract the eggs. Still, they uncovered enough of the nest to realize what it was before taking out the egg clutch, and surely stood back and gazed in awe at it when it was fully exposed. The bulls-eye pattern caused by the egg clutch, dead center in the structure and circumscribed by the rim, left little doubt of its significance. This was a nearly perfect example of a dinosaur ground nest, and must have been made by one or both *Troodon* parents before the eggs were laid.

Before making a nest, *Troodon* parents must have first scouted their surroundings to find suitable locations. For example, bad places to nest would have included underwater (eggs drowned), in places with abundant scavengers or predators (eggs eaten), areas with high traffic from other dinosaurs strolling through the area (eggs smashed), or on solid rock (eggs exposed, perhaps fried). Consequently, these parents probably sought out areas with well-drained and crumbly soils that were also well above the local water table, and away from other animals that might have accidentally or purposefully harmed the eggs. The fortuitous fossilization of the eggs and nest must have happened because a nearby river overflowed its banks and buried both before the eggs hatched. Thus even though the dinosaurs had made a nest rim high enough to prevent most rising waters from enveloping the eggs, it wasn't high enough in this instance. In other words, what was unlucky for the expectant *Troodon* parents turned out to be fortunate for the paleontologists about 75 million years later.

Although we still don't have sufficient information to say for sure, *Troodon* nest building probably involved their using a combination of rear feet, hands, and snouts to excavate underlying soft sediments. Making a circular nest probably required digging in a circular pattern, and the easiest way to make a doughnut-shaped pile of soil outside a depression would have been by moving either clockwise or counter-clockwise from a central point inside of the nest. The width of the nest, rim included, was about half the total body length of an adult *Troodon* (3 m, about 10 ft), implying that only one adult at a time could have fit comfortably in the center to dig. Two at a time would have gotten in each other's way, but both parents also might have taken turns.

The soil piles defining the nest exterior also likely required its maker(s) to pat these down and otherwise mold the rim of the nest, firming it so that it later hardened, perhaps cemented by precipitating minerals. A firm exterior would have been essential, as loose mounds of dirt would not have offered much protection to the eggs: one hard rain or windstorm would have easily eroded such inept attempts to shield their brood. The gentle slope on the interior of the nest also suggests this area was somewhat compacted, perhaps by trampling in a grape-stomping sort of way. However, this might have been the final action of nesting before actual brooding, happening after the eggs were laid. Because the eggs were oriented vertically, they must have been stuck into soft sediment first; a parent *Troodon* then may have compressed the surrounding soil to reinforce and stabilize them. Regardless of when the nest interior was finished, the nest rim must have been sculpted—like hand-building with wet clay—an action that could have been done from inside or outside the nest.

How long did it take a *Troodon* (or pair) to make a nest? The answer depends on whether just one or both parents were making it. If both were involved, the time might have been halved, but as many human couples know all too well, this would have depended on their level of cooperation (or lack thereof). Nevertheless, one can easily imagine cooperative digging and molding, with one parent on the inside and the other on the outside, making a nest together, and thus decreasing their labor. Just to compare, though, crocodilians and sea turtles normally take just a few hours to dig a hole nest. However, this isn't the best model for how dinosaurs constructed a ground nest, as these hole nests are less complex structurally than what the dinosaurs made.

A more realistic estimate might be gained by looking at times required by alligators and mound-nesting birds to make their egg-protecting structures. For alligators, constructing a nest normally takes about 10 to 15 days. On the other hand, mound-nesting birds may need months to build the massive, vegetation-laden structures they use for burying and incubating their eggs. Some of these mounds, such as nests of the Australian brush turkey (*Alectura lathami*), can be 2 m (6.6 ft) tall and 20 m (66 ft) wide. Based on the comparatively modest proportions of *Troodon* nests, the bowl-like structure itself could not have taken more than a few weeks to dig. These nests have no evidence of having been topped by vegetation, but if they were, their construction would have taken a bit longer.

Troodon egg-laying itself probably took a couple of weeks, assuming that each pair of eggs took at least a day to emerge from the mother's two oviducts. These anatomical traits of *Troodon* mothers are interpreted on the basis of how eggs are paired. Mentioned earlier as "statistically significant," you actually don't need statistics to see the pairing, though, as it is obvious in every nearly complete *Troodon* egg clutch studied thus far. As noted before, the eggs also must have been aligned vertically and stuck in the ground by one or both of the *Troodon* parents. The most parsimonious way to accomplish this sort of interior decorating would have been to:

- Lay two eggs;
- Dig two shallow holes next to each other, each slightly more than the width of and half the length of an egg;
- Place the eggs in their respective holes, pointy ends down;
- Compact the soil around the eggs;
- Repeat with each new pair of eggs until done;
- Sit above eggs until hatched.

Another intriguing trait of the eggs, when oriented like this, is that they had a greater concentration of pores toward their exposed tops. More pores were there so the rest of the embryo living underneath the eggshell could breathe easier in that position. This, along with all of the other trace and body fossil evidence, implies that the eggs were not buried under a thick pile of vegetation or soil, but more likely were protected by a *Troodon* parent sitting above the egg clutch. This is a reasonable assumption, as the relatively small size of these dinosaurs meant they would not have crushed the eggs. Additional evidence from body fossils of related theropods also suggests that *Troodon* and its close kin had feathers, which would have provided some insulation for the eggs, too. Even better evidence of brooding behavior is the adult itself, as is seen with *Citipati* and *Oviraptor*: the bones of an adult *Troodon*, found directly in contact with a partial clutch of *Troodon* eggs.

Further study showed that this *Troodon*, as well as *Citipati* and *Oviraptor*, were probably male, providing an important clue related to how dinosaur parents took care of an egg clutch. For instance, in modern birds, large egg-clutch sizes correlate with paternal care, whereas small clutches correspond with both parents helping with the brooding. Hence large clutches for *Troodon*, *Citipati*, and *Oviraptor*, along with their skeletons on top of these egg clutches, implied they were male. Also, large modern ground-nesting birds—such as emus, ostriches, and rheas—females develop *medullary bone*. This bone has a distinctive texture, imparted because the mother birds extract calcium from their bones to help build calcium carbonate in the eggshell microstructure. In contrast, male birds lack medullary bone, as did the bones of *Troodon*, *Citipati*, and *Oviraptor*. As a result, this absence suggests that the adult dinosaurs on top of each nest were not the mothers, but the fathers.

Not coincidentally, this same sort of paternal protection is a modern behavior likewise seen in nesting emus, ostriches, and rheas. Emu fathers, for instance, sit on or otherwise stay close to egg clutches laid by the mothers for 50 to 55 days, even eschewing food and water during that time. Once their babies hatch, they consume the leftover eggshells, which not only gives these new fathers much-needed sustenance but also decreases the likelihood that a predator will smell freshly hatched eggs and make its way to the nest for some easy snacks. Did *Troodon* have the same behaviors, cleaning up a newborn mess while also getting a long-awaited meal? So far we don't know, and the only trace fossil evidence that could support this would be a nest structure with scraps of eggshells but also holding abundant footprints by at least one adult and hatchlings.

What were ecosystems like in the same area of the *Troodon* nests? Much study on the sedimentary rocks there, including their geochemistry, revealed that this part of Montana was probably warm and semi-arid 75 to 80 million years ago. Furthermore, *Maiasaura* and *Troodon* nested on river floodplains that were cut by meandering rivers and dotted with small lakes. Relatively little vegetation was in the immediately surrounding landscape, based on how paleontologists have not found many fossil trees or plant-root trace fossils directly associated with the nesting grounds. Nonetheless, some root traces in nest rims show that at least a few plants may have taken root in these after having been built. Forests were farther away from the nesting grounds, as was an active volcano that occasionally erupted and helped to preserve both the trees of these forests and dinosaur bones. Some sedimentary deposits near, above, and below the *Troodon* and *Maiasaura* nests further indicate periodic flooding, which along with other geological evidence suggests monsoonal environments with annual dry and wet seasons. However, the area must have been suitable enough for nesting to motivate *Troodon* mothers to keep coming back, as paleontologists found at least ten egg clutches from three separate levels in the rock.

I was reminded of this previous observation during a visit at Egg Mountain on a field trip led by Varricchio and one of his colleagues, Frankie Jackson. On this visit, I noted how the distinctive light-brown micrite composing the rim of the undoubted *Troodon* nest was also apparent in at least three horizons of the hillside there. These may have been partially preserved nests, but ones not accompanied by eggshells or bones: trace fossils in the purest sense, unadulterated by body fossils. If so (although this requires more testing), this would be another way to find out whether *Troodon* had what modern biologists have called *site fidelity*. This is a behavioral trait shared with nesting sea turtles, crocodilians, and birds, in which mothers come back repeatedly to the same nesting site for brooding, or their offspring return to the same site where they hatched.

To better fill out this picture of the dinosaur nesting environments, and as an example of how dinosaur trace fossils are best interpreted with trace fossils made by other animals living with the dinosaurs, I later did a study with Varricchio on fossil insect nests near the *Troodon* nesting sites. Over years of field work there, he and other paleontologists in the area had noticed that several strata contained concentrated zones of fossil insect cocoons and burrows. I was astounded the first time I visited these sites, as the fossil cocoons were so abundant that handfuls of them could be scooped up from the ground where they eroded and fell out of the outcrops. The cocoons were also sometimes directly associated with dinosaur nests, having been reported previously by Horner, Varricchio, Jackson, and others. In fact, during several subsequent visits to this area, I noticed cocoons embedded in the rims of a few of the *Troodon* nests, showing that insects may have been nesting at the same time and same place as the dinosaurs.

The cocoons were exquisitely preserved, some with impressions of spiraled silk weaves still apparent, and a few burrows bore scratch marks from the legs of their makers. These trace fossils were apparently preserved through filling of the burrows and cocoons by fine-grained sediment, followed by microscopic precipitation of micrite. This process was further aided by a semi-arid climate, which would have caused rapid evaporation of water in between sediments, forming calcium-rich soils called *calcisols*.

What we also eventually figured out about these insect trace fossils was a little surprising. Based on the appearances of the cocoons and burrows, we surmised that burrowing wasps probably made most of them, with perhaps only a few made by beetles or other insects. Because modern ground-nesting wasps also burrow into and pupate in well-drained soils, we presumed that these trace fossils were made in similar soils. These insect trace fossils thus affirmed that *Troodon* probably picked its nesting locations for the same ecological reasons as the wasps and other insects. Horner originally conjectured that these cocoons, which were also near the *Maiasaura* nest sites, were those of carrion beetles, which he imagined fed on hatched eggs, or dead eggs and hatchlings. In the light of the new interpretation of the cocoons, his more gruesome scenario is now doubted. In contrast, dinosaurs and wasps may have peacefully co-existed, just like most people and wasps do today, with only a few sting-induced exceptions.

In summary, because of a wonderfully complete amount of both body and trace fossil evidence—but especially trace fossils—*Troodon formosus* of Montana is arguably the best understood of nesting dinosaurs, only rivaled by its neighbors in the same field area, *Maiasaura*. This is not to say that we have stopped learning about *Troodon* and its nesting, but the holistic approach taken toward its study, including ichnological perspectives, gave paleontologists a superb model for comparing with other nesting dinosaurs.

Big Dinosaurs Scratching Out Big Nests

Thanks to *Maiasaura* and *Troodon* for their tracemaking abilities, as well as exceptionally good fortune in both fossil preservation and discoveries, paleontologists had a new search image for dinosaur ground nests as trace fossils going into the 21st century. Thus it was not surprising that one of the same people who studied the *Troodon* nests and eggs with Varricchio in Montana, Frankie Jackson, later noticed and defined similar structures in Late Cretaceous rocks of Patagonia, Argentina, in 2004. However, what was astonishing about these nests was that they were not from a mere 3-m (10 ft) long dinosaur like *Troodon*, but were made by titanosaur sauropods, which were among the largest animals to ever walk the earth.

The wealth of evidence supporting nest building in these sauropods included rimmed nest structures, but also complete egg clutches, embryonic titanosaur bones inside the eggs (some preserved with skin impressions), and at least four horizons containing nests or eggs. The latter showed that the titanosaurs returned to this nesting site over years and perhaps generations, suggesting site fidelity in these dinosaurs too. Further supporting this idea, paleontologists working at this site figured it held thousands of egg clutches, varying from eggshell fragments to entire eggs. The number of eggs was 15 to 35 per clutch, and each individual egg was big, about 13 to 15 cm (5–6 in) wide. Also, eggs were shaped like slightly squashed balloons: round ones, that is, not the cylindrical ones used to make funny balloon animals (including sauropods). Unlike the *Troodon* egg clutches, the clutches did not show pairing or any post-laying arrangements by the sauropods, and looked more haphazardly placed.

Paleontologists working the site were also excited to find six definite nest structures directly associated with egg clutches. Five of these six nests were adjacent to one another and less than a meter (3.3 ft) underneath a bed with abundant adult-sized sauropod tracks. All of this trace fossil evidence suggested that these nests might have been made at about the same time, and that adult sauropods came along later in the same place as the nests, perhaps to start another generation of titanosaurs.

Probably the most surprising trait of these titanosaur nests, though, was their size, which did not match the proportions of their tracemakers. Instead, some of these nests had about the same width and depth of the *Troodon* nests, measuring about 1 to 1.4 m (3.3–4.6 ft) wide and 10 to 18 cm (4–7 in) deep, respectively; a few were only half the size. Some of these nests—like those of *Troodon*—also had partially preserved rims, although these were made of sandstone, not micrite. The nests cut into an underlying well-stratified sandstone that was otherwise nearly identical to the sandstone composing the rims, implying that the rims were originally derived from soft sand below. These were shockingly delicate works for creatures of their size, like an elephant molding a peanut bowl out of clay.

However, where these nests diverged the most from the *Troodon* nests was in their overall shape. Whereas the best-preserved *Troodon* nest was nearly circular when viewed from above, the titanosaur nests were longer than they were wide, giving them oblong or kidney-bean profiles. This made the titanosaur nests look less like kiddie pools and more like troughs. Moreover, *Troodon* egg clutches were tightly packed in a relatively small central area of the nest, whereas the titanosaur eggs were spread throughout their nests, filling most of the interior space. Complicating matters further, the expansion and contraction of soils around the eggs fractured and moved them after burial. This meant that some eggs laid by different sauropod mothers at different times were mixed together. The egg arrangements of *Troodon* and these titanosaurs were as different as the rooming habits of Felix and Oscar (respectively) in *The Odd Couple*.

So although these titanosaur nests shared some of the same traits as the *Troodon* nests, they also differed enough to raise more questions about how they were made. For instance, one can more easily imagine a small theropod using its hands, rear feet, and maybe its snout to dig into loose sediment, neatly composing a ground nest. This same creative exercise becomes more challenging with massive sauropods. For one, their small heads don't look as if they would have been much help in excavating a nest, and their snouts seemed ill suited for doing much more than grabbing plants. Looking elsewhere on a sauropod body, a quick cursory glance at sauropod feet also disappoints, as these do not have any obvious adaptations for scratching out a nest. Or do they?

Let's take a closer look at sauropod feet. One might initially think that the feet of sauropods and elephants were similar, considering that, just like elephants, sauropods were large terrestrial herbivores that greatly outweighed all of their contemporaries. Yet when their feet are compared, striking differences emerge, with the most obvious being their toes. Elephants tend to have blunt *unguals* ("nails"), whereas most sauropods have pointed unguals, which are especially apparent on their rear feet. Even more odd is how the ungual on the first digit of the rear foot of a sauropod—homologous with our "big toe"—is significantly larger than all of the others. Then the unguals in that same foot are all oriented so that they point away from the center of the body, instead of straight forward.

This same skewed orientation shows up in well-preserved sauropod tracks, so we know that this was a real trait in living sauropods and not something inferred mistakenly from putting together their foot bones the wrong way. All of this evidence thus led paleontologists to ask a simple question: What's with the weird feet?

Interpretations of these pedal oddities, which paleontologists had noted since the 19th century, have been quite varied and include: (1) better traction on uncertain terrain, kind of like how snow tires or treads on running shoes are used for the same function; (2) use in defense, in which a kick from a massive leg with huge curved claws would devastate a theropod attacker or sauropod rival; (3) use in mating (more about that later); and (4) digging out nests. Fortunately, with regard to digging, no one has seriously offered the idea that sauropods burrowed, in which their long necks might have served as periscopes. Instead, the phrase "digging sauropods" implies that they merely scratched a surface and never got too deep.

To test the "scratch-digging" hypothesis, two paleontologists, Denver Fowler and Lee Hall, followed up on Jackson's and others' discovery of sauropod nests in Argentina by more critically examining the rear feet of sauropods, as well as how living animals make ground nests. In their paper published in 2009, they proposed that many sauropod feet were perfectly adapted for digging out nests. A creative aspect of their study was to completely reject the "elephant foot" comparison, and instead look at the feet of much smaller modern tortoises. Once size is discounted, the foot anatomy of some sauropods and tortoise species are remarkably similar. This was an intriguing clue, because some tortoises, such as the gopher tortoise of the southeastern U.S. (*Gopherus polyphemus*), are magnificent burrowers and can make twisting 5 to 10 m (16–33 ft) long burrows. More important, though, females use their rear feet for scratching out their ground nests in which they lay their eggs.

To get a better idea of how sauropods might have dug out their nests, Fowler and Hall read many accounts of tortoises digging nests and watched videos of tortoises making nests, noting the following sequence of motions:

- Finding the right spot with suitable sediments for digging a nest, which sometimes was aided by the tortoise urinating onto the sediment to loosen it;
- Digging a small "starter" hole with the front feet, which does not always happen but can be part of the process;
- Placing her cloaca over the potential nesting spot, rear feet on either side;
- Digging with the rear feet, with each foot alternating, in which she moves her foot to the midline of her body and pulls up a dollop of sediment on the bottom of her foot, then places sediment outside of the hole. On the other hand (or foot, rather), she may just fling sediment out of the hole, or she might have saved enough urine to soften the sediment just a little more;
- Digging continues until she has made a hole about the same depth as the length of her rear limbs. The holes differ in overall shape, but most have been described as "flask-like," wider at the bottom and narrower at the top;
- Once egg laying commences, she might move the eggs a bit by shifting them laterally or pressing them down;
- After the entire clutch is laid, she buries the eggs with the excavated sediment on the sides of the hole, again using her rear limbs. The mother tortoise then walks away, and the eggs incubate under the enveloping sediment.

What they found out was that tortoises produced nests in a way that would explain sauropod nest building. Although the sizes and overall shapes of the tortoise and sauropod nests are quite different—flask versus trough, respectively—their origins are both explainable by scratch-digging. However, the whole sequence of events just described for tortoise nesting is not a perfect fit for nesting sauropods, either. For example, paleontologists who studied the Argentine sauropod nests do not think these eggs were actively buried by the sauropods. Still, other sauropod eggshells have the right kind of pore structure for having been buried and incubated under a layer of sediment, just like in tortoises. But we also presently have no trace fossil evidence about whether any sauropods urinated on the ground first before nest building, even though one sauropod urination trace fossil has been interpreted from Late Jurassic rocks in Colorado, explained in a later chapter.

The scratch-digging hypothesis was then further tested with sauropod eggs and nests in Late Cretaceous rocks in Spain. Bernat Vila, Frankie Jackson, and others looked at nine stratigraphic horizons with fossil eggs, which they classified as megaloolithid, an egg type affiliated with sauropods. Using some computer-aided 3-D modeling, they were able to reconstruct the original undistorted forms of the egg clutches, which also outlined the forms of the nests. These nests, which on average held clutches of 25 eggs, apparently lacked the rims of the Argentine nests, but their shapes had the same sort of asymmetry: shallow, oblong, wider at one end, and a little curved, like a kidney bean. These paleontologists also attributed such shapes to scratch-digging, in which the wider end was where the sauropod started digging with the rear feet, and the narrower end was about where the foot pulled out. They further proposed that the size and form of the enlarged unguals on the rear feet of sauropods were very likely responsible for making the widest and deepest part of the nest, in which the majority of eggs were laid.

So to make a 70-plus-million-year-long story short, the discovery of sauropod nests as trace fossils helped paleontologists test the hypothesis that sauropod feet were not just meant for walking, but also for digging. Nonetheless, perhaps alert readers noticed a wee problem with this hypothesis, which is that it does not take into account gender differences. In other words, did male or female sauropods dig nests, or did they do this cooperatively? If done either by males or cooperatively, this would have been a radical departure from reptilian behavior, in which only female tortoises, turtles, lizards, and crocodilians dig nests. Alternatively, many male–female pairs of birds share the burden of building a nest, so why not dinosaurs?

Another question remains in thinking about scratch-digging sauropods: Why would both genders have the same scratch-digging traits in their feet? One might also ask why both male and female mammals have nipples, and yet these are only functional in one gender (although I have not thoroughly nor personally tested this). Regardless, it is not such a big deal to have a similar anatomical trait related to reproduction in both genders of the same species.

All the same, Fowler and Hall thought about this problem too, and conjectured that large unguals in the front feet of sauropods may have been used for getting a grip—on their mates, that is. This certainly would have applied to male sauropods, in that they needed some way to hold on to a female while copulating. After all, the negative consequences of a 20-to-30-ton sauropod falling off its mate in mid-coitus would have caused much more damage than just humiliation in front of sauropod peers.

This is where the claws of a male sauropod's front feet, such as the first on each foot that pointed in toward their midlines, would have come in handy, acting as grappling hooks to hold on to the forward sides of a female when mating. However painful this action might have been for the female, it would have prevented both of them from toppling over, a mishap that would have had far worse consequences for both dinosaurs. (Nevertheless, this would have made for a magnificent trace fossil, perhaps combined with the body fossils of both sauropods that died on impact.) Trace fossil evidence that might test the "get a grip" hypothesis (also known as the "hang on, it's going to be a bumpy ride" hypothesis) would be broken or otherwise damaged front-foot unguals, caused by struggles between male and female sauropods trying to get into the right position, or gouge marks in bones near the front of a sauropod skeleton, such as on its scapula, ribs, or vertebrae just behind the neck. However, titanosaurs lack these front-foot claws, an absence that calls into question the "get a grip" hypothesis for these sauropods.

These nests and their egg clutches inspire yet another question, which is how sauropod mothers laid their eggs without occasioning onomatopoeia such as "crunch" and "splat." For instance, from a full standing position, a mother titanosaur's cloaca might have been as high as 4 to 5 m (13–16 ft) above the ground. An egg that fell from such a height would have become instantly scrambled upon impact—no matter how soft the substrate—and a very messy multi-egg pile-up would have resulted from nonstop laying. Clearly these sauropods had to reduce the height from which their eggs dropped, or at least slow down their descent.

This could have been done in several ways. One would have been for the sauropods to squat low enough that eggs had less distance to fall and hence generated less force. Alternatively, these mothers may have had an anatomical attribute, such as an extendable tube that projected from the cloaca and thus brought the eggs closer to the ground without the sauropod mother having to squat. Another method would have been to use mucus, which would have acted like a slimy envelope to keep each egg from going into freefall, instead gently lowering it to the nest. Sadly, sauropod soft tissues or mucus produced by these animals are extremely unlikely to have been preserved in the fossil record, so these are very difficult hypotheses to test properly.

In any case, an idealized suite of trace fossils documenting sauropod egg-laying would show her hind-foot impressions on either side of the nest structure, pointing forward and perhaps overlapping as she laid eggs from the rear of the nest to the front. If all things were equal in the mud or sand preserving the tracks, though, evidence of sauropod squatting should have generated more depth or deformational structures in and around her tracks, which would also help to explain whether they got close to the ground or stood up high. Also, however thrilling it would be to find a trace fossil of an egg impact, complete with splatter pattern caused by the explosion of yolk and eggshell, this is rather doubtful when one considers how quickly such an inept reproductive strategy would be lost from a sauropod lineage.

Based on what we now know from a blend of body fossils and trace fossils, nesting titanosaurs should have had the following sequence take place, from doing the deed to making the nursery, so to speak:

- Successful mating which resulted in fertilization (let's not forget that important step);

- A female and/or male sauropod, but probably the female, found a suitable spot for digging;
- The sauropod used one or both of its rear legs to scratch down and behind her, but added an outward twist to remove the loosened sediment;
- The mother stood over the elongated and slightly asymmetrical trough to deposit the eggs, probably starting at the rear and moving forward;
- The sauropod(s) may or may not have buried the eggs.

We can be relatively sure that mother or father sauropods did not sit on their egg clutches afterwards, or if they did, they did not pass on those genes. (See previous caveats about dropping eggs from great heights.)

Knowing about the nesting behavior of these titanosaurs and *Troodon* also makes one wonder if they stayed in the same place after laying their eggs, however temporarily that might have been. In other words, did they have any post-hatching care, like that seen in modern crocodilians and birds, or other dinosaurs like *Maiasaura*? We don't know an answer for sure yet, although some sauropod tracksites show a wide range of sizes in their tracks—from very small to extra large—suggesting that young sauropods and parents may have been in the same place and time. A few instances of adult sauropod skeletons preserved with much smaller, juvenile ones, such as in a few instances with *Camarasaurus*, also argue for young sauropods sticking close to their parents for a while after hatching. One thing is for sure: This question could be better answered by studying a combination of trace fossils, such as nests and tracks together.

The Oldest Dinosaur Day Care

Dinosaur body fossils make up the bulk of evidence for documenting the earliest years of dinosaurs, consisting of eggs, embryos, hatchlings, and half-grown juveniles. Although this wealth of information provided by unborn and baby dinosaurs has contributed to our understanding of dinosaur life cycles, paleontologists still have to face the reality that these all represent lives that were snuffed out before they behaved much. As just discussed, what do we know about the lives of hatchlings freeing themselves from their enclosing eggshells? How would we know through trace fossils whether newborn dinosaurs were altricial (depending on their parents for extended care), or precocial (able to leave mom and dad soon after hatching)? Furthermore, what clues can trace fossils give us about when such behaviors first evolved in dinosaurs?

One of the most impressive dinosaur nest sites with other trace fossils to address these questions is in Early Jurassic rocks of South Africa from about 190 *mya*. Because of their antiquity—these are the oldest known dinosaur nests in the world—this site also tells us much about the early evolution of nesting in dinosaurs, and is unparalleled for the sheer amount and quality of body- and trace-fossil goodness in a relatively small place. At this site are nest structures, eggshells, embryos, and juvenile footprints, all identifiable to the sauropodomorph *Massospondylus*. (Sauropodomorpha is a clade that includes sauropods and their relatives, among which were "prosauropods" like *Massospondylus*, lineages that only lived during the Late Triassic through the Early Jurassic, about 230 to 180 *mya*.) *Massospondylus* was not a massive dinosaur, probably weighing about 130 to 150 kg (285–330 lbs); this made it much smaller than *Maiasaura* or the nesting titanosaurs, but about three times bigger than *Troodon*. All in all, these fossils provided a rare and intimate look at dinosaur nesting from a time when this behavior might still have been relatively new in dinosaurs. To put this in perspective, nearly all other nests interpreted thus far, such as those of *Maiasaura*, *Troodon*, and titanosaurs, are from about 80 to 65 *mya*, meaning the South African nests are more than twice as old as these.

Paleontologists had known about *Massospondylus* embryos and eggs since at least the 1970s, but the exact source of these fossils had not been documented until paleontologists followed up on a few clues much later, in 2006. The results of this exploration and discovery, summarized in a report published by Robert Reisz and four other paleontologists in 2011, were spectacular. The first embryos and eggs were discovered at this site in 1976 in loose rocks on the ground, next to where a highway crew had cut into the strata to allow better passage of a road in Golden Gate Highlands National Park. (Such exposures, which geologists worldwide revere and covet nearly as much as easy access to beer, are known simply as roadcuts.) This meant that people visiting this national park—unaware of what world-class paleontological secrets resided there—had driven by the undiscovered nests, tracks, and other fossils there for forty years before the rocks were scraped again.

The paleontologists working on this roadcut were elated to find that it held a mother lode of mother-dinosaur data. Contained within a vertical span of siltstone about 2 m (6.7 ft) thick and across a horizontal distance of 23 m (75 ft) were ten egg clutches; eight of these were apparently unmoved, implying these groupings of eggs marked the original sites of the nests. (Keep in mind, though, that this is only a two-dimensional sample, and many more nests might be inside the roadcut awaiting geological unveiling.) The nests were on four horizons and close to one another, which the paleontologists interpreted as good reasons to infer site fidelity and nesting colonies. If so, these were the oldest known examples from the fossil record for dinosaurs, or any other vertebrates for that matter.

One of the clutches had as many as 34 eggs, with perhaps a few more lost to erosion, a significant number of eggs compared to other dinosaurs. From an ichnological perspective, the paleontologists could not find any other evidence of actual nest structures that fulfill all of the previously mentioned criteria for dinosaur nests, but they inferred their presence from depressions that held the tightly packed eggs. The eggs, which formed discernible rows within each clutch, also may have been organized thusly by one of the parents after laying. Invertebrate trace fossils (burrows) in the sediments surrounding the eggs show that the clutches may have been partially buried in soil by one or more of the parents before being buried in a more lethal way by river floods. This assumption is strengthened by the close-fitting arrangement of the egg clutches.

If these eggs had been left in an open nest when a nearby river flooded, they surely would have been scattered about like balls in the opening break of a billiards game.

Nonetheless, it was not only the nests and eggs that made this suite of fossils truly extraordinary, but also some startling dinosaur trace fossils in the rocks around the nests: teeny footprints. These four-toed tracks, preserved as natural casts on blocks of rock that fell from the roadcut, just happened to match the front and rear feet of a sauropodomorph like *Massospondylus*. Yet these tracks were only about 15 mm (0.6 in) long, just smaller than a U.S. dime. When compared to embryonic *Massospondylus* foot bones, the tracks were made by sauropodomorphs about twice as big as expected for hatchlings, and showed they were walking on all fours, in a manner similar to that interpreted for the adults.

This trace fossil evidence, along with the proximity of the tracks to the nest sites, suggested two points: the hatchlings had enough time to grow larger, and they must have been receiving parental care near the places they were born. These tracks, then, extend the evolution of dinosaurs as good parents more than a hundred million years farther into the geologic past than previously supposed, thus providing yet another example of why trace fossils tell us much more about dinosaur lives than just bones, eggs, or other body fossils.

CHAPTER 5

Dinosaurs Down Underground

Discovering a Digging Dinosaur

Big surprises often arrive in small packages; in this instance, it was in an e-mail message. On August 26, 2005, while preparing to teach my first classes of a new semester, I was glad to see the message from my friend Dave Varricchio. Although we had known each other since 1985, chatted in person at professional meetings, and had kept in touch via e-mail, at best we got together only once a year and our exchanges were infrequent. This seemingly unsocial behavior was not because we were less cordial with one another, but instead could be attributed to distance, time, and personal demands. Dave was busy earning tenure in his assistant-professor position at Montana State University in Bozeman, Montana, while also serving as a staff paleontologist at the Museum of the Rockies there. Meanwhile, I was at Emory University in Atlanta, Georgia, having just finished rewriting the latest edition of a dinosaur textbook and teaching full-time.

The longest span of time we had spent together since our graduate-school days—which was in the late 1980s—was a week of field work in northwestern Montana in 2000. While there, he and I took a close look at the Late Cretaceous fossil insect cocoons and burrows near the *Troodon* nest sites. Yet the resulting research paper from that investigation was dormant and nowhere near ready for submitting to a journal. So what could be the purpose of his message?

The message subject gave me no clue whatsoever. It simply said "specimen." Dave's message text was similarly cryptic, albeit enlivened by an injection of Freudian humor:

> *Hey, What do you think of this feature in the photo? Yes, it looks somewhat phallic, but that's not what I'm asking. The central structure ranges from sandy mudstone to sandstone and crosscuts claystone. Beds dip to the left. Brush is one inch wide. DV* [Dave Varricchio]

I opened the photo attached to the message and studied it. The top part showed a red mudstone overlying a gray mudstone. The latter had a big chunk removed—probably by shovels and picks—which freshly exposed a darker-gray rock underneath its weathered surface. There was indeed a "central structure" within that gray mudstone (what Dave called a "claystone"), so I focused my attention on that. It protruded slightly compared to the rock around it, so it was made of more resistant material, and was a lighter gray.

These observations and inferences of mine synced with Dave saying it was a sandy mudstone or sandstone, which would have contributed to its bas-relief appearance. It also had a horizontal segment toward its top, which turned abruptly downward into a vertical segment. In my imagination, I converted this part into three dimensions, and a spiral shape came to mind. The vertical part connected directly to a white oval near the bottom of the photo, which had a piece of shiny metal poking from it. Even though I almost never worked with fossil vertebrate skeletons, I nonetheless recognized the white oval as a plaster jacket, which must have been placed around the remains of a fossil vertebrate. The metal was likely aluminum foil the field crew used to cover part of the rock around the skeleton before slapping burlap strips soaked in wet plaster of Paris around the fossil specimen, the latter a time-honored practice in vertebrate paleontology since the 1880s.

And yes, there was a small field brush as the only scale in the photo, off to the left. Assuming a width of about 2.5 cm (1 in), I used that to extrapolate its length as about 15 cm (6 in). I then applied this brush length to figure the approximate length and width of the plaster jacket. Using this seat-of-the-pants reckoning, it came out to 55 to 60 cm (22–24 in) long and 40 cm (16 in) wide. The central structure kept a consistent width throughout its length, and was also about 30 to 35 cm (12–14 in) wide, flaring some as it connected with the rock holding the plastered specimen.

I waited until the next day to write and send Dave my assessment of what was depicted in the photo. It went like this:

> *Hypotheses:*
>
> 1. *Large burrow and terminated with the burrowmaker (the latter inferred from expansion at the end and your nicely jacketed specimen)*
> 2. *Large burrow and terminated with some critter that was not the burrowmaker*
> 3. *Collapse feature, but with some hapless critter at the end of it that fell into/drowned in it*
> 4. *Impact structure, with aluminum from the alien spaceship still preserved (and the spaceship safely concealed from prying eyes of the guvmit [government])*
> 5. *Not enough information, you speculative ichnologist!*

I ended my reply with a summary interpretation: ". . . it would most likely be a bank burrow (adjacent to a channel) going into a fluvial [river] overbank deposit . . . I'll guess that you got a crocodilian at the end of its burrow."

My thinking at the time was influenced by what I knew about big modern burrows I had been studying on the Georgia barrier islands, namely alligator dens. Alligators are represented by two species that live in widely separated places: the southeastern U.S., which has the American alligator (*Alligator mississippiensis*), and eastern China, with the Chinese alligator (*Alligator sinensis*). Both are burrowers, especially the Chinese alligators, which dig tunnels as long as 20 m (67 ft) into soft sediments next to water bodies. The American alligator burrows are not so extensive, but still impressive structures. These can be more than a meter (3.3 ft) wide at their entrances, are 4 to 6 m (13–20 ft) long, and, like their Chinese relatives, are made next to ponds or intersect the local water table.

Dens are used by alligators for a variety of reasons, such as keeping their skin from drying out from sun and heat, maintaining more reliable "indoor" temperatures during cold winters or hot summers, or providing safety from droughts or fires. Most important, though, these burrows serve as a refuge for young, newly hatched alligators, which are watched closely for more than a year by the ultimate of overprotective mothers. In my studies on the Georgia barrier islands, I had already experienced a few too-close-for-comfort encounters with alligator mothers and babies near their dens and learned to proceed with extreme caution near them, treating all such large burrows as if they were loaded guns.

In their evolutionary history, Chinese and American alligators shared a common ancestor not too far back in the geologic past, but they also connect to all other crocodilians, a lineage that extends well into the Jurassic Period. However, despite this ancient heritage, no crocodilian burrows had been interpreted from the fossil record. This made the structure in Dave's photo potentially exciting, but not knowing the age of the rocks in the photo, or even their location, I could not allow myself to get too enthused about it. Still, I knew that Dave often prospected Cretaceous-age rocks in Montana and other parts of the western U.S., so I held some hope that he had found a Cretaceous example of a crocodilian burrow, and one with a crocodilian at the end of it. Such a discovery would have been very pleasing to me, because it would have linked directly with my Georgia-coast interests in alligator dens, while also documenting a much deeper history of burrowing behaviors in crocodilians.

My mild curiosity concerning whether I was right or not about my crocodilian-in-a-bank-burrow hypothesis was addressed later the same day, but in a completely unexpected way. It turned out that I was both right and wrong. Dave's next message had an intriguing title: "hoping you say 'yes'." Here are the first three sentences of text from his note:

> *You have correctly inferred that there is a vertebrate at the end of this structure. But it's not a [ancient] crocodile. It's a dinosaur, a hypsilophodont.*

Never had I been so happy to be mistaken. It was a dinosaur in a burrow. Sure, lots more scientific questions had to be asked before jumping to any more conclusions, but this revelation warranted a happy dance (as demonstrated by Snoopy in *A Charlie Brown Christmas*), followed by popping the cork on a champagne bottle, and concluded by a happy dance while swigging vigorously from that champagne bottle. Lacking champagne, though, I merely pried off the caps of whatever beers were in my fridge at the time and toasted this discovery with my wife Ruth, who was promptly sworn to secrecy about Dave's discovery.

Just to give you an idea of the depth of my wrongness in not thinking about a dinosaur as a possible burrowmaker, as well as the consequences of Dave and his field crew finding a dinosaur in a burrow, the following is a definitive statement I had made in the second edition of my then-newly revised dinosaur textbook. As you read this statement, keep in mind that I had just extensively rewritten the book, after blistering peer review and editing, and it was already on its way toward publication a few months later:

> *Furthermore, no convincing evidence has revealed that dinosaurs lived underground because no dinosaur has ever been found in a burrow, nor have any burrows been attributed to them.*

By the way, did I mention I was wrong? Of course I did not think "dinosaur" when looking at the photograph earlier in the day, as all of my ideological blinders had been switched on. Was I studying modern alligator dens at the time, biasing me to think "crocodile" instead of "dinosaur"? Check. Had I previously thought about burrowing dinosaurs but rejected their existence on the basis of what was then known? Check. Had I let these biases overrule the fact that Dave studies dinosaurs almost exclusively? Check. Was I an idiot? Check. But I was pretty damned delighted to be one in this instance.

Dave's message had more to say, including a brief confession of his long-held suspicion that other small dinosaurs such as "hypsilophodonts"—a non-clade grouping of small ornithopod dinosaurs—might have burrowed. This informed guess was based on his observation of the small Late Cretaceous hypsilophodont *Orodromeus* near his *Troodon* nest sites in northwestern Montana. In more than a few specimens, Dave had noticed that the *Orodromeus* skeletons were unusually complete and concentrated in compact masses, encased in well-cemented rocks. The simplest explanation for this unusual preservation, he thought, was that these small herbivorous dinosaurs might have been in chambers at the ends of those burrows, then either were buried after dying or buried alive in their burrows. Nonetheless, he had no other evidence for this idea, such as an actual burrow connecting to these masses of bones. So it remained just a nagging conjecture awaiting more persuasive evidence. And now he had it: a probable burrow, attached to a probable burrow chamber which held the body fossil of its possible maker, which just happened to be a hypsilophodont dinosaur. The last paragraph in Dave's message neatly summarized and resonated with my feelings at the time:

> *Obviously . . . some interpretation is dependent upon the actual specimen within . . . but still . . . a cool specimen. I think that really only hypothesis 1 and 2 are suitable. The sequence includes paleosols not too unlike the Egg Mountain area.*

When he referred to "hypothesis 1 and 2," he meant what I had enumerated previously: (1) a large burrow that had the burrow-maker at one end, which in this instance was a dinosaur; and (2) a large burrow with a dinosaur at its end, but one that was not the maker of the burrow. In this second scenario, the dinosaur could have lived in the abandoned burrow of another unknown animal—dinosaurian or otherwise—or had its remains washed into the burrow where they were subsequently buried and fossilized. This would have been quite a geological coincidence, but it still had to be considered as a possibility. Regardless, Dave was right about one thing: it was a cool specimen.

My reply to him the same day was understandably animated, and I wrote my responses in between his quoted text to approximate a call-and-response dialogue. Following his quotation "It's a dinosaur, a hypsilophodont," I said:

> [Colloquial expression—censored here—denoting excitement and enthusiasm, containing several profane words, and comparing the discovery to the application of a foot to one's posterior.]
>
> *What I meant, of course, was that is a most interesting observation. Congratulations on a noteworthy find, perhaps warranting further investigation.*
>
> [Repetition of colloquial expression—also censored here—denoting excitement and enthusiasm, containing several profane words, and comparing the discovery to the application of a foot to one's posterior.]

Discoveries are only the beginning, though. It was time for us to do a lot more science.

I Have No Shovel, and I Must Dig

Only in the past fifteen years or so have paleontologists realized that a few dinosaurs dug into the ground underneath them. As we learned previously, scratch-digging sauropods were one such example, in which titanosaurs used their specially adapted feet to gouge the earth before laying their eggs. *Troodon* was another, and it might have used both its hands and feet to excavate and shape its rimmed nests. A few other dinosaurs had anatomical attributes that could have helped them to dig, such as the small Late Cretaceous theropod *Mononykus* of Mongolia, which had weird shortened and pointy forearms that looked like pickaxes. These appendages have been inferred as tools for breaking into termite or ant nests, gouging these nests to gain access to their goodies, sort of like a dinosaurian anteater. Moreover, one specimen of the Early Cretaceous ceratopsian, *Psittacosaurus*, was found with 34 juveniles just like it, all collected in a bowl-like depression. Paleontologists speculated that all of these animals may have died together in a collapsed burrow, but no one could tell for sure if this is what happened, nor whether or not the adult made the depression.

More recently, in 2010, trace fossils attributed to predatory dromaeosaurs—such as *Deinonychus*—were interpreted as dig marks made by claws on their rear feet as they tried to unearth small mammals, whose preserved fossil burrows were just underneath these dig marks. But these and other dinosaurs were only scraping the surface, so to speak. What about dinosaurs that went deeper and actually spent extended periods of time underground?

Before jumping into our time machines and going back to the Mesozoic Era, a much better idea to ask ourselves first is what modern vertebrates might tell us about burrowing behavior. After all, the variety of today's burrowing vertebrates is astounding, ranging from lungfish to spadefoot toads to skinks, alligators, naked mole rats, puffins, aardvarks, and more. Furthermore, just limiting your list to mammals that live part or most of their lives underground will result in hundreds of species, and they range in size from Oldfield mice (*Peromyscus polionotus*) to grizzly bears (*Ursus arctos*).

There is no mystery about why so many vertebrates burrow, either, with a number of perfectly fine evolutionarily based reasons to excavate a home and live in it. For one, parents can raise their young in a safe, quiet place, away from the prying eyes and noses of predators. In upland environments, where undergrowth and trees constitute fuel for wildfires, burrows act as shelters from those natural hazards. Burrows also maintain equitable temperatures year-round. Thus, in places with extreme daily or seasonal fluctuations in temperatures—especially in deserts or polar regions—burrows are not too hot, not too cold, but just right. For a few animals, such as lungfish or some toads and turtles, these burrows serve as aestivation or hibernation chambers in which they conserve energy and water during lean times, sometimes for months or years.

Given this burrowing norm in modern vertebrates, including the living relatives of dinosaurs—birds and crocodilians—along with the incredible diversity of dinosaurs during their 165-million-year history and the myriad complex behaviors paleontologists had inferred from their bones and trace fossils, one would think that someone, somehow, had thought of dinosaurs living underground. Turns out that, yes, a few people had thought of subterranean dinosaurs, although not in the formal sort of way that stood a chance of acceptance in the face of fierce skepticism by the paleontological community.

Indeed, the case for burrowing dinosaurs was not helped in that one of the first people to publicly propose it was the bombastically flamboyant and iconoclastic gadfly of dinosaur paleontology, Robert ("Bob") Bakker. More P.T. Barnum than Barnum Brown, Bakker is well known as a great populist of dinosaur lore, with many of his views summarized in his still-intriguing and never-dull 1986 book, *The Dinosaur Heresies*. The popularity of this book stemmed at least in part from his gleefully sticking fingers, toes, and other appendages into the eyes of stodgy paleontologists who viewed dinosaurs as up-scaled, dull, and cold-blooded variations on lizards. As a result, Bakker is often credited for rebooting dinosaurs in the public imagination as active, hot-blooded animals that acted much more like birds or mammals. (Bakker was not original in this concept, though. Instead, paleontologists rightfully laud his Ph.D. advisor, John Ostrom, as the person who built the foundation for the "birds are dinosaurs" argument in the early 1970s.) Despite the hit-and-miss nature of Bakker's assertions and conjectures about dinosaurs in his book, it nonetheless ensured him a near-permanent presence on dinosaur documentaries from then on, in which he disagreed with nearly every other paleontologist. After all, conflict sells.

Anyway, Bakker speculated that *Othnielia* and *Drinker*—both of which were small hypsilophodont dinosaurs that lived during the Late Jurassic Period—must have been living in burrows of their own making. He did not present much data to support this supposition—like, say, trace fossils of their burrows—but instead based it on how these dinosaurs were little and some of their carnivorous contemporaries were big. Thus, burrowing would have been a great way for these cute, innocent herbivores to avoid getting chomped by nasty, ravenous predators. Bakker mentioned this idea of burrowing ornithopods in a 1996 paper, and promoted it in talks afterwards.

As mentioned earlier, digging was proposed as early as 1985 to explain sauropod feet, which was confirmed about two decades later by their nests. *Troodon* nests found in the 1990s also showed that some dinosaurs could dig. Based on their anatomies, the small theropod *Mononykus* was mentioned as a possible burrowing dinosaur, as was the small ornithopod *Heterodontosaurus*. However, for *Mononykus*, its forearms seemed too short to do much more than break up soil.

Sure, all of this talk about whether dinosaurs lived underground or not is very interesting, but runs the risk of becoming an esoteric subject that is only appreciated by the most dedicated of dinosaur fan-boys or fan-girls. Ultimately, one has to ask that most underused but instructive of questions in science: So what? Other than being able to exclaim "We got ourselves a burrowing dinosaur! Yee haw!" what meaning did such a discovery have in a much larger sense, relating to the Mesozoic world and modern times alike?

Here is one word that should warrant sitting up and paying attention: extinction. As discussed earlier, most scientists now accept that a large meteorite hit the earth about 65 *mya*, landing in the area of what is now the Yucatan Peninsula. Apart from those that died from the direct hit, this impact likely had dire consequences worldwide for life in the oceans and on land, including a dust cloud—mixed with soot from forest fires—that would have blocked out sunlight for several years. Take away sunlight and terrible things happen to most ecosystems. For example, many photosynthesizing algae in Cretaceous oceans were adapted to full sun year-round, as well as land plants that grew and reproduced best when illuminated. Most of these species would have died within a year or two. Plummeting global temperatures also would have contributed further to this massive die-off of photosynthesizers, as the dust cloud would have acted like the ultimate sunscreen, preventing infrared radiation (heat) from making it to the earth's surface.

Far fewer photosynthesizers meant less food was being produced at the base of nearly all ecosystems, which meant animals that ate these photosynthesizers died too. After that, animals that ate the animals that ate photosynthesizers died. (Although I also imagine scavengers very briefly having enjoyed an end-of-the-world party, greedily chowing down on a sudden bounty of dead animals.) Think of the food web holding together an entire ecosystem having one thread after another ripped apart. We witness such trophic cascades today in ecosystems that undergo a precipitous shock such as a major forest fire, volcanic eruption, or tsunami.

Yet the end-Cretaceous was far worse because it was global in scale, affecting the food supply for animals in the seas and on the continents alike. Also, animals living at the end of the Cretaceous with the misfortune of having large body sizes had greater caloric needs, meaning bigger was not better for them in this situation. So large animals likely died the soonest after the impact, and yes, I am talking about dinosaurs, including much beloved ones such as *Triceratops*, *Tyrannosaurus*, *Hadrosaurus*, and *Ankylosaurus*, to name a few. All non-avian dinosaurs had no way to survive such terrible conditions, even the ones living in polar environments. Thus they went extinct, while some of their somehow luckily adapted descendants—birds—survived and thrived, giving dinosaurs a second chance in a radically transformed world. Among land-dwelling vertebrates, mammals, crocodilians, turtles, lizards, and amphibians also got tickets to ride on the Cenozoic Express; their lineages evolved and had their own more modest extinctions since.

Now throw a post-Cretaceous monkey wrench into this scenario in the form of burrowing dinosaurs. Burrowing was a strategy that at least a few small non-avian dinosaurs could have used to survive an end-Cretaceous "meteorite winter." Just recall a few of the advantages of a burrowing lifestyle, and suddenly an apocalypse becomes ever so slightly more optimistic: equable temperatures year-round, protection from hungry predators (made more desperate from a lack of food), a room to hibernate, and—most important from a delaying-extinction standpoint—a safer place to raise offspring.

It is not a coincidence that people of the 1950s and 1960s, to cope with nuclear-tinged fears of the Cold War, built underground bomb shelters in an attempt to assure themselves they would survive a nuclear exchange between the Soviet Union and the U.S. More realistically, though, such shelters only would have delayed the inevitable, and perhaps small burrowing dinosaurs lived only a few years longer than their larger, non-burrowing relatives before succumbing to the wretched after-effects of a massive meteorite hitting their planet. Nonetheless, knowledge that at least a few dinosaurs had the means to live just a few more years bolstered credence to a concept of dinosaurian resilience, defying long-held and unfair stereotypes of maladapted ineptitude summarized in the phrase "dead as a dinosaur."

Lastly, burrowing is a behavior present in every major group of vertebrates that made it past the mass extinction: birds, mammals, crocodilians, turtles, lizards, and amphibians. Could burrowing have been one of the essential tools in a Swiss Army knife of adaptations possessed by terrestrial animals, which came in handy once disaster struck? Is it possible that an ability to burrow has allowed certain lineages of animals, such as crocodilians, to make it past multiple mass extinctions in earth's history? How might burrowing aid future generations of modern animals to survive human-influenced extinctions? All such questions swirled around the revelation that at least maybe one dinosaur burrowed well before the last breaths of the rest of its kin.

Digging up the Diggings of a Dinosaur

In September 2005, one month after learning about this dinosaur and its possible burrow, my friend Dave thought it would be a good idea to have me come out to Montana to look at the burrow with him. Despite my hesitation—thinking about it for as long as three seconds—I found no reason to disagree with him. Fortunately, he had research funds available to buy me an airline ticket, so I flew from Atlanta to Bozeman, and within a day of my arrival he drove us to the field site. It was about three hours away in southwestern Montana, an area of the country I had never before seen.

The nearest town of any notable size near the field site was Lima, which has a population of just more than two hundred. Travel from the claustrophobic overpopulation and traffic of metropolitan Atlanta to the unpeopled expanses of "Big Sky" country requires some mental adjusting for me, but is always a welcome kind of different. During the drive, Dave and I caught up on what had been happening in our lives the past few years, but he also filled me in on some of the details about what led up to my seemingly improbable presence there with him, united by an ichnologically significant find.

This region of Montana had outcrops of the Blackleaf Formation, which was composed of mudstones and sandstones formed by rivers during the middle of the Cretaceous Period, about 95 *mya*. Few fossils of any kind had been reported over the years from the Blackleaf, but among those were some scrappy and vaguely identified dinosaur bones. Apparently, this was enough incentive for Dave to decide it was worth prospecting for more dinosaur material. So during the summer of 2005, he gathered a group of experienced dinosaur-bone spotters and pickers to accompany him to places with extensive exposures of the Blackleaf Formation, some of which were on U.S. Forest Service land.

During one of their searches, one of the crew, Yoshi Katsura, spotted a few bones protruding from a hillside. Once they determined that these were from a small ornithopod and that more remains were probably just underneath the surface, they decided to collect them. They excavated around the potential skeleton, not knowing exactly what was there, but leaving plenty of rock around the exposed bits so that the specimen would be as complete as possible. The field crew was happy about the find because it meant that the Blackleaf might have dinosaur skeletons after all. Even better, because it had been explored so little, it might even hold species previously unknown to science.

It was during their digging around the specimen, and Dave taking notes about the sedimentary rocks immediately around the bones, that he began to notice something peculiar. What he experienced is a feeling shared by many field-oriented scientists, expressed succinctly in a quote by famed science fiction writer Isaac Asimov: "The most exciting phrase to hear in science, the one that heralds new discoveries, is not 'Eureka!' but 'That's funny . . .'" Dave noticed that the bones were concentrated in an egg-shaped mass of sandstone, which also was better cemented than the surrounding and overlying mudstone. He looked above where the bones were located, and saw that the sandstone continued upward as a vertical structure, and then turned to the right. *That's funny,* he thought. This structure had to have been a significantly sized hole that was later filled by sand from above, making a natural cast of it. What would have made such a hollow form during the Cretaceous Period, and one that also had a dinosaur at the end of it? Even better, the dinosaur was a small ornithopod. Although Dave could not yet identify its species, it reminded him of *Orodromeus,* another small ornithopod dinosaur he had seen often in northwestern Montana. As mentioned before, he had suspected *Orodromeus* was a burrowing dinosaur, but lacked the key supplementary evidence for it: namely, a burrow. But now it looked like he might have both a trace fossil and its tracemaker together.

Fortunately for him, and for vertebrate paleontology in general, Dave is a careful observer and cautious with his interpretations. He knew that much more evidence needed to be gathered to address a few key questions. For example, was this odd structure really a burrow, or just a hole in the ground made by some other process, such as erosion along a riverbank? Even if it were a burrow, was the dinosaur in it the same one that made it? If this dinosaur was not the burrowmaker, what was it doing there, and how did it become entombed in the burrow? Did it die in there, or did it die somewhere else and was later placed in the burrow, whether by river currents or a predator that used the hole as a cache?

Dave shared a few of these thoughts with the rest of the field crew at the time, and once the specimen was jacketed and detached from the earth, they carried it out on a stretcher, a victim of Cretaceous circumstances undergoing its last journey. He and everyone else then had to wait for the skeleton to be prepared at the Museum of the Rockies in Bozeman, with the hope that its bones would tell Dave more secrets.

However, before this specimen was separated from its entombing sediment in a preparatory lab, Dave had used me—in an intellectually sordid way, I might add—to test his hypothesis. His exclusion of much information about the find, such as its age, location, interpretation of the original environment, and the oh-so-minor detail that it involved a dinosaur, was purposeful and sneakily effective. Still, it reduced the possibility of bias in my assessing the photo and ensured that I would interpret it solely on its face value as a possible trace fossil and not as something I was hoping to see.

Once in Lima, Dave had prearranged a meeting with a friend and his wife who were passing through Montana that week. Both were non-geologically inclined but able-bodied and eager to help us, so the four of us went to the field area. Like many dinosaur discovery sites in the western U.S., getting there involved getting off pavement and going down a dirt road, with cow pastures on either side. Once on U.S. Forest Service land, we parked, unloaded shovels, picks, and other human-powered earth-moving tools from Dave's car, and began walking across a sparsely grassed expanse with rolling hillocks and minor gullies.

I scanned the ground as we traveled, looking for pieces of sandstone weathering out of the mostly weathered mudstone bedrock in the area, stopping to pick up and look at the more interesting samples. Some of these sandstone bits contained small holes that led to longer tubes, which I recognized as burrows. These trace fossils were made by invertebrate animals—probably insects—living with the dinosaurs in this area about 95 *mya*. Other than these ichnologically motivated pauses, the stroll to the discovery site was surprisingly short, taking only about twenty minutes. We stopped when Dave said "This is it" and started digging.

In accordance with their permit for working on U.S. Forest Service land, the field crew had followed a "Leave No Trace" edict by burying the discovery site and restoring it to an approximation of what it looked like before their arrival. I found this amusing on many levels. For one, the disturbed soil there betrayed the crew's former presence, as did the dried flecks of white plaster on the ground, which had persisted despite whatever weather had taken place between their jacketing the dinosaur in July and our visit in September. I also pondered the metaphysical implications of the Forest Service imploring us to minimize our tracemaking, which involved digging, while we investigated a trace fossil from a digging dinosaur. I further wondered whether such thoughts might generate a self-absorbed poem, read aloud later at an urban coffeehouse and accompanied by clicking fingers of appreciation. Fortunately for much of the world, this ambition slipped away as soon as our shovels chopped the ground and we started to pile weathered mudstone around us.

At some point in our careful excavation, we reached the surface where Dave and the others in the field crew had placed plastic sheets above the possible burrow and the not-so-final resting spot of the dinosaur skeleton. From here we became even more vigilant and began to clear away some of the softer reddish-green mudstone that enveloped the more solid whitish sandstone filling the presumed burrow. However, this work was cut short as the sun thwarted our objective by falling toward the western horizon. We hastily covered our previous efforts with a shallow layer of sediment. Dave's friends had to leave the next morning, but he and I would come back to see what lay beneath. That night, we enjoyed a meat-and-pie-laden dinner at a restaurant in nearby Dell, Montana—The Dell Calf-A, with its name written cursively with a neon "lasso"—and we slept at a cheap motel with running water, electricity, and an intact roof overhead. In other words, we were experiencing the height of luxury for field paleontologists.

In the morning, Dave and I said good-bye to his friends, who were off to see other parts of Montana, and he and I returned to the field site to resume our handiwork. The fresh familiarity of the previous day made the walk seem shorter, although we also moved a little more quickly in anticipation of learning something new that day. Once at the site, we first used shovels to uncover the thin layer of sediment dumped on top of our objective, but soon set these aside once the white sandstone emerged. Replacing the shovels with rock hammers—pointy ends first—and small hand picks, we began breaking the mudstone closest to the sandstone. Our goal was to reveal this mystery structure in all of its full three-dimensional glory, which would answer many of our questions or, more likely, generate a lot more.

While we worked, I kept comparing the whitish structure in front of us to the flat digital image Dave had sent me the previous month. It definitely was a big, sandstone-filled tube, cutting across the mudstone around it. Starting from above, it turned into a horizontal segment, and then twisted down to where the dinosaur bones resided until just recently. Despite this torsion, its width stayed constant throughout, at about 30 to 35 cm (12–14 in). If emptied of its sandstone fill and turned into a hollow tube, it would have been too tight of a fit for Dave or me to crawl into (especially after the previous night's meal). But it would have been perfect for a five-year-old child, or a small dinosaur.

What was its exact form? Before we started, I told Dave that I thought it was semi-helical, descending with regular turns and not quite making a complete spiral: more like a water slide rather than a corkscrew. But this was mostly a hunch based on what I had seen in much smaller invertebrate burrows with similar forms. Was it wrong for me to apply the same reasoning to something many orders of magnitude larger? I wasn't sure, considering that this was the first time I had seen something like this in the geologic record.

I also speculated about trace fossils of other animals that might be there and connected directly to the supposed dinosaur burrow. For example, modern gopher tortoise burrows play host to 200 to 300 species of other animals, making them more like underground menageries and not just holes in the ground. Some of these commensal animals include burrowing insects and mice, which readily punch through burrow walls to add their own homes, like adding extra rooms to a house. The day before, I'd said to Dave, "If this is a burrow, it should have commensal burrows, too," followed by an explanation of what little I knew then about gopher tortoise burrows and their cohabitants.

Using the stark textural and color contrasts between the sandstone and mudstone as our guide, we chipped away at the mudstone around the sandstone, feeling like sculptors collaborating on an expression of our inner visions. In that vein, as the fuller form of the structure began to emerge from the mudstone, Dave stopped and with a nervous laugh asked, "Are we sure we're not just making this up?" It was a good self-doubting question to ask. Ichnologists have heard variations of this inquiry for years, often posed by people (including other paleontologists) who know absolutely nothing about ichnology. Yet it is stated confidently because, after all, their skeptical ignorance trumps our expertise. Still, ichnologists are also self-effacing enough to acknowledge the possibility that a supposed "trace fossil" might later turn out to be a random inorganic structure, blemish, break in a rock, or other such oddity that has nothing to do with traces of life.

With this possibility in mind, Dave and I sat back and reassessed the situation. After some back-and-forth deliberation, we soon agreed that the differences in sediment between the sandstone and mudstone were too sharp, too distinct to be something we were imagining. Reality checks are good, as is documentation. As a Russian saying goes, "Trust, but verify." We got back to work.

With the removal of each piece of mudstone, the structure became better defined and took shape. Once done with our handiwork, we stood up and then walked around it to gain a fuller perspective. It was a curious object. For starters, the horizontal segment joined with a gently inclined one above, and another inclined one below, making it just more than 2 m (6.7 ft) long. The last segment joined with the enlarged part of the sandstone where the dinosaur had been. Sure enough, just as I had visualized earlier, the whole thing was semi-helical, close to an "S" or a "Z" lying on its side. Looking at it from above, we could see it made an abrupt right turn, and then a left, ending at the spot where the dinosaur had been taken out. The first two segments were about equal in length, 60 to 70 cm (24–28 in), and their widths were within a narrow range of 30 to 38 cm (12–15 in), a little bit taller than wide. The gentle inclines for each of the segments added up to a vertical drop of about 50 cm (20 in), which was about the length of my lower leg.

The field crew had truncated the lowermost part of the structure when they took out the expanded mass of rock and its dinosaur in the supposed burrow chamber at its end. Despite the absence of this chamber, we drew its approximate outline on the ground, which helped complete our understanding of the entire structure. This was a real burrow all right, and with a form and dimensions that made sense as a burrow. We started to talk about how this geometry would have worked for a small dinosaur moving through it, tail included. We also wondered whether the dinosaur collected from here, once prepared, would turn out to be too large, too small, or just right for this structure as the burrowmaker.

But this is what really got me excited: the structure had small projections of sandstone poking from each of the two corners of the bending tunnel. On the upper turn, a bundle of a half-dozen horizontally oriented narrow sandstone cylinders—each about the width of a pencil—suggested that something else had been added on to the main structure. A horizontal sandstone cylinder about the width of a baton stemmed from the lower turn. "There they are," I said to Dave. "Commensal burrows." These were likely small burrows that also had been hollow spaces connected to the main burrow and filled in by the sandstone. Insects could have made the upper cluster of burrows, whereas the lower burrow was more appropriately sized for a small Mesozoic mammal. The study of modern traces—also known as *neoichnology*—had been kind to us, with the gopher tortoise burrows of Georgia supplying a sensible explanation for these seeming aberrations.

We took as many measurements and photos as we could manage, and each of us drew labeled sketches in our field notebooks. Neither of us planned to ever come back there, so we made sure to gather enough information to analyze what was there and later write a coherent story about it. Once satisfied with our data collection, we placed plastic sheeting over the structure and buried it as best we could. By interring the structure instead of leaving it out in the open to weather, we increased our chances of it being there for us to study again someday if necessary. Furthermore, given the GPS coordinates and our descriptions of how to find it, future generations of other paleontologists could investigate it, no doubt aided by new technologies or knowledge that might lend further insights on what we so crudely described in 2005. "Leave No Trace" took on a duplicitous meaning as we covered this fossil burrow and prepared to say good-bye to it.

During the drive back to Bozeman, Dave and I discussed and debated more about what we had just seen, felt, and studied. By the end of the day, though, we were nearly convinced that this sandstone-cast feature was indeed a large burrow. Yet we also knew that we needed much more information before declaring that this was the world's first known dinosaur burrow. It was time to take a closer look at the dinosaur found in the burrow, and find out whether it had died in a burrow of its own making or not.

Documenting the Discovery of a Digging Dinosaur

Meanwhile, the jacketed dinosaur was still in its presumed burrow chamber, surrounded by plaster and sitting in the basement of the Museum of the Rockies in Bozeman. After I left Montana and went back to life in Georgia—which did not involve excavating dinosaur burrows—I inquired every month afterwards about when the dinosaur skeleton was supposed to be freed from its jacket and sandstone matrix. My impatience was understandable. I felt as if Dave and I had done a thorough job diagnosing the burrow and we really needed the dinosaur to complete the study. Nevertheless, the Museum of the Rockies is a very busy place for dinosaurs, and our little ornithopod was sitting in line behind a number of other, much larger dinosaurs, which accordingly take much longer to prepare.

Finally, at one point, Dave told me that only a *Triceratops* skull preceded our specimen in the queue. Having this big, sexy, and geologically younger dinosaur prepared before ours seemed a little unfair to me, like the doorman to an exclusive club lifting a rope to let it in while all of the not-so-hot dinosaurs, including our little ornithopod, stood out in the rain. Also, the preparator assigned to our dinosaur was well known for her precise and meticulous craft in separating dinosaur bones from rocks. Thus we began to talk about the amount of time it would take her to complete the extraction of a *Triceratops* skull, which changed the meaning of "*Triceratops* skull" to a unit of time in our minds.

Our patience was tried but rewarded, and big-time. Preparation of the dinosaur yielded a few shocks, but the good kind. As I told people later, it was like celebrating Christmas on your birthday. The first shock came when Dave sent me an e-mail message in which he relayed the news from Montana:

There's another dinosaur in there, a juvenile.

The adult dinosaur had a youngster in the presumed burrow chamber with it, and it was of the same species. Although not as complete as the adult, some of its limb bones were there and they matched the adult's bones in form. Based on relative proportions of their limb bones, the juvenile was about 55 to 60% the size of the adult, the equivalent of a growing teenager who was still hanging out in the burrow with Mom or Dad. This discovery thus had huge implications related to dinosaur behavior. An adult with a half-grown juvenile in a burrow was not only evidence of denning—a previously unknown behavior in dinosaurs—but also of extended parental care. The latter synced beautifully with previous work done by Dave's advisor, Jack Horner, who had proposed that *Maiasaura* raised its young in their nests until they were large enough to fend for themselves.

That was exciting just by itself. So you can imagine how it felt to receive additional news a week later from Dave that said this:

There's a second *one [juvenile].*

This called for another happy dance. It is hard enough to find one dinosaur, let alone three of the same species, and possibly from the same family, all together. Although this specimen was also incomplete, enough of its bones were there to distinguish it from the other, and their matching bones were the same size. These dinosaurs could have been brother and sister to one another, and might have been entombed together with one of their parents in a burrow. This revelation strengthened our denning hypothesis, albeit while also conjuring sad thoughts of parental loss.

More pertinent information came in as Dave studied the bones, looking for anatomical clues that might tell him whether the adult dinosaur was capable of burrowing. Once he finished his study, three relevant traits stood out. First, one of the bones integral to its shoulder—the scapulocoracoid—had attachment sites for large muscles. This meant it had the right musculature for using its forelimbs in burrowing, similar to that of modern burrowing mammals like armadillos. Second, the bone on the front of its snout—the premaxilla—was fused, an unusual feature in a dinosaur. This reinforced its nose, making it easier to use as a small spade to augment the dinosaur's busily digging hands. Third, when compared to its closest relatives (other small ornithopods), its hip had an extra vertebra. For example, *Orodromeus* had six vertebrae in its hip, whereas this newcomer had seven. Dave surmised that this additional vertebra strengthened its hip, which the dinosaur used as a brace while digging. As anyone might notice while watching his or her dog dig under a neighbor's fence, dogs anchor themselves into digging position with their rear legs while scratching enthusiastically with their front feet; only later are the rear feet used for excavating. It was now easy to imagine that this dinosaur, with three adaptations for a burrowing lifestyle, would have behaved similarly.

So, like a burrowing dinosaur cutting through substrata, we uncovered even more evidence that connected the remains of this dinosaur to its probable trace fossil. Using the bones to reconstruct the size of the dinosaur, Dave calculated that it was just more than 2 m (6.7 ft) long, with much of that taken up by its tail. Its torso (from hips to shoulders) length was about 70 cm (28 in) long; its body was 26 to 30 cm (10–12 in) wide, and its mass (weight) somewhere in the range of 20 to 30 kg (44–66 lbs). Take away its tail, and it was about the size of a collie. When documenting the burrow at the field site, Dave and I had noted the regular lengths of tunnel segments, and that it was slightly higher than it was wide. This got us to thinking about "burrow fit." In other words, how easily could this dinosaur squeeze into it and turn in the tunnel?

If the burrow was too small for this dinosaur, then we had to come up with a very complicated (but still possible) scenario in which one adult and two juveniles of the same species had their bones deposited and buried together in some other unknown animal's burrow. If the burrow was too big, then that suggested the adult and its young were squatters, taking over a burrow made by another, larger animal: a Cretaceous version of the "Occupy" movement, 95 million years ahead of its time.

So let's look at those numbers again. The torso length of the reconstructed dinosaur was about 70 cm, and the tunnel lengths were 70 cm. The width of the burrow was 30 to 38 cm, and the width of the dinosaur was 26 to 30 cm. So far, so good. But was there a way to calculate the weight of an animal based on its burrow? Why, yes, there was. In a study published the same year the dinosaur was discovered (2005), a zoologist compared the cross-sectional areas of burrows made by a wide variety of modern burrowing animals to the weights of their makers and found a positive correlation between these. He had even provided a handy formula, which we applied to the cross-sectional area of our burrow. Taking into account the slight variations of tunnel width along its length, we came up with a range of 22 to 32 kg (48–70 lbs), which overlapped with the mass Dave calculated for the adult dinosaur on the basis of its bone. The burrow dimensions matched the dinosaur.

Granted, the burrow was a close fit, but this situation also applies to many modern burrowing and denning mammals, such as coyotes (*Canis latrans*), striped hyenas (*Hyaena hyaena*), and aardwolves (*Proteles cristatus*). It also makes good evolutionary sense for burrows to be narrower rather than luxuriously spacious. For one, tight spaces limit who else can go down the burrow, including predators of either the adults or juveniles. Also, slender burrows are better able to retain heat and humidity, maintaining better climate control that is less influenced by surface conditions and making it more like a cave. The twisting Z shape of the burrow would have helped both of these factors, while also discouraging or confusing predators.

For the latter, put yourself in the place of a predator and imagine looking down the burrow entrance and only seeing the short tunnel segment to the first turn, not the next segment burrow, let alone the final burrow chamber with its potentially yummy, tender juveniles. Gopher tortoise burrows display exactly the same sort of strategy, in which these at first run straight down, but then turn abruptly to the right or left within a meter or so from the entrance.

What made everything even more groundbreaking than the burrowing nature of the dinosaur was that the dinosaur itself was a new species. Based on its anatomical traits, it was sufficiently different from other hypsilophodont dinosaurs that it warranted its own unique name before we revealed it to the rest of the world. This took much deliberation of Greek and Latin roots, which we discussed via e-mail with the discoverer of the dinosaur and our co-author, Yoshi Katsura (who was in Japan), and a paleontologist friend of mine who was an expert on the rules of nomenclature, Andy Rindsberg (who was in the faraway and exotic land of Alabama). Eventually, Dave, Yoshi, and I agreed on the name *Oryctodromeus cubicularis*, which literally translates to "running digger of the den," a unique name that would signify how this dinosaur was capable of digging and denning, but also could run about on two legs.

So just to summarize, we had the following:

- A new species of dinosaur;
- Anatomical evidence in that dinosaur, showing it was adapted for burrowing;
- A burrow that fit the reconstructed dimensions of the dinosaur;
- A burrow that had smaller burrows attached to it, showing a commensalism like that seen today in large vertebrate burrows;
- Two half-grown juveniles of the same species, indicating denning and extended parental care.

In our estimation, then, this discovery did indeed fulfill the earthy expression of applying a foot to one's posterior. Clearly it was time to do the logical next step in the scientific method, which was to hold a press conference.

Just kidding. What scientists are supposed to do after a momentous discovery, well before scheduling interviews on talk shows, dating Hollywood stars, or uploading self-promoting videos to YouTube, is to put all of our observations and analyses into a coherent report that states our argument. The report is then sent to a scientific journal, where our peers—other paleontologists—give it a beady-eyed critical review. The very first step of this peer review happens with a journal editor. She or he decides whether the report deserves to be sent on for peer review, or whether it should be kicked back immediately to the authors with a pithy and dismissive "not worthy of our journal" notice. If this report gets past the editor, it is sent to at least two experts in the appropriate field for review.

In the initial submission of the report, authors are welcome to suggest potential reviewers who they think would give fair, thorough, and relatively impartial assessments of your work. Authors can also state who should *not* review it. This is normally because of a conflict of interest, such as a potential reviewer being a cantankerous pedant who has never agreed with a single word written in any of the authors' previous reports, including "and," "the," and especially "but." If you're really unlucky, though, the editor will pick just that person, either out of a sense of being "fair and balanced" (in a cable-news sort of way) or because the editor enjoys watching intellectual fireworks. The editor may even pick the much-dreaded third reviewer, who always seems to have an opinion that diverges wildly from those of the other two reviewers, presenting editors and authors alike with head-scratching dilemmas.

Once the reviews are in to the editor, she or he makes a decision about the paper. Choices are: accept it into the journal with minor revisions; accept it but with major revisions; or outright rejection. Most journals also have an "anonymous reviewer" policy in which they keep the reviewers' identities secret, unless they make themselves known to authors after the review, which allows for some dialogue. To make matters even more challenging, editors and reviewers are volunteers, performing these duties for free on top of their other professional obligations. This means that reviewing a scientific paper can easily become a low priority, perhaps eking out a spot ahead of, say, taking out the garbage or cleaning the kitty litter at home. Then, if the article is accepted, authors must revise it, which includes formatting each and every cited reference to the exasperatingly exact and idiosyncratic standards of that specific journal. Sometimes authors get so discouraged that they give up during this part of the process, and the article never gets revised and resubmitted. All of this means that science is not just about making discoveries and announcing them to the world, but also is about jumping through hoops and hopefully being rewarded with publication, or slinking away in abject failure.

With all of this in mind, Dave, Yoshi, and I wrote our report about this new species of burrowing dinosaur, its young, and its burrow, and we gave it a simple title: *A Burrowing, Denning Dinosaur.* If you had asked us then, we would have told you that our paper was astonishing, a true masterpiece of paleontological literature that would surely earn the academic equivalent of an A++++++ (evoking Ralphie's daydream from the movie *A Christmas Story*). With supreme confidence, in mid-2006 we submitted it to an elite scientific journal, one in which Dave had published papers before.

But then peer review struck. Remember in the preceding description where the editor can say "not worthy" and kick it back to you without further review? That happened to us not once but twice. An editor from the first journal sent a terse reply within just a few days of submission, informing us that it was "not appropriate for the journal at this time," a response that made us wonder if the previous or following week would have been better. A little daunted and confused, we regrouped and revised the article, then submitted it to another elite scientific journal. This met a similar fate, with the rejection notice again coming only a couple of days after submission.

We were confounded. What had gone wrong? It was hard to say, considering that neither journal editor addressed anything about the content of our manuscript. Perhaps it was just a matter of bad timing, as both journals may have had too many other dinosaur-discovery articles in review or being published then and those other specimens were more attractive or from more exotic places. Or as one paleontologist acerbically commented to Dave afterwards, "Well, of course they didn't accept it. It [the dinosaur] didn't have feathers and it wasn't from China." Oh well. That's the nature of science.

With egos properly deflated, we decided to try one more time, and with a journal that would be more receptive to the notion of a burrowing dinosaur from Montana. We had to hurry, though, as our concept of burrowing dinosaurs was starting to take on form through other people. For example, in a paper published in late

2006, its author (David Loope) proposed small dinosaurs as possible burrowmakers for large structures he found in Early Jurassic rocks of the western U.S. Although Loope did not have any bones in these burrows, his analysis of the burrows was very well done and perfectly credible. The secrecy we had kept around our discovery was also starting to unravel. In mid-2006, I stumbled onto an ostensibly innocent online discussion about burrowing dinosaurs, but one that had been prompted by someone associated with Dave's original Montana field crew. It was only a matter of time before more information got out, and peer review might be compromised by any such rumors (or facts). Editors and reviewers tend to frown on papers in which authors seek pre-publication publicity, and we did not need to risk this study any further.

For the proverbial "third time's the charm" attempt, toward the end of 2006 we sent our report to a venerable British journal, *Proceedings of the Royal Society of London*. This time, the editor kindly gave us a chance and sent the article to two reviewers, both of whom signed their names to their honest and in-depth assessments of the research. One agreed with nearly everything we said, but the other thought we had placed too much emphasis on biologically based arguments and needed to include more geological evidence. Oddly, the editor then told us the paper was "rejected," but encouraged us to resubmit a revised article. It was like being told by a date that he or she thinks you're ugly, smelly, and stupid, but would like to go out with you again, just as long as you lose some weight, take a shower, and start playing Sudoku. Nonetheless, we did as told, revised and resubmitted, and the paper was finally accepted.

In February 2007, the online version of the paper was finally released, which coincided with our sending out press releases, and a good amount of media attention followed. Granted, this was not a feathered dinosaur from China, nor was it a close relative of *Tyrannosaurus rex*. Nevertheless, after 95 million years, *Oryctodromeus cubicularis*, a dinosaur buried in a burrow of its making and with its offspring, was now known to the rest of the world. The words "denning," "burrowing," and "dinosaur" could be used in the same sentence, and thanks to a fortuitous combination of trace and body fossil evidence, dinosaurs had entered yet another dimension in our Mesozoic imaginations: underground.

Dinosaurs Down Under

We were lost. As a result, this otherwise fine fall day of May 10, 2006 had turned into an unexpectedly long one for our group of eight while we hiked atop the high cliffs of Cretaceous rocks along coastal Victoria, Australia. We were looking for an auspiciously named locality—Knowledge Creek—that continued to elude us despite our maps, GPS units, and field-savvy participants, including a few Australians who knew the surrounding area. Having already taken two wrong turns down toward the shore, only to double back and climb up, we were all becoming a bit tired and frustrated. Below us, ocean waves burst onto the rocks, a dull, rhythmic booming carried to us by a strong, cool sea breeze, imploring us to try again.

In the early afternoon, we finally found the proper route, a grassy path that cut through the scrubby coastal forest parallel to a ravine cut by Knowledge Creek. Two of our group went back to fetch our vehicles, a sacrifice that would shorten the travel time for the rest of us at the end of the day and supply our field lunches, which we had dumbly left behind. The steep incline down to the coastal outcrops promised a long, slow fight against gravity that would coincide with diminishing light, as the sun began arcing toward the horizon. Underestimating the length and difficulty of the hike in, some of our group had not brought enough water, and those of us who did shared what little we had left. On the off chance that any dinosaur bones would be found at the site, one of our party was hauling a portable rock saw on his back, and another was carrying fuel for it. I did not envy their journey back up the slope we were now descending for a third time. None of us had ever been to Knowledge Creek, and two paleontologists who had visited before us had only gone once, firmly vowing to never come back. We were starting to understand why.

Australians have a long tradition of ill-fated expeditions, and I felt like I had unwittingly instigated one of these. Yet our motivations were paleontological, and with good reason. First of all, I was in Victoria on a semester-long sabbatical from my university to work on a science-education project with Patricia (Pat) Vickers-Rich. In 1980, she and her husband Tom Rich trekked to Knowledge Creek; yes, they were the two aforementioned paleontologists who had been there before us. They were prospecting for dinosaur bones in the Early Cretaceous (~105 *mya*) rocks there, which had been found at other coastal outcrops just east of Knowledge Creek. While there, they failed to find any bones, but they did manage to discover what was then the only clear example of a dinosaur track in all of southern Australia. Only about 10 cm (4 in) wide and long, with three stout and well-defined toes, the track was attributed to a small ornithopod dinosaur, probably a hypsilophodont.

By the time I arrived in Australia—more than 25 years later—it was still the only undisputed dinosaur track in all of southern Australia. I had found some not-so-clear dinosaur tracks only a few months before at another spot, but it was time to find more. Thus, the purpose of our troublesome foray that day was to revisit the source of that track and look for more dinosaur tracks. Pat and Tom, still filled with the wisdom imparted by Knowledge Creek the day they went there in 1980, had declined our invitation to come along.

Dinosaur bones were relatively rare in this part of the world and, most interesting, represented a polar dinosaur assemblage. Based on plate-tectonic reconstructions, the rocks in this part of Australia were originally formed near the South Pole, when southern Australia was connected to Antarctica during the Early Cretaceous (130–100 *mya*) before drifting north to its present location. Nearly all of the dinosaur bones and teeth in strata there were from small dinosaurs, and most of these were hypsilophodonts. Only a few pieces were from theropods, such as one bone that came from a dinosaur similar to the Late Jurassic *Allosaurus* of North America.

Hence, Pat and Tom often asked themselves, "Why hypsilophodonts?" and wondered how these ornithopods, which were probably too small to migrate, had adapted to long, cold, dark winters of polar environments during the Cretaceous. In a paper they and other co-authors published in the journal *Science* in 1988, they proposed that hypsilophodonts and other dinosaurs in the region were likely endothermic, or "warm-blooded," generating their own body heat. At that time, warm-blooded dinosaurs were still being hotly debated (no pun intended), and recall that Robert Bakker's *The Dinosaur Heresies* had only been published two years before then. Not all paleontologists were so accepting of the idea that dinosaurs were less like reptiles and more like birds, and others thought that maybe dinosaurs represented something entirely different from either group of modern animals.

Pat and Tom, along with many other contributors to their paper, fed this debate further by documenting the best-known polar dinosaur assemblage in the Southern Hemisphere. This added support to the then-remarkable idea that dinosaurs lived in frigid places, a hypothesis later backed up by discoveries of thousands of dinosaur bones and tracks in Cretaceous rocks of Alaska that were also in formerly polar environments.

However, a single sentence in this 1988 paper later shouted at me, showing some remarkable prescience by the authors. It was a simple sentence, and one that very easily could have been cut by the authors, peer reviewers, or editor for being too speculative. Here is what they said:

> *With the possible exception of the* Allosaurus *sp., all of the animals were small enough to have found shelter readily by burrowing.*

That particular day, while winding down into the valley carved by Knowledge Creek, I was not aware of that sentence, nor would I have cared. We were doggedly trying to get to the shoreline to look at the exposed rocks along the marine platform to see whether any more dinosaur tracks were there. Once the path exited the forest, we crossed the trickle of fresh water that was Knowledge Creek, and most of us successfully avoided the leeches there, wriggling excitedly at our warm-blooded presence. Our boots met Cretaceous rock surfaces, which extended out as a platform and met the sea. Successful in reaching our goal, we paused to catch our breath but also found ourselves gasping in another way at the spectacular cliffs of bedded sandstones and conglomerates looming above us. Because this part of the world was entering winter, the waves at the shore were noticeably higher and more vigorous than usual, encouraged by offshore winds that had just blown over and around Tasmania.

With little time left before we had to meet our two vehicle-driving companions, we started looking for dinosaur tracks, bones, or anything else paleontological that caught our attention. One of our party, Mike Cleeland, was a long-time volunteer with Pat and Tom in their Victoria-dinosaur endeavors, and despite his great height, he was legendary for his uncanny ability to spot tiny scraps of bone. Hence, he and most others there had their "osteo-eyes" switched on, looking for bones and teeth. Alternatively, my wife Ruth (who had enthusiastically persevered with us through the day) and I were using our "ichno-eyes" to search for trace fossils ranging from small invertebrate burrows to dinosaur tracks.

What I was not expecting to see there was a dinosaur burrow. Even less likely, it matched the form and dimensions of the fossil burrow that Dave Varricchio and I had unearthed in the Cretaceous rocks of Montana only eight months before. Yet there it was in the outcrop, a gently dipping structure with a sediment-filled tunnel that twisted right, then left, and connected with an expanded chamber. It was naturally cast with sandstone toward its top and conglomerate at its bottom, but clearly cut across the bedding surrounding it. I stood transfixed, barely breathing, and gaped at it long enough to prompt Ruth to come over to stare with me.

"What is it?" she asked.

"I think it's a dinosaur burrow," I whispered.

At the time, she was one of only a few people in the world who knew about the fossil burrow in Montana. In fact, Dave, Yoshi, and I had not yet named *Oryctodromeus*, whose remains, along with the two juveniles, were still encased in plaster and rock. So we were still doing all we could to keep its discovery quiet until it had gone through a scientifically proper vetting process. Consequently, I had not yet told any of my new Australian friends about it and was not about to start now.

Nonetheless, the problem with outwardly expressing curiosity and wonder is that it attracts other people who then want to join in with you. As a result, only a few minutes elapsed before one of several geology graduate students with us, Chris Consoli, saw us standing there and wandered over to ask the same question as Ruth: "What's that?" I answered truthfully but evasively, "I'm not sure, but it's an interesting structure." I then asked him if he would mind being the scale in my photographs, and quickly snapped three shots in succession. He soon strolled away once I supplied no other information about what was there. Ruth then helped me to take a few quick and surreptitious measurements of the enigmatic structure. Mike Cleeland also stopped by and asked about it, and I was similarly nonchalant with him.

These were the only photos I took of the suspected burrow that day, as I really did not want to call much attention to it and we had other work to do in our short time there. All of us wandered about, looking down on bedding planes, glancing above us on the outcrops, and in between at nearby sections of the strata. This perusal was mildly successful, as during our short time there Mike and I found a few faint and incompletely expressed dinosaur tracks, including one that was likely from a large theropod. This meant that it would be worth coming back to look for more and better tracks. I also noted many fossil invertebrate burrows that were about the right sizes for ones made by insect larvae. These were valuable indicators of the original environments there, such as river floodplains that formed from the run-off of spring thaws following polar winters. Yet we saw no bones.

All told, we were at the site for only a little more than an hour before slogging back up the hill to meet our two friends, who wondered how things had gone. Amazingly, we still had enough time that afternoon to stop at nearby Dinosaur Cove, a world-famous locality for polar dinosaur bones that many of these same Australians accompanying me had quarried in the 1980s and 1990s. We departed just as the sun set over Dinosaur Cove, a long and satisfying day of field work completed.

Later, I studied the three photographs I had taken of the odd structure at Knowledge Creek and could not get over the eerie sameness it held compared to what I had seen in the Cretaceous rocks of Montana. Not knowing what else to do, I sent a photo to Dave Varricchio for him to assess. As expected, he was non-committal about what it meant, but it felt good to share this coincidence with someone else in the know. Nevertheless, one thing was for sure: I had to go back to Knowledge Creek.

So I returned the next year, in July 2007, only a few months after the publication of the paper on *Oryctodromeus* and its burrow. The day that paper came out, I wrote to Pat Vickers-Rich and Tom Rich to finally disclose my secret that they had a possible dinosaur burrow there in the Cretaceous rocks of Victoria. Of course, they were excited about this prospect (albeit understandably skeptical) and encouraged me to come back to Victoria. During this second trip, Mike Cleeland was along for the ride again, but joining us was Lesley Kool and five others. Lesley was a long-time dig-site manager for Pat and Tom, and like Mike is an extraordinarily keen-eyed finder of vertebrate bones.

Mike and Lesley looked at the possible burrow with me and confirmed what I had surmised initially, which was that it contained no skeletal debris. Thus this was going to be pure ichnology, dinosaur burrows without bones. However, it was on this foray that I noticed a second twisting, sandstone-filled structure above the one that had originally grabbed my attention. It was not as completely expressed as the other, with only a former tunnel and lacking a chamber at the end, but was the same width and filled with the same type of sandstone. This time, with much less urgency than the previous time, I carefully sketched, measured, and photographed both structures, documenting them enough so they could be summarized in a paper and evaluated by my peers.

Because nearly everyone in our group was exhausted by the hike down but still faced a climb up, they took off as I did this tedious but necessary work. Roger Close, a young geology graduate student, stayed with me and assisted where needed before we also walked out. At the top, Lesley, her husband Gerry, and several others who had made the round trip greeted us with big grins and happily announced that they never needed to visit Knowledge Creek again.

Life got in the way throughout most of 2008 for me to do anything with these hard-earned data, until a burst of writing during the 2008–2009 holiday break from teaching duties between semesters resulted in a manuscript. I decided to author the paper by myself, because if it turned out to be totally wrong, I would own all of the failure. My interpretations of the field observations were risky because, unlike the Montana burrow, the structures did not hold any accompanying dinosaur bones.

However, in the paper I made sure to thoroughly explain the reasons why some polar dinosaurs *should* have burrowed as a behavior that would have allowed them to overwinter during cold, dark, harsh times in polar environments. For one, the Cretaceous rocks of Victoria abounded with evidence for hypsilophodonts, small ornithopods morphologically similar to *Oryctodromeus* and *Orodromeus* in North America. For another, such dinosaurs were too small to have migrated long distances between seasons. Thirdly, many modern polar vertebrates, from puffins to polar bears, burrow into dirt or snow to take refuge in those harsh environments. Consequently, in January 2009, I sent the paper to the journal *Cretaceous Research* and kept fingers, toes, and thumbs crossed that it would be received favorably.

The paper was reviewed in a timely way and fortunately was not rejected outright. Still, the reviewers and editor expressed concern about its content, requiring responses to specific points. As mentioned before, this is where many scientists might give up on both revising and resubmitting a manuscript, without the guarantee of acceptance. However, I had already planned to go back to Australia and promptly put Knowledge Creek on my itinerary to answer a few questions, such as the exact nature of the sedimentary rocks surrounding the structures I claimed to be dinosaur burrows.

Thus my third (and perhaps last) visit to Knowledge Creek took place in May 2009. This time only three of us went: my wife Ruth (her second visit); Mike Hall from Monash University of Australia, an experienced field geologist who also was an expert on sediments and sedimentary rocks; and myself. Mike had never been there, and it had been long enough for me that we became disoriented and displaced—some might call it "lost"—on the way down to the site. This was a disheartening déjà vu for Ruth, echoing her experience from three years previous with me and other geologists, and justifiably shaking her faith in our navigational abilities. Her unease grew more pronounced when Mike proposed that we circumvent the nearby sheer sea-cliffs by following a wallaby trail through the thick coastal scrub forest. Travails notwithstanding, we made it to the site okay, and once more Ruth and I stood together on the nearly flat marine platform of Cretaceous strata of Knowledge Creek. This time we had plenty of daylight ahead of us, though, and Mike's help as a geologist to double-check my field results.

This trip was well worth it, as we noted what might have been yet a third partially preserved dinosaur burrow. It was shorter than the other two but almost the same diameter and had a similar sandstone fill. This implied that all three had once been hollow structures and had likely been filled by the same sedimentary processes at the same time. Elsewhere on the outcrop, in strata above the supposed dinosaur burrows, we also measured and otherwise documented dozens of invertebrate trace fossils, all small-diameter burrows. With only three of us there, we even had enough time after my data collection to explore our gorgeous surroundings.

It was a magical experience, a sense of wonder evoked by walking on the remains of polar rivers from more than 100 million years past, as waves from a present-day sea churned violently against these same rocks just behind us. All paleontologists who have done field work end up having favorite places where they feel lucky to learn and realize something new about the history of our planet. Knowledge Creek had become one of mine.

Once back in the U.S., I finished the revisions of the manuscript, wrote a rebuttal to the reviewer remarks that nearly eclipsed the length of the manuscript itself, inserted a mention of a possible third burrow at the same site, and resubmitted it. Gratifyingly, the editor accepted the paper the following week, which had the title "Dinosaur Burrows in the Otway Group (Albian) of Victoria, Australia, and Their Relation to Cretaceous Polar Environments." Unexpectedly for me, though, the accepted manuscript was also posted online only a week after that, which meant that science reporters began calling me, asking about these dinosaur burrows in Australia, the oldest interpreted from the geologic record.

My university media folks and I scrambled to put together a press release, and we used edited video footage from Knowledge Creek to accompany it as a sort of modern-day newsreel. As of this writing, the video had the most views of any in the history of my university, and the press attention was flattering. So two places in the world—Montana, USA and Victoria, Australia—had been proposed as sites with dinosaur burrows, with the hope that more would be added to this list in upcoming years, as paleontologists now knew what to look for.

"Mythbusting" a Dinosaur Burrow

Many scientists welcome media attention or other forms of interacting with the public, whether through lectures or writing popular-outreach pieces such as magazine articles or blog entries. Yet a mantra I often preach to my students and try to put into daily practice is that these attempts at public outreach and communication do not necessarily make our science truer. As I mentioned previously, science does not prove, it disproves. Hence, paleontologists who live up to this ideal by testing their own results—treating their work to the same degree of scrutiny and skepticism as they would their rivals' research—always impress me. It is an intellectual honesty we all need to practice, such as in interpreting the indirect evidence represented by dinosaur trace fossils.

Along those lines, Dave Varricchio assigned an undergraduate student of his, Cary Woodruff, to do just that. Woodruff's job was to test whether the *Oryctodromeus* bones could have been buried while in the den (that was the original hypothesis) or whether they were carried in from outside (that's the alternative hypothesis). If they were moved from outside of the den, this would be a strike against these dinosaurs having lived and died there and would have cast doubt on their having been the original burrow inhabitants. Instead, their dead remains may have been tossed into a big hole, burrow or not, that just happened to be in the area when a river overflowed back in the Cretaceous.

So in an experiment straight out of the popular TV show *Mythbusters*, Woodruff joined PVC piping and set up scaffolding to make a half-sized version of the burrow. He then made up mixtures of sand, mud, and water, and used rabbit bones as a proxy for the dinosaur bones. Thirteen times he ran the experiment, in which he filled a plastic bucket with sediment and poured it down the hole. Sometimes rabbit bones were added to this sedimentary stew, but other times he placed the bones inside the artificial burrow chamber first, then decanted. Out of the thirteen times he did this, six resulted in the same sort of jumbled distribution of bones in the burrow chamber that Dave had observed with the original *Oryctodromeus* bones and their sedimentary matrix. Four of these arrangements were made with the rabbit bones already in the den, and two came from outside, with the bones in the bucket.

Four to two: we win! Except not really, because science is not a game in which simple scores decide which hypothesis is the better one. These results still meant the bones feasibly could have been deposited into the burrow and that the *Oryctodromeus* bones could have come from somewhere else outside of it. Woodruff then tried to disprove this hypothesis by scrutinizing the original *Oryctodromeus* bones, seeing whether they held any nicks, scratches, or dents from having been bounced along the bottom of a stream.

Many dinosaur bones bear such evidence, telling how those parts may have traveled far from the spot where a dinosaur died. However, in this instance the bones had no such marks. This lack of evidence implied that the bones of this probable parent and offspring did not travel far at all. They either died together just outside of the burrow, or in it.

But wait: Could the burrow have belonged to some unknown burrowing predator that scored a super-sized meal one day by taking down an adult and two juveniles of the same species? Then maybe it dragged these bodies into its burrow to nosh on them, left its partially consumed dinner in the burrow, and was conveniently somewhere else when its prey's bones were buried by the flooding of a nearby river? For that idea to have more support, though, more trace fossils were needed, such as toothmarks on the bones. Yet none were to be found. Thus, considering the happenstance of bones from an adult and two juveniles of the same dinosaur species being together, no signs of long-distance transport of those bones, no evidence of anything chomping on the bones, and all of the aforementioned ichnological evidence showing a match between the burrowmaker and the burrow, the hypothesis that this *Oryctodromeus cubicularis* was a burrowing, denning dinosaur still stands. So far, so good.

However, this experiment also served as a reminder that for scientists to stay authentic, we should never rest on our laurels, however hard-won those might be. For example, the effort expended to make three trips and back from Atlanta, Georgia to Knowledge Creek in Victoria, Australia, and discovering and documenting possible dinosaur burrows in Cretaceous rocks there, does not give me the inviolable right to omit the word "possible" when discussing them. After all, someday someone with more knowledge, experience, technology, and luck than me may reinvestigate those strange structures and decide that, no, they are something else entirely. On the other hand, someday someone somewhere else might find far better examples of dinosaur burrows, and that dinosaurs we never expected to have burrowed made them. (Still, I am not holding my breath about sauropods or tyrannosaurs as probable burrow dwellers.)

All of this would be perfectly fine, a happy circumstance of how paleontology, like any science, progresses through the slaying of old ideas and the inclusion of newer, better-tested ones. Burrowing dinosaurs are no different in this respect, but thanks to ichnology, a previously obscure idea about dinosaurs is now there for us to consider, poking its head out of the ground for a look around.

CHAPTER 6

Broken Bones, Toothmarks, and Marks on Teeth

Dinosaurs and Their Embodied Trace Fossils

This is where I cheat. Take a look at the title of this book again, and you will see an absence—"without bones"—as a central theme. So in defiance of that dictum, I will now turn to dinosaur bones and other body parts, such as teeth, for whatever wisdom these can provide.

This concession to body fossils is necessary because many dinosaur trace fossils are in their bones and teeth, or in bones and caused by teeth. A few of these trace fossils announce themselves as broken or otherwise injured bones, some of which could only have been inflicted by other dinosaurs. Sometimes this evidence of dinosaur-on-dinosaur violence suggests that it originated from within a species and perhaps was inspired by competition, whether over territory, a mate, or food. In other instances, injuries were caused by another species of dinosaur that attempted to kill and eat the assaulted dinosaur, although most trace fossils in bones are marks made after their owners perished. In short, the bodies of dinosaurs or other vertebrates—not sediments—were the places where these marks of dinosaur behavior were recorded.

How do you imagine such trace fossils? You already have. Recall that this book began with a piece of fiction set about 70 million years ago, opening with two rival male *Triceratops* that squared off in combat, and with one losing face. A few other dinosaurs in this tableau—small anonymous feathered theropods—were chewing on pterosaur bones while this ceratopsian drama played out nearby. A tyrannosaur, initially interested in the outcome of the ceratopsian battle but ultimately disappointed with its results, switched her ravenous attention to a herd of *Edmontosaurus*. With these dinosaurs, she bungled her ambush of a young male *Edmontosaurus* but managed to get a snack by biting a chunk of flesh and bone out of his tail. The hadrosaur escaped with only a wound, and he lived happily ever after until dying of some other cause and getting fossilized. For the rest of his life, though, he ate plants near the ground, which were more likely to include grit that scratched his teeth. Although this scenario was imagined, the trace fossil evidence for all such behaviors is real.

This realization that bodies were substrates for dinosaur behaviors enables us to discern much more than just looking at the rocks surrounding and entombing dinosaur bodies. For example, how can one tell whether or not dinosaur toothmarks on a dinosaur bone were punched into or scraped against it while the prey was still alive, dying, or dead? How does one go about identifying dinosaurs that left toothmarks, especially if they were not kind enough to leave a calling card in the form of a dislodged tooth?

Dinosaur toothmarks are not just recorded in the bones of other dinosaurs but also in bones of vertebrates that lived at the same times and places as dinosaurs. Even dinosaur teeth, which in some places might be the only body fossils reflecting a dinosaur presence in Mesozoic rocks, can host trace fossils, too. Some of these traces are extremely subtle, such as the microscopic marks imparted when dinosaurs chewed plants with grit on their leaves or with mineralized parts. Other traces are more overt, such as broken teeth of carnivorous theropods which must have bitten off more than they could chew.

By looking at these body parts and their trace fossils, many questions about dinosaurs become more answerable. For instance, how did they relate to one another, especially if they became upset with another of their own species? Were predatory dinosaurs picky eaters, only going after one species of prey, or were they more opportunistic, eating whatever appealed (and was available) at the time? Did they normally eat big adults, or did they primarily seek out juveniles, or the old and weak? How did they eat: daintily, with a furtive nibble here and there, or with the gluttonous bad manners of a Renaissance-festival banquet? Did some dinosaurs prey at all, or did they live their lives mostly as oversized vultures, relying on the kills of other dinosaurs or other already-dead, ready-to-eat meals? Did any dinosaurs ever succumb to the evolutionary taboo of cannibalism by eating—or at least biting into—their own kind?

All of these are behaviors about which we might otherwise speculate if not for trace fossils becoming more pieces in the puzzle and completing pictures of how dinosaurs lived their daily lives.

It's Only a Flesh Wound

Through the power of trace fossils, paleontologists can tell by looking at a dinosaur's body that another dinosaur attacked it. Yet, as is typical in science, such dramatic interpretations can be equivocal. Such a quandary is exemplified by a 2013 discovery of fossilized dinosaur skin from South Dakota. Although this "skin" was actually just a natural cast of the original skin, it was closely associated with the bones of *Edmontosaurus annectens*, a Late Cretaceous hadrosaur from the western U.S. and Alberta, Canada. Yet the skin itself is not the trace fossil. Instead, a probable trace fossil is *in* the skin. Whether this trace fossil was made by the hadrosaur or inflicted by another dinosaur, though, is subject to vigorous debate, as explained here.

Owing to their rarity, dinosaur skin impressions give paleontologists good reason to celebrate such finds. These unusual body fossils were normally formed first as impressions against soft sediment, and then naturally cast in sandstone, similar to how many dinosaur footprints were preserved. (Along those lines, the best-preserved dinosaur tracks have scale impressions, but these are the marks of living skin, not the skin itself.) This particular patch of skin had an irregular feature which paleontologists interpreted as an apparent healed injury. This interpretation provoked a spirited discussion about what constitutes good trace fossil evidence for a dinosaur attacking another dinosaur.

The skin impression and the irregularity on the skin are surprisingly small. The patch of skin measures only about 12 × 14 cm (4.7 × 5.5 in), about the size of a notebook that can fit in a shirt pocket. The oddity on that patch stands out from the main scaly pattern, looking like a partly closed human eye and coincidentally about the same size as one. It has a raised exterior and indented interior, as if something sharp had punched through the skin, then the skin annealed around the wound.

So here are two scenarios for the given evidence, in which everyone accepts the basic premise that this is indeed a healed injury:

> Scenario A: Hadrosaur was walking peacefully through a forest, minding its own business, while unknowingly stomping on and otherwise terrorizing insects, amphibians, and small mammals. Suddenly, a ferocious tyrannosaur bursts out of hiding, ambushes the hadrosaur, and gets in a good bite. At least one tooth punctures the hadrosaur's skin and damages the bone underneath. Luckily for the hadrosaur, but sadly for the tyrannosaur, the injured dinosaur escapes and lives long enough for its skin to heal. The end.

> Scenario B: Hadrosaur was walking peacefully, but awkwardly, through the forest; perhaps it was a teenager.

> Suddenly, it stumbles into a tree with prominent, sharp spines. At least one spine punctures the hadrosaur's skin and damages the bone underneath. Fortunately for the hadrosaur, it extracts itself from the tree and lives long enough for its skin to heal. No one knows, or cares, what happened to the tree. The end.

So which hypothesis do you want to be true, especially if you absolutely adore the notion of huge and horrific theropods chomping on hapless ornithopods? Which do you think would make for a sexier story in a news release written about the study? Which one is most likely to be recreated through the latest CGI technology and show up in a cable-TV special on dinosaurs in the next few years, narrated by semi-comatose celebrities or overexcited paleontologists? Why, yes, A, A, and A are all correct answers. Yet Scenario B has not been disproved, nor have all sorts of variations on the interpretation that this fossil malady was not caused by a ravenous predator. How about a sharp rock? Another hadrosaur in a fight over a mate or meal? Or, most ignoble of all, a festering sore caused by a bacterial or fungal infection?

Just like the hadrosaur, you get the point. Trace fossils of injuries preserved in dinosaur bodies become difficult to interpret when we can't distinguish whether these wounds were caused by another dinosaur, a non-dinosaur animal, or self-inflicted (albeit accidentally). Also, some of these marks may have been made through other means, such as infections, which are not trace fossils at all. This is where the study of dinosaur health problems enters a realm of forensics that gets contentious, just like evidence for a court case going to trial.

Acting as the prosecuting attorney, however, I purposefully withheld some crucial evidence until the last part of the trial so that it would have a maximum effect on the jury. You see, the skin impression from this hadrosaur was directly next to its skull, and the skull had toothmarks. Furthermore, the toothmarks were widely spaced, and most were from an animal with pointy teeth and large enough to attack an adult *Edmontosaurus*. Even better, these toothmarks on its skull also show signs of healing. So, ladies and gentlemen of the jury: Is it not true that these toothmarks came from a carnivorous dinosaur? Is it not true that they match the lineup of teeth in the mouth of a tyrannosaur? Is it not true that tyrannosaurs liked to dine on *Edmontosaurus*? Is it not true, then, that the puncture mark on the skin came from the same attack as the skull?

Well, maybe. After all, even with healed toothmarks on its skull and a nearby patch of skin with a healed puncture wound, this specimen of *Edmontosaurus* may have had two mishaps separated in time, with only one inflicted by a large theropod. However, if I were a betting ichnologist, I would wager that the toothmarks in the skull comprise the real evidence of a tyrannosaur attack. In contrast, the healed wound in the skin impression is more circumstantial and might be from a separate injury that the hadrosaur did to itself.

So why did these trace fossils get reversed in importance by the paleontologists who studied them? Probably because dinosaur skin impressions are rare, evidence of wounded skin is rarer, and evidence of healed skin is exceptional. Hence it was the skin that justifiably became the focus of the study, whereas the multiple toothmarks in the bone below it became additional information, shuffled to the background because they seemed so ordinary—which they most assuredly are not.

Old Ailments, Traces, and False Traces

The discernment of dinosaur disorders falls under the category of *paleopathology*, which is the study of ancient ailments. Although not all of paleopathology concerns itself with dinosaurs—much of it centers on evidence for diseases in pre-historic human remains—this science is being applied enthusiastically to dinosaurs. In the practice of paleopathology, physicians and veterinarians sometimes collaborate with paleontologists in a neat mix of old and new.

Still, we must always return to basic principles of ichnology before allowing ourselves to be dazzled by other sciences. For any paleopathologic evidence in fossil bones, skin, or teeth to qualify as a dinosaur trace fossil, it must have behavior behind it. This criterion is necessary regardless of whether a trace fossil was made by the dinosaur with the injury, from another dinosaur, or from a non-dinosaur animal.

For example, one specimen of *Psittacosaurus* from Early Cretaceous rocks of China also, like the *Edmontosaurus*, has a skin impression associated with its body, and the skin impression has two apparent puncture marks. But these marks show no sign of healing. Accordingly, paleontologists would simply say these traces are from predation or scavenging by another animal that intended to eat this already-dead *Psittacosaurus*. Possibly the tracemaker was a theropod, or maybe it was another animal that also had pointed teeth, such as a lizard or crocodilian. So rather than showing how this dinosaur escaped an attack, with its punctured skin later closing up and leaving a noticeable scar, this was evidence that its inert body was included on some animal's meal plan.

Thus basic ichnology is combined with basic paleopathology whenever a paleontologist looks at a dinosaur skin impression or bone and notices an abnormality: holes, dents, breaks, enlargements, or other traits that stand out as something extra, something that was not part of its original anatomy. Even chipped teeth fall into this category, in which a theropod lost part of a tooth. When this happens, the most basic of questions a paleontologist can ask is "Dead or alive?" As in, was the dinosaur out of commission for good, or was it still moving under its own power when the feature was added? If living, the trace fossil might be attributed to the dinosaur itself, although in some instances another dinosaur might have caused it while the two were tangling with each other. If dead, though, another tracemaker was entirely responsible for whatever trace fossil is preserved in a dinosaur's bones or teeth, which in some instances might have been a dinosaur, too. However, a composite trace fossil also could have been made from the behavior of the dinosaur with the preserved injury combining with the behavior of the dinosaur that dealt the injury.

Paleontologists and paleopathologists alike also must remember to be good scientists by asking a follow-up question, "How could I be wrong?" This is when they contemplate the possibility that their supposed paleopathologic evidence is anything but that. After all, holes, dents, breaks, and other marks in Mesozoic bones (dinosaurian or otherwise) could be trace fossils from other vertebrates, trace fossils from invertebrates, or—most upsetting of all—not trace fossils at all.

Among vertebrate trace fossils on dinosaur bones, these could be toothmarks made by animals that lived in environments completely separate from dinosaurs. We know that because, for example, some dinosaur bones ended up in shallow-marine sediments. Yet because they were land-dwelling animals, their decaying gas-filled bodies must have washed out from land and floated along until they had a burial at sea. (This explanation for how dinosaur parts got into marine sediments is nicknamed the "bloat-and-float" hypothesis.) Amazingly, a few of these bones have toothmarks and embedded teeth from scavenging sharks, which evidently could not pass up an exotic (albeit rotting) carcass passing through their neighborhood.

Meanwhile on land, lowly mammals imparted their distinctive incisor incisions on a variety of Late Cretaceous dinosaur bones from Alberta, Canada. These toothmarks were not made by mammals feeding on dinosaurs, exacting revenge for all of their consumed relatives. Instead, they were trace fossils like the toothmarks left by modern rodents that chew bones to wear down their constantly growing teeth and get more calcium into their diets.

Insect borings are other trace fossils in dinosaur bones that, at first glance, might be mistaken for dinosaur toothmarks. Carrion beetles, which dine on dead bodies, or termites, which make small pits or tunnels in bones, could have made these. Carrion beetles would have chipped bones with their strong mandibles while stripping flesh for food, just like they do today. Not all termites restrict themselves to wood or soil, and some modern species drill into bone, making themselves at home. As many people might know from watching lurid crime shows on TV, specific insects and their lifecycles are directly associated with colonizing dead bodies. Hence these insect trace fossils tell dinosaur researchers much about what happened after the dinosaurs died, such as whether their bodies were exposed for long or buried quickly.

So between sharks, small mammals, and insects, recently dead dinosaur bodies were valuable sources of nutrition and potential homes, with the bones lending themselves to lots of tracemaking. What about marks left on bones that were not traces of animal behavior? This is where paleontologists and overzealous ichnologists (guilty as charged) must exercise caution. For instance, plant roots can invade any available spaces in bones and push them apart, making trace fossils that could be mistaken for insect borings. Bones also can gain nicks, scrapes, and dinks from being dragged along a stream bottom. There's also something about having several kilometers of sedimentary rock overhead that tends to crush bones. All of these non-living phenomena could have wrought false traces, the bane of ichnologists everywhere.

Bang Your Head, Wake the Dead

Once in a while, two dinosaur partners of the same species contributed to making one type of trace fossil on bones. Among my favorite examples of these are big holes in the head shields of *Triceratops*. When paleontologists first noticed these holes, they were a bit mystified. For one, these were not normal parts of head-shield anatomy, which admittedly can become quite holey in some ceratopsians such as *Chasmosaurus* or *Torosaurus*. Although most had roundish outlines, these holes also varied in size and location on the head shield. Most important, though, the holes showed signs of injury and healing of the bone, meaning they were lesions. This implied they were wounds acquired during the lifetimes of the dinosaurs, and thus not post-death artifacts caused by erosion, dissolution, or fracturing of the bones and surrounding rock over tens of millions of years.

So paleontologists Andrew Farke, Ewan Wolffe, and Darren Tanke, in an attempt to make sense of these holes, took a closer look at the sizes, shapes, and placements of them on *Triceratops* and another ceratopsian, *Centrosaurus*. In *Triceratops*, most of these injuries were toward the rear and bottom part of the head shield, having been registered on the squamosal and jugal bones; almost none were in other bones. This clustering suggested that some type of behavior was behind them. After all, if these lesions had been from bone infections, they more likely would have been evenly distributed throughout the head. So were they toothmarks, delivered by their theropod contemporaries and presumed archenemies *Tyrannosaurus rex*? No, because *Centrosaurus* head shields showed almost no holes. Yet because *Centrosaurus* lived at the same time as *Triceratops* and *Tyrannosaurus*, it also should have been on a *T. rex* menu. (Paleontologists currently have no evidence of *T. rex* being a fussy eater.) Granted, we also know through trace fossils that *Triceratops* was eaten by *T. rex*, and with much gusto. But that ichnological point will have to wait until later.

Left without disease and predation as explanations, the paleontologists who studied these injuries identified *Triceratops* itself as the culprit. What anatomical traits did *Triceratops* possess that could impart such grievous traumas? Were these not mere herbivores, possessing no real means of defense other than through herding together as big happy families? Did their means of meanness ever show on their faces?

As many dinosaur fans can relate—especially those under ten years old—the genus name *Triceratops*, assigned in 1889, translates as "three-horned face." For generations, paleontologists and laypeople alike have looked at those three horns, imagined them pointing forward on a charging *Triceratops*, and thought, "I would hate to be skewered by those." Two of these horns, the longest, are positioned one above each eye and thus are nicknamed "brow horns." These horns give it a bull-like appearance, but remember they were on an animal that weighed about ten times as much as a modern bull. The third horn is more centrally located above the beak (prefrontal) that defined the top of a *Triceratops* mouth. Paleontologists are also now sure that *Triceratops* horns changed in size and shape throughout their lives, becoming more formidable with age.

These horns took considerable energy to grow and were very much a part of these dinosaurs' lives. Hence it makes sense, evolutionarily speaking, that these horns had some useful purpose, whether to prolong life to reproductive age ("Back off, predatory theropod!"), ensure reproduction ("Hey, baby, check out my horns!"), correctly identify others of your species ("We can do this, right?"), or fend off rivals within your species ("I found her first!").

At first, people weaponized these horns, imagining them as deadly counterpoints to ravaging tyrannosaurs. Paleontological artist Charles Knight (1874–1953) most famously depicted such a scenario in a 1927 mural, in which a lone *Triceratops* stands defiantly in front of a *Tyrannosaurus*, its paired brow horns pointing suggestively at the predator's soft underbelly. Since then, children and adults alike have imagined *Triceratops* fatally goring its attacker with those horns, or *Tyrannosaurus* somehow slipping past this armature to get in a triumphant killing bite.

Sadly, ichnology has little to tell us about whether or not *Triceratops* ever practiced such self-defense arts against *Tyrannosaurus* or any other predator. For instance, if *Triceratops* horns ever did pierce tyrannosaur skin, we have no record of it. *Tyrannosaurus* skin impressions are thus far unknown, and even if such spectacular finds are uncovered, these may or may not have recorded any wounds, let alone ones attributable to *Triceratops*. One would also think that the force generated by a jousting *Triceratops* would have left distinctive marks on tyrannosaur leg bones or other skeletal parts, but these are likewise unidentified.

Only one tantalizing *Triceratops* skull tells of a living battle with a *Tyrannosaurus*, in which one of its brow horns and a squamosal bone were chomped and later healed. But this evidence is just enough to provoke far too many questions. Who started the fight, the tyrannosaur or ceratopsian? Was this an argument over food—as in, the *Triceratops* was potentially food for the tyrannosaur, but objected—or was it over territory, in which one of two powerful animals needed more space? And of course, who ended the fight? Did the tyrannosaur walk away with no injuries, some injuries, or worse? All we know is that the *Triceratops* survived this testy encounter. Otherwise, without the trace fossils so helpfully left by a *Tyrannosaurus* on a *Triceratops* skull, we might not have known about it at all.

Lacking any clear evidence that horns were used as implements of self-defense against large theropods, paleontologists turned away from violence and reached for sex. That is, *Triceratops* horns combined with head shields are now regarded more as sexual advertisements than battle gear. In this role, their large heads would have been quite useful for recognizing the same species and perhaps gender, thus neatly avoiding two egregious mistakes sometimes made whenever the urge to mate takes over nearly all reason. Moreover, once these ceratopsians correctly recognized the same species and opposite gender in another ceratopsian, these horns and head shields then may have served a second overlapping purpose as implements for combat. These contests most likely would have been over mates, but also could have been inspired by a need for more territory, scarce food and water resources, protecting young, or an ill-tempered ceratopsian simply deciding to take out its frustrations on another. As a result, colliding heads and piercing horns left their marks as healed holes in *Triceratops* skulls, leaving us with intriguing scenarios of ceratopsian conflict.

Pachycephalosaurs, like *Triceratops*, are also great candidates for having made trace fossils caused by their own species and having those trace fossils left in their skulls. Pachycephalosaurs, such as *Pachycephalosaurus*, *Stegoceras*, and others, are relatively rare dinosaurs and only found in Cretaceous Period rocks. They are distant relatives of ceratopsians, having shared a common ancestor before the Cretaceous. Although their limbs are poorly known, they were bipedal, and their teeth look perfectly adapted for a plant-eating lifestyle. So far, no one has interpreted pachycephalosaur tracks, which is completely forgivable as their feet are unknown. Fortunately, pachycephalosaurs are well represented by their opposite ends, which are the tops of their heads. These skullcaps are incredibly thick, and in *Pachycephalosaurus* can be nearly 25 cm (10 in) thick. They are composed of parietals and closely associated skull bones, which are sometimes accompanied by crowns of spikes and horns. In many instances, such bones are the only evidence of pachycephalosaurs in a given time and place.

Paleontologists have long wondered why these dinosaurs were such boneheads. Growing bone is energetically expensive and must have meant this trait had some adaptive advantage during the evolutionary history of pachycephalosaurs. How could this have helped them, whether to survive, to have sex, or survive to have sex?

Nearly everybody agrees that these robust skulls must have been used for butting, as in, these were the Mesozoic equivalent of crash helmets used to protect their heads as they smashed them against something. What reasonable people disagree on, though, is what kind of butting? Did pachycephalosaurs use their heads for defense against predators, such as running full-tilt into the flank of a theropod, leaving it to limp away and make small mental notes not to attack pachycephalosaurs? Did they use them to drive off other herbivores from their favorite plants? Or did they turn against one another and, like their ceratopsian cousins, hit each other, whether for mates, establishing territory, or both?

The consensus is that pachycephalosaurs' thick skulls were most likely used for violent confrontations with one another. But figuring out why or how they used their heads is worth lots of discussion. Fortunately, explanations for their skulls can be reduced to just two. One is that pachycephalosaurs were head-bangers, knocking into each other directly with skull-to-skull contact. The other is that pachycephalosaurs went for softer targets, such as torsos, because breaking ribs might have been less risky for an attacking pachycephalosaur than taking on a skull like its own.

Let's think ichnologically, then, about what trace fossil evidence would be needed to figure out which of the two scenarios actually happened. Ideally, two pachycephalosaur trackways made at the same time would do the trick. For head-to-head combat, each trackway would have long stride lengths (running at high speed) and along the same line, but directly opposed (pachycephalosaurs ran toward each other), and abruptly ending, with one or perhaps both trackways connecting with much shorter and irregular steps off to the side of the previous trackways (staggering away dizzily). For flank-butting combat, the trackways would be nearly the same except one of them might be at right angles to the other, ending where they intersect. However, as mentioned before, not one pachycephalosaur track has been identified, let alone a trackway, let alone two trackways, let alone two intersecting trackways made at the same time. So as much as this ichnologist hates to admit it, we must rely on bones to resolve this problem.

In 2011, two paleontologists, Eric Snively and Jessica Theodor, had head-butting in mind when they looked at skulls of *Stegoceras* and *Prenocephale*, both Late Cretaceous pachycephalosaurs from North America. Using a combination of CT (computer tomography) scans and computer modeling of stresses and strains that would have been transmitted to the head and neck by this behavior, they tested whether the additive impacts of two pachycephalosaurs were feasible or not. After all, any such behavior, if performed regularly, might have resulted in permanent head or neck injury, and hence would not have lasted very long in a pachycephalosaur lineage. They then repeated their analysis on three species of modern mammals that habitually head-smash one another: white-bellied duikers (*Cephalophus leucogaster*), which look much like deer; giraffes (*Giraffa camelopardalis*) of Africa; and musk oxen (*Ovibos moschatus*) of North America. Other mammals they studied as control groups were: bighorn sheep (*Ovis canadensis*), which head-butt when younger but stop such shenanigans once older; pronghorns (*Antilocapra americana*) of North America, which hit each other with their antlers, but not their skulls; and three mammals that eschew head-knocking altogether, llamas (*Lama glama*), elk (*Cervus canadensis*), and peccaries (*Tayassu tajacu*).

Their conclusions were that a few modern head-butting mammals and *Stegoceras* shared the same types of skull tissue and other cranial traits for absorbing head-to-head blows. Having a more rounded, dome-like head was an advantage, as well as having more compacted (*cortical*) bone supported by more porous (*cancellous*) bone, like a motorcycle helmet with cushioning just below its hard exterior. Somewhat surprisingly, duiker skulls were the most similar to those of the pachycephalosaurs, despite these animals being much smaller than pachycephalosaurs. This led these paleontologists to suggest that we should look at these hoofed mammals in particular as models for pachycephalosaur behavior.

But wait! Don't give up on trace fossils just yet. If pachycephalosaurs were knocking noggins with one another, surely this left marks on their skulls where they impacted. Alternatively, if pachycephalosaurs went for flank butting, this behavior would have broken ribs in the one receiving the blow, and in a definite, localized pattern. Just to put their bone-breaking potential in perspective, the largest of pachycephalosaurs, *Pachycephalosaurus*, may have weighed as much as 400 kg (880 lbs), or about three times that of the largest linebackers in the NFL. Imagine, then, this bipedal dinosaur running at full speed, head down and pointed forward, and the consequences of its head meeting any other solid object, such as another head, and with the rest of another *Pachycephalosaurus* behind it. Adaptations or no adaptations, traces of this behavior should have been imparted on the bodies of the "butter" and "buttee" as trace fossils.

In 2012, in a bit of good timing as a follow-up to Snively and Theodor's 2011 study, two other paleontologists—Joseph Peterson and Christopher Vittore—scrutinized a *Pachycephalosaurus* skull and found exactly the sort of damage one would expect from a head-on collision. Two big closely spaced depressions, each about 5 cm (2 in) wide and 1.6 cm (0.6 in) deep, are on the top surface of this skull, together making a figure-8 pattern. Closely associated with these are about twenty pits, ranging from 1 to 10 mm (less than 0.4 in) wide and deep. The paleontologists discounted post-death erosion of its skull, such as breaking or dissolving. Likewise, they did not find any reason to think that bone-boring insects made any of these marks. Instead, they concluded these were lesions caused by some sort of skull trauma that did not result in the death of the pachycephalosaur but left it with a dented dome and a terrific headache.

What could have caused these marks? Quite reasonably, these researchers concluded that it was a result of the pachycephalosaur smashing its head into a solid object: perhaps another pachycephalosaur head, but certainly something that did not yield easily, like a tree trunk. The only problem with this evidence was that it came from a single specimen, and in science, we always like to see results repeated before confidently stating an idea might actually have some truth behind it. In a follow-up study published in 2013, Peterson and two other paleontologists found many more examples, verifying the previous study. As a result of these studies, dinosaur paleontologists may now look more carefully at dinosaur skulls for similar trace fossils, testing whether this head banging happened in other species.

Tell Me Where It Hurts: Distinguishing Dinosaur Trace Fossils from Other Injuries

Do trace fossils show that some species of dinosaurs hurt other species of dinosaurs? Yes. But this is also where ichnologists and paleopathologists alike urge caution when looking at formerly broken dinosaur bones that healed. For example, it is awfully tempting for paleontologists to look at the smashed (but mended) leg bones of a Late Jurassic *Allosaurus* and say, "Look, it's the trace fossil of an ankylosaur or stegosaur!" Such leaps of logic are connecting the formidable tail clubs of an ankylosaur or tail spikes of a stegosaur contemporary of *Allosaurus*, and one that successfully whacked its tail against this particular theropod marauder's leg. Alternatively, the *Allosaurus* might have tripped over a log and broken its leg from the fall. Again, just like the *Edmontosaurus* skin injury discussed earlier, paleontologists are at their best as scientists when they consider many possibilities for the origins of dinosaur wounds. Then, in a Sherlock Holmes sort of way, each possibility is contemplated, tested, and eliminated until only one, however improbable, remains.

Fortunately, two types of trace fossils recorded in the bones of both *Stegosaurus* and *Allosaurus* tell tales of their tails. One is of injured but healed bony tissue in a few tail spikes of *Stegosaurus*. These wounds are consistent with damage it would have suffered from striking solid objects. Another is of a hole in a tail vertebra of an *Allosaurus*, but one that probably healed around a piece of a *Stegosaurus* tail spike, an unwanted souvenir. *Stegosaurus* had four pointy spikes on the end of its tail, held horizontally and paired on each side, which surely became lethal weapons when whipped by powerful muscles attached to the tail. Paleontologists since the first half of the 20th century had figured that stegosaur tail spikes were used for self-defense, but lacked further confirmation until these trace fossils were described in 2001 (broken spikes) and 2005 (a healed hole in the bone of a predator).

This was not the only example of *Allosaurus* having had a bad day or two. One *Allosaurus* in particular had its rough-and-tumble life recorded all over its body. Nicknamed "Big Al" and studied by paleontologist Rebecca Hanna in the 1990s, this *Allosaurus* was unusual in two ways: Its skeleton was about 95% complete, and it had 19 instances of bone injuries, including breaks that healed. Among the bone breaks suffered by "Big Al" were two of its ribs, and its right foot was badly hurt, meaning that its tracks would have reflected a pronounced limp. These breaks may be trace fossils from an interaction with another dinosaur, whereas other breaks may have been self-inflicted, which would also qualify as trace fossils. However, where bone ailments would not count as trace fossils is if diseases caused them, because these would *not* be related to dinosaur behavior.

Unlike *Tyrannosaurus*, *Allosaurus* is one of the best represented of all dinosaurs in the fossil record, with thousands of its bones identified. This means that paleontologists are seeing a more complete range of its bones, which accordingly more closely reflect the overall health of an *Allosaurus* population at a given time. In that respect, *Allosaurus* was more likely to get hurt than another copiously represented theropod dinosaur, the Late Triassic *Coelophysis* of the southwestern U.S. Hundreds of specimens of *Coelophysis* have been studied, and very few of these show signs of healed bone injuries. One of the reasons for this may be an easy one: *Coelophysis* was a much smaller dinosaur than *Allosaurus*, measuring about 3 m (10 ft) long and weighing about 45 to 50 kg (100–110 lbs). In contrast, an adult *Allosaurus* could have been as much as 10 m (33 ft) long and weighed about 2.5 tons (more than 5,000 lbs). Yet another factor to keep in mind is that *Allosaurus* (on average) probably lived longer than *Coelophysis*, meaning it had more opportunities to rack up injuries.

Bigger was not better for a bipedal dinosaur, owing to the effects of gravity. As elaborated in a previous chapter, if a large two-legged dinosaur tripped, both its greater weight and height would conspire against it once it hit the ground, fulfilling the old saying "the bigger they are, the harder they fall." Big theropods also may have gone for accordingly larger prey items, which would have objected to a theropod's invitation to dinner and fought back. Last but not least, big theropods, just like ceratopsians or pachycephalosaurs, may have competed with other large members of their own species, whether over real estate, sustenance, or desirable mates. All of these possibilities add up to a greater likelihood of a hefty theropod suffering physical damage during its lifetime, especially if compounded by aggressive behaviors.

Inconveniently, other than the few examples here, most healed bone breaks in dinosaurs have not been interpreted more narrowly as trace fossils made by other dinosaurs or by the dinosaur itself. In most instances, paleopathologists are rather conservative in interpreting these abnormalities, offering a basic diagnosis—such as "tibia fractured, later became infected"—and leaving it at that. Perhaps more precise attributions to bone injuries—such as "tibia fractured by *Ankylosaurus* tail club, later became infected"—will be made in future studies. But in the meantime, one can always turn to one of the most unambiguous of dinosaur trace fossils known: toothmarks.

When Tooth Met Bone: Dinosaur Toothmarks as Trace Fossils

Dinosaur toothmarks are lovely trace fossils, simply because most leave little doubt about the dinosaur's motivation, which was eating. Other than the few exceptions of toothmarks interpreted in skin impressions mentioned earlier, almost every example of a dinosaur toothmark described thus far was registered in bone. However entertaining it might be to imagine dinosaurs gnawing random, non-living items in their surroundings, such as sticks, mud, or rocks, no trace fossils of this behavior are known. Similarly, dinosaur toothmarks have not yet been interpreted in fossil plants, despite our surety that sauropods, ornithopods, ankylosaurs, stegosaurs, ceratopsians, and some theropods enthusiastically consumed vegetation. One of the most obvious problems with such a discovery is that plant material, such as leaves, stems, roots, and so on, may not have readily preserved such evidence.

When a dinosaur bit into another dinosaur and its teeth reached bone, the resulting punctures or scrapes are trace fossils of the dinosaur that did the biting. Just like mentioned before, though, a good question to always ask about toothmarks as trace fossils is whether that bone belonged to a living or already-departed dinosaur. For instance, if a bitten dinosaur happened to be alive when chomped by another dinosaur, it very likely would have reacted to this assault and may have added a trace of its own behavior to the wound. Think about how the first reaction of a person bitten by their pet dog, cat, parrot, snake, or Komodo dragon would be to pull away from the source of pain. Unless the biter has really clamped down on its victim, this defensive motion should cause the wound to elongate, making the trace a composite one that is a result of both the pet and pet owner's behaviors. The same principle should then be applied to dinosaur toothmarks. When paleontologists look at a bone with dinosaur toothmarks in it, they most often assume that the bone belonged to an already-dead animal. Nevertheless, they could also look closely for signs that an animal may have still been breathing, and possibly adding its traces to that of its attacker.

More good questions to ask include "Who did it?" which relates to the tooth anatomy of the dinosaur that made the toothmark, and "How did this dinosaur bite?" For example, did the dinosaur rake its teeth across the surface of a bone as it was tugging meat from it? Or did it punch past skin, muscle, and other soft tissues and go directly into the bone? Did it mostly use teeth from its upper jaw (maxilla), or did it also use its lower jaw (mandible)? Which teeth were used in the biting: ones toward the front of the mouth, ones more toward the back, or a combination of the two? Did the dinosaur bite multiple times in the same general area, inflicting an overlapping array of toothmarks? And in a related question, did more than one dinosaur snack on this delectable treat, adding its toothmarks to a varied collection?

Before going on any further with toothmarks, though, it is probably a good idea to learn about dinosaur teeth, and with a focus on theropod teeth, because theropods included a large number of dinosaurs that ate other dinosaurs. Like many other groups of toothed vertebrates, theropods had a wide variety of tooth shapes and sizes appropriate for their varied uses. Although paleontologists now know that not all theropods were meat eaters, and some were even toothless, we are also aware that "meat eating" is too broad a category for toothed dinosaurs to have evolved "one-shape-fits-all" tooth forms, too.

For example, a few smaller theropods, such as the Late Cretaceous theropods *Velociraptor* and *Saurornitholestes*, had thin, curved, and beautifully serrated teeth. In contrast, larger Late Cretaceous theropods, such as *Daspletosaurus* and *Tyrannosaurus*, had thick, stout teeth with only minor serrations. Serrations on theropod teeth were formed by numerous raised and bladed bumps called *denticles*, which were evenly spaced along the narrowed edge of a tooth. These functioned just like the serrations on a typical steak knife, helping the teeth to saw through flesh. How this helps ichnologists is that toothmarks can also act as a "fingerprint" for identifying a dinosaur species when examining any bones bearing long, parallel series of grooves. In such marked bones, each groove corresponds to a denticle. Hence the spacing and numbers of these traces should match those of known theropod teeth.

Tooth forms even varied within the same dinosaur, a condition called *heterodonty* ("different teeth"). Humans, by having incisors, canines, bicuspids, and molars all in the same mouth, are well acquainted with heterodonty, which is a typical condition for mammals with varied diets. However, heterodonty can be subtler in other vertebrates, such as theropods. For instance, if you were to glance at a live tyrannosaur's open mouth, all you would see at first are a large number of its big pointy teeth, which understandably might all look alike as you quake in fear. But if you looked deeper into its mouth, say, while being eaten, you would see that the teeth in the front curve more toward the rear of the mouth, are more incisor-like on their ends, and have D-shaped cross sections, with the rounded part of the "D" facing the front of the mouth. Then, just before you pass down its gullet and become a bolus, you would note that its rear teeth are rounder and blunter than those in the front. This will be explained further in just a minute, but in the meantime, just digest the preceding.

Theropod teeth also formed distinctive patterns in how they lined up on both the upper and lower jaws. This row of teeth is appropriately known as a tooth row. Certain theropods had an exact number of teeth per tooth row, but that number could also be divided into two parts, a front (anterior) tooth row, and a rear (posterior) tooth row, which are divided by a gap called its inter-tooth spacing. Regardless, you should mind the gap, as the front teeth and rear teeth served different functions in a theropod mouth. Obviously, the front teeth were what mattered most when it was love at first bite, whether performed on a living or dead animal.

If a theropod's potential food item, such as a small ornithopod, was still alive and having issues with a proposal that it should devote its life to feeding a theropod, then the front teeth were the most persuasive tools used by that theropod. Depending on tooth forms, these would have been used initially to slash, grip, or crush live prey, then later nip and rake flesh from the body. Conversely, the rear teeth were more likely to have been used well after any disagreement had ceased between the diner and its meal, as these were better suited for chewing meat and crunching bone. These variations in forces within a theropod mouth were a function of physics. For example, think of how a nutcracker is used more effectively when a nut is placed closer to its hinge rather than at its ends.

All in all, theropod toothmarks in bones can be connected quite reasonably to recognized theropod teeth and their intended functions. These trace fossils are also interpretable by applying just a little imagination and keeping in mind how meat is separated from bone. For instance, toothmarks could have been made by a dinosaur scraping its teeth along the length of a bone, trying to get every last little tasty bit of skin, fat, sinew, and muscle off it. (Think of people eating ribs or chicken wings, using just their incisors.) These marks would have been made by a pulling, lateral movement, so they would be shallow, long grooves on a bone surface, maybe with parallel lines made by denticles.

A theropod also could have punctured a bone while accessing deeper sources of meat in a carcass. Because this action would have been more perpendicular to the bone surface, these puncture toothmarks will be deeper and limited to small areas that more or less reflect tooth cross-sections. A combination of these two actions—puncture and pull—would have inflicted deep toothmarks followed by long, shallower grooves, made when a dinosaur bit into its meal, hit bone, and then pulled meat off those bones with its teeth, like a carnivorous version of raking leaves.

So through a combination of knowledge about tooth anatomy, toothmarks in bones, and experiments on bite forces—including those of living animals—paleontologists can indeed figure out which dinosaurs ate other dinosaurs, what parts they ate, how they ate, and other such gory details. Although dinosaur toothmarks were first interpreted as trace fossils near the start of the 20th century, the first studies to pull together all of these lines of evidence were not done until the mid-1990s. This was when paleontologist Greg Erickson and a few of his colleagues decided it was time to start describing and interpreting what were likely *T. rex* toothmarks in *Triceratops* bones.

What made them think these toothmarks belonged specifically to *T. rex* stemmed from a combination of size, shape, place, and time. Very simply, these toothmarks were big, they matched the shapes of *T. rex* teeth, and this massive theropod lived in the same places and times as *Triceratops*. Because some of the toothmarks were deep punctures in *Triceratops* hipbones, the paleontologists were able to identify the biter by stuffing putty into the holes, extracting the putty, and presto: they had beautiful molds identical to the shapes and sizes of *T. rex* teeth.

Naturally, just identifying the toothmark maker was not enough to satisfy these curious paleontologists. They also wanted to answer a basic question: How much force was generated by a *Tyrannosaurus* bite? This potentially could have been answered through complicated computer models that took into account muscle attachments, jaw mechanics, pivot points, and other factors. Yet the toothmarks were an actual record of a tyrannosaur biting into bone, so all the paleontologists had to do was try to replicate the toothmarks.

Disappointingly, they did not reanimate a long-dead *T. rex*, but instead set up a "biting" mechanism with proxy tyrannosaur teeth, which were made of aluminum and bronze. These artificial teeth were then used to imitate a tyrannosaur bite, but one in which the force exerted by each "bite" could be measured. Force is measured in newtons (named after Sir Isaac, not the confection), which is equal to a kilogram (2.2 lbs) moving a meter (3.3 ft) per second squared. The formula for force is:

$$F = ma \text{ (Force = mass} \times \text{acceleration)}$$

However, force for a biting tyrannosaur did it no good if it just bit air. Once applied to an area on whatever it was biting, this force then translated into pressure. To better understand the difference between force and pressure, gently place a heavy book on the upper parts of both of your feet. Not so bad, is it? Now imagine what it would feel like if you dropped this book from waist height, and onto just the big toe of one foot. Dropping the book would certainly increase the force exerted by the book, but pressure would also increase dramatically because of this force striking a smaller area, and even more so if the corner of the book was the first point of contact. Now think about a *Tyrannosaurus* mouth, the muscles connected to its jaws, those jaws opening and then closing, with the huge force of the bite transmitted into the small areas represented by points of teeth.

Still, the researchers' biting mechanism needed something to actually bite, such as real bone. This would effectively test whether depths of the original toothmarks in bone could be produced artificially. Seeing that *Triceratops* bones were both too valuable and too fossilized to use, the researchers turned to a more easily procured supply of fresh hipbones from domestic cattle. So with this biting machine and "victims" in place, the researchers were ready to start making their own traces.

The results of this experiment were astounding. Although nearly everyone suspected that *T. rex* was a terrific chomper, no one had been able to put numbers to this presumption. Bite-force estimates came out to 6,400 to 13,400 N, which at the time were greater than those known for any living animal, and confirmed that *T. rex* could have easily crunched its food, bones and all, living or dead. In later experiments done on modern alligators and crocodiles, Erickson and other researchers found the largest American alligators (*Alligator mississippiensis*) had bite forces within the same range as *T. rex* (9,400 N). But most impressive of all were estuarine crocodiles (*Crocodylus porosus*)—known warmly by Australians as "salties"—which had maximum bite forces of 16,400 N. These findings further implied that *Deinosuchus*, a gigantic Late Cretaceous crocodilian, probably had a bite far more powerful than that of *T. rex* or any other land-lubbing theropods. (It was not too surprising, then, when paleontologists later found toothmarks attributable to *Deinosuchus* in dinosaur bones.)

Despite a usurping of tyrannosaurs by crocodilians as champion biters, other *T. rex* toothmarks invoked awe-inspiring scenarios of this dinosaur occasionally saying "Off with their heads!" This gruesome idea came about when Denver Fowler and several other paleontologists noticed, while looking at *Triceratops* bones from the Late Cretaceous of Montana, that some of the outer frills of *Triceratops* skulls had toothmarks on them, which once again matched those of *Tyrannosaurus rex*. The most perplexing aspects of the toothmarks, though, were their specific locations. Some were along the edges of *Triceratops* head frills, whereas others on the same skeleton were on the neck vertebrae. The toothmarks on the skull showed no signs of healing and were from teeth on upper and lower parts of the jaws puncturing opposite sides of the frill: no scraping or pulling, just clamping. For feeding, these traces did not make much sense, because there was very little ceratopsian meat to be enjoyed on its skull. In contrast, toothmarks on the neck vertebrae were more puncture and pull, definite traces of where the tyrannosaurs bit into and pulled off scrumptious hunks of ceratopsian flesh.

To answer this mystery, take another look at a *Triceratops* skeleton, and you will note that its huge frill covers its neck quite thoroughly and effectively. Hence, it is intuitively obvious to even the most casual observer that this dinosaur not only must have been already dead for a tyrannosaur to feed on its neck muscles, but also must not have had a frill in the way. Accordingly, the easiest way for a tyrannosaur to access that flesh would have been to remove the frill. For *T. rex* to do this, given their jaw strength and body mass, all it would have needed to do was bite down onto the edge of a *Triceratops* frill, tug, and yank the head from the body to expose the good stuff below. In such a scenario, it also may have used one or both of its feet to stabilize the *Triceratops* body while pulling.

So these trace fossils tell us that *Tyrannosaurus* ate dead dinosaurs, which is good to know. But some of us want something more to satisfy our bloodlust. In our minds, *T. rex* was not a big vulture, however noble vultures might seem to be in their own way. Instead, we imagine *T. rex* or other tyrannosaurs as apex predators of their environments, among the greatest that ever lived. So do toothmarks ever show where a tyrannosaur bit into a living dinosaur?

Recall again the story at the start of this book, in which a *Tyrannosaurus* took a bite-sized chunk out of the tail of an *Edmontosaurus*. It turns out this is a story based on fact, and the trace fossil evidence backing it up can be viewed publicly. In the dinosaur hall of the Denver Museum of Science and Nature is a beautiful mounted skeleton of the Late Cretaceous hadrosaur *Edmontosaurus*, flawless in nearly every way save for one small blemish. About halfway down its tail vertebrae, on the top surface, its vertebral spines look as if they were clipped, forming a semi-circular pattern, almost as if someone used a cookie-cutter on them. When paleontologist Ken Carpenter took a close look at this oddity, he saw signs of healing around the bone, which meant this pattern was from an injury and not from, say, vertebral spines snapping off after death. The curvature and width of the wound was also intriguing, as it was the right size and shape for the tooth row of a large theropod dinosaur. The only theropod large enough to own such big jaws and that lived in the same area and time as this *Edmontosaurus* was *Albertosaurus*, a tyrannosaur closely related to *T. rex*, or *T. rex* itself.

This was among the first firm indicators that tyrannosaurs did indeed go after live prey, refuting naysayers who have put forth the case that the *T. rex* was just a lowly scavenger versus a fearsome predator. The reality, like most realities, is much more nuanced. Since Carpenter's study, other healed toothmarks found in *Edmontosaurus* bones, either attributed to *T. rex* or *Albertosaurus*, have both affirmed and restored the original reputation of tyrannosaurs as predators. Both trace fossils tell us that tyrannosaurs ate both dead and live dinosaurs; hence, these theropods probably used a mixture of predation and scavenging to feed, much like many modern top predators today, such as African lions, grizzly bears, and hyenas.

Fine Young Cannibal Dinosaurs

Consider being armed with knowing a theropod's individual teeth sizes and shapes, the number of teeth and spacing within its jaws, and how to interpret toothmarks. All of these bits of knowledge then become handy forensic tools for figuring out who ate whom. So this is exactly how three paleontologists—Ray Rogers, David Krause, and Kristi Rogers—discerned that *Majungasaurus*, a large theropod from Late Cretaceous rocks of Madagascar, was a cannibal. In one specimen of *Majungasaurus*, its ribs and vertebrae had toothmarks on them, and rather distinctive ones. The toothmarks were a series of thin, evenly spaced grooves caused by serrations, ones that perfectly matched those on the teeth of *Majungasaurus*. These grooves were also separated by a gap that corresponded with the intertooth distance of—you guessed it—*Majungasaurus*.

Based on similar toothmarks these paleontologists had seen on sauropod bones, they knew that this theropod normally ate those dinosaurs. But at least one decided its dead relative looked too appealing to pass up as a snack. Based on the bodily locations of the toothmarks, the *Majungasaurus* must have been dead when they were made, so these trace fossils are not only of cannibalism but also of scavenging. Such evidence is a bit perplexing, leading paleontologists to wonder if there was a time when ecological conditions during the Late Cretaceous in Madagascar became bad enough that large theropods resorted to eating their own.

From an evolutionary standpoint, one might think that cannibalism is uncommon in modern large carnivores. After all, any long-term reliance on consuming your own species could lead to eventual extinction. Yet cannibalism happens. Komodo dragons, crocodilians (including alligators), big cats such as lions and tigers, and bears are among the large predators that will eat their own species. Some cannibalism is a consequence of competition, in which a male lion kills and eats the cubs of a rival male, or bad timing, such as when a baby alligator swims too close to a hungry adult. But cannibalism also can be more opportunistic, such as during hard times; when you're starving and nothing else is around, you might as well eat your brother. This situation is especially more likely to take place during times of ecological stress, when food supplies tend to shrink. Think of droughts, which cause many plants to wither and die, negatively impacting herbivores, which would in turn affect carnivores. Hence, the biological taboo of cannibalism becomes less of a barrier when food becomes scarce.

So let's apply this idea to the geologic past. Did Late Cretaceous ecosystems of Madagascar undergo any sort of stresses, such as droughts? Apparently so, as its rocks show the region was semi-arid while dinosaurs lived there, but with pronounced wet-dry cycles.

Such environments were more susceptible to droughts than, say, a tropical rainforest. One of the more compelling pieces of evidence for droughts in this area during the Cretaceous comes from other trace fossils, namely lungfish burrows.

Modern lungfish, to keep from drying out during times of low rainfall, make burrows. They then reside in these while also enveloping themselves in slimy cocoons, thus doubling their protection. In an ichnologically wonderful paper by paleontologists Madeline Marshall and Ray Rogers published in 2012, they interpreted more than a hundred lungfish burrows in a Late Cretaceous river deposit in Madagascar. These burrows, which were nearly identical to modern ones, pointed toward this as a behavioral response to a long, dry time that required hunkering down. So just like insect cocoons next to *Troodon* nests in Montana, these trace fossils tell us something about how animals around dinosaurs were adapting to their climate. Did such dry periods also affect food supplies of Late Cretaceous dinosaurs in Madagascar? Perhaps, although paleontologists need more details before they can say for sure.

Regardless of reasons why, finds of dinosaur-cannibal trace fossils from so long ago are remarkable enough to prompt paleontologists to reexamine toothmarks on theropod bones, checking to see whether or not these came from the same species. From there, they could then test whether these behaviors were anomalous or more normal than originally surmised.

Along those lines, trace fossils tell us that another large theropod, our old friend *Tyrannosaurus rex*, decided to count on its relatives for some of its meals. In a 2010 study conducted by Nicholas Longrich and three other paleontologists, they examined *T. rex* bones in museum collections at the University of California (Berkeley) and the Museum of the Rockies (Montana) and realized that not only did some of these bones have toothmarks, but big ones. Even before measuring these marks, the paleontologists knew that their depths, lengths, and spacing between teeth limited them to the largest land carnivore that lived at the same time as *T. rex*, which would have been *T. rex*. Unexpectedly, bones from four separate specimens held these toothmarks. Considering the rarity of *T. rex* specimens in general, and to have four with toothmarks also coming from *T. rex*, cannibalism in that species might have been more common than originally thought. Still, such acts might have been done in desperation, as three of the affected bones were from feet, and one was a humerus. Let's just say that if you were the world's largest land carnivore and you were resorting to not only eating your own species but also the least meaty parts, you were much too hungry. Also, no reasonable person can imagine a live tyrannosaur quiescently accepting another tyrannosaur stripping flesh from its toes or arms. In other words, a prone, dead tyrannosaur would have made for a much easier snack for one of its kin.

Trace fossils also tell us that big theropods chewed on one another's faces, but while their faces were still alive. This idea, first proposed by paleontologists Darren Tanke and Phil Currie in 1998, was based on healed bite marks they noted in skulls of *Sinraptor* (a Late Jurassic theropod from China), *Albertosaurus*, *Daspletosaurus*, and *Gorgosaurus* (all Late Cretaceous tyrannosaurs from Canada). Interestingly, some of this unruly face-biting happened while these theropods were still relatively young. For example, subadults of *Albertosaurus* and *Gorgosaurus* have face bite marks, as did another unidentified tyrannosaur from the Late Cretaceous Hell Creek Formation of Montana that got its nose out of joint (literally) caused by a bite to its nasal and maxilla that bent its face to the left. Because these wounds healed, these were probably from competition or territorial aggression and do not really qualify as cannibalism. Although, considering that these were adolescent dinosaurs, premating "love bites" cannot be discounted, either. If so, these were hickeys from hell.

Dinosaur Dentistry: Looking Closely at Wear on Teeth

Pretend you are a dentist, or, if you already are a dentist, just be yourself. Your patient, who just walked into the waiting room, is a 7-ton hadrosaur, and he is complaining of a toothache. You've had hadrosaurs stop in before for checkups, and unlike most theropods, they're usually great patients. Nevertheless, you dread taking a look inside any of their mouths because of their dental batteries. Dental batteries are tightly packed arrangements of small teeth, and in a typical hadrosaur, they could have hundreds. Consequently, finding a problem tooth among many healthy ones becomes a time-consuming task. It also doesn't feel all that necessary, considering that another, newer tooth directly underneath the bad one will replace it anyway.

Using a mirror, directed light, and a magnifying lens, you see thin, shallow scratches on its teeth. These scratches tell you that the hadrosaur moved its upper jaws out and to the sides while the lower jaw stayed put. This motion differs greatly from that of nearly all humans, who can move their lower jaws forward and laterally while their upper jaws are fixed. For this and other hadrosaurs, though, this action is normal because of how its jaw is hinged, and such an arrangement provides the grinding action needed to chew its food, which are various near-the-ground plants. Seeing this evidence, you admonish your patient for eating too many of those succulent low-lying plants along a nearby river floodplain—especially after the river has flooded—and for chewing too much. Your patient is both surprised and impressed that you somehow figured out what, where, and how he had been eating, and he promises to do better. You then schedule him for a cleaning six months from then, and plan your vacation to coincide with that appointment.

Again, thanks to trace fossils on teeth—these tiny scratch marks—we can tell that certain species of hadrosaurs and other herbivorous dinosaurs were grazers, eating plants that grew close to the ground, rather than being browsers, which meant going for greens hanging high in trees. Whole, healthy teeth certainly can have shiny polished surfaces, but if you look at them with a magnifying glass or microscope, you will also see tiny grooves and scratches. You have such wear on your teeth, too, especially if you are in the habit of eating mineral-laden plants.

These marks, called *microwear,* were scored on dinosaur teeth when they chewed plants containing silica or plants with grit on them. The grit also would have consisted of silica-rich minerals such as quartz. As any geology major will gleefully tell you, quartz is harder than the mineral apatite, and the latter is what composes vertebrate teeth and bones. Hence, silica in plant tissues or silica-rich grit on plants, combined with dinosaurs chewing those plants, would have caused more scratches than if the dinosaur had either swallowed those plants whole without chewing or eaten plants with less silica. On the other hand, browsing on plants growing well above the ground should have resulted in fewer scratches.

To better understand how silica was included in some plant tissues, we look to evolution. Plant-eating dinosaurs may have left few bite marks on fossil plants, but plants certainly "bit back." Thorns or spikes, toxins, or indigestible parts became common in some plant lineages as a consequence of dinosaurs treating them like indiscriminate items on a salad bar. Among these defenses were *phytoliths*, tiny grains of silica precipitated in plant tissues. Phytoliths represent a war of attrition, which, through sheer numbers and high abrasiveness, slowly wear down herbivore teeth. Hard seeds and nuts are more overtly offensive, inflicting breakage in teeth and thus forming pits.

Phytoliths are common in many plants today, especially monocotyledons, which include all grasses, orchids, bamboo, palm trees, and many others. Monocotyledons also got their start in the middle of the Mesozoic Era. Coincidence? Maybe, but the evolution of the largest land herbivores of all time surely resulted in land plants responding ferociously; after all, plants have no moral compulsion to fight fair, and they want to survive and reproduce, too. Thus, it comes as no surprise that sauropods, hadrosaurs, and other herbivorous dinosaurs also evolved fast-replacing teeth (in sauropods) and dental batteries (in hadrosaurs) to more easily replace teeth worn down by insidiously vicious plants.

Plant-eating dinosaurs also might have had a double threat to their dental health posed by plants with phytoliths: silica-rich grit on their surfaces. How did this grit get on plants? Take a close look at any vegetation alongside a stream that experiences frequent flooding in a place with silica-rich rocks and you will likely see clay, silt, and fine sand adhered to the leaves, branches, and stems of plants there. As any flood subsides, suspended sediment carried by a stream during a flood settles, and some of it sticks to the plants. I look for such residue whenever tracking animals along stream banks, and wherever noted, it informs me of former flood heights in that stream valley. Of course, wind can also put some grit on plants, but it adheres more easily if already wet. Nonetheless, also think of how dust clouds were likely kicked up by herds of dinosaurs, suspending plenty of fine-grained sediment near the ground and likewise adding these grains to any low-lying flora.

Paleontologists who studied microwear in *Edmontosaurus* found out that this dinosaur—when it was not breaking off tyrannosaur teeth—chewed its food through a definite series of movements, and that it was grazing. Microwear on its teeth consists of four sets of scratch marks, with each set showing different orientations on tooth surfaces. For the set with the deepest scratches that also cut across the others, the paleontologists defined these as having formed during the "power stroke" phase of chewing. This was when the hadrosaurs put the most effort into grinding down their food, moving their jaws vertically and together.

Microwear is also visible on the teeth of other herbivorous dinosaurs, such as ceratopsians, ankylosaurs, and sauropods. Most of these trace fossils, while telling us how these dinosaurs chewed their food, also suggest a grazing habit, instead of their cropping tall tree tops. For sauropods, this may seem to contradict interpretations of how their long necks were used to do just that, reaching higher to sample some tasty canopies. But anatomical studies done on some sauropods now imply that some of these lengthy necks were maybe better suited for sweeping large areas—back and forth—across fields of low-lying vegetation.

Trace Fossils in Bones and Teeth: Greater than the Sum of Dinosaur Parts

Hopefully by now it should be absolutely clear that taking a gander at dinosaur body fossils for their trace fossils was a good kind of ichnological cheating. For paleontologists, these skin impressions, bones, and teeth with their trace fossils are two-for-one specials that not only are the bodily remains of dinosaurs but also give us valuable insights on dinosaur behavior. Again, just looking at these body fossils without noting their trace fossils is not enough. When viewed through an ichnological lens, these parts unfold into much more than just a piece of a dinosaur, including complex relationships between their former owners with other dinosaurs and many other facets of their Mesozoic ecosystems.

CHAPTER 7

Why Would a Dinosaur Eat a Rock?

Your Inner Gastrolith

The dark rock stuck out in the pinkish mudstone, looking like an errant raisin tossed into a strawberry mousse. My curiosity properly piqued, I walked over to pick it up and held it in my left hand, admiring its fine qualities. It was a black, polished chert, which is a hard, flint-like mineral. It was beautifully rounded and about half the size of a baseball, but slightly longer than wide. I bounced it up and down to better feel its heft. A closer look revealed a few irregularities on its otherwise flawless surface, little chatter marks where it must have impacted against some other solid object, such as another rock.

Somehow I identified with this stone, feeling like a small oddity pasted onto a homogenous background, with an apparently polished exterior but also bearing blemishes for all to see. It was the summer of 1983 and I was well into the second week of geology field camp in northwestern Wyoming. About a hundred other geology students were there too, all of us being directed by three geology professors. Yet I was an earth-science neophyte, having just finished my first year of an M.S.-degree program in geology, but entering it with a background in biology and art. I was also a Midwesterner who had been confined to flat, cornfield-dominated landscapes east of the Mississippi River for all of my life. So this was my first trip to the western U.S., and my first time walking on Mesozoic rocks.

Underneath my feet, the weathered mudstone—streaked with white, gray, or maroon, and occasionally interrupted by beige sandstone lenses—was the Late Jurassic Morrison Formation, renowned worldwide for its massive fossil bones. This was dinosaur country, and being in it was absolutely thrilling. I had read about these giant extinct animals all of my life, but at the time my home state of Indiana had no museums with mounted skeletons of dinosaurs, nor did it have any rocks of the right age holding either dinosaur bones or trace fossils. I was a transgendered Alice in Wonderland, albeit with more amiable and slightly saner companions. After studying the rock for a few more minutes, I set it down and walked away, but held on to its memory.

Later that day, the professor in charge of our geology field camp, Dr. Wayne Martin (no relation), casually remarked on these anomalous rocks in the fine-grained mudstone. "Gastroliths," he informed us with his distinctive slow West Virginian drawl. Geologists regarded these blemishes in the otherwise smooth, fine-grained Morrison Formation as evidence that dinosaurs were there. Sure, you could also find dinosaur bones if you looked hard enough, and the Morrison Formation was rightly celebrated elsewhere in the western U.S. for yielding some of the best-loved of all dinosaurs: *Apatosaurus*, *Diplodocus*, and *Allosaurus*, to name a few. But where we were, these large stones were far more common than bones. Despite nearly two hundred eyes looking at the ground that day, not one person spotted a dinosaur bone, but we found plenty of these strange rocks.

I remember being intrigued by this vague, indirect evidence of a dinosaurian presence. Filled with naivety and impressionable enthusiasm, I continued to wonder about that black rock from my field camp and gastroliths in general. Was it, along with similar rocks there in the Morrison Formation, really connected to dinosaurs? And if so, how? I wanted to believe they were. Yet when I read more about gastroliths much later in my graduate studies, doubt began to erode my faith. Indeed, in a class the year after this field camp, Dr. Martin seemingly contradicted himself, saying that such rocks should not be called "gastroliths" unless found in the ribcage of a dinosaur. It later seemed that many geologists and paleontologists told me, to paraphrase Freud, "Sometimes a rock is just a rock," saying that these rounded stones could be readily explained by non-dinosaurian means, such as rivers, wind, or ocean waves. This creeping uncertainty was fueled by later trips to the western U.S., in which I would find similar rocks in Mesozoic strata and ask if these were gastroliths. In response, other geologists and paleontologists would scoff at the suggestion, as if I had asked about Bigfoot, aliens, or Bigfoot aliens.

Were these skeptics completely or partially right, or should they have been laughing at themselves for being gastrolith deniers? But maybe they had a point. After all, what exactly were gastroliths? How could a mere rock tell us whether a dinosaur had been on or near a given spot in the western U.S.? Moreover, if gastroliths were real, how could these curios actually be trace fossils, supplying deep insights on dinosaur behavior and evolution?

Between a Rock and a Soft Place

Starting nearly 200 million years ago, well before our ancestors were attracted to certain rocks, chose them with care, and began using them as tools, dinosaurs had their own rock collections. Some of these rocks even had fossils in them, which meant that dinosaurs might have been unwittingly the first fossil collectors, too. The biggest difference between these dinosaur- and human-made assortments of stones is that dinosaurs carried them on their insides. "Gastrolith" literally means "stomach stone" (in Greek, *gastros* = stomach, *lithos* = stone), and they thus refer to rocks that somehow made it into the digestive tract of an animal. Because many modern animals—from invertebrates to vertebrates, marine to terrestrial—have rocks in their innards, no one questions the physical reality of gastroliths. Nevertheless, with dinosaurs of the long-lost past, disagreement stems from exactly how those rocks got there in the first place, whether they served any purpose, and if so, their intended functions. Thus it becomes easy to see why these are among the most controversial of dinosaur trace fossils.

If one can stomach it, a little bit of jargon goes a long way in better understanding these enigmatic trace fossils. For one, they can be divided into two categories: *bio-gastroliths* and *geo-gastroliths*. As these hyphenated prefixes imply, biological processes make one of them, whereas the other is formed by geological means. However, the latter type certainly has to involve an animal at some point in a gastrolith's lifecycle by making its way into a gut. So these could be called geo-bio-gastroliths, or bio-geo-gastroliths. But either of those terms would be needlessly pedantic and cause science editors everywhere to flinch and twitch uncontrollably, especially if these bio-geo-gastroliths are touted as "missing links" in a "living fossil." And we wouldn't want to do that.

With bio-gastroliths, the animals make these themselves, secreting minerals internally and forming concretions in their body cavities. Sound strange? Not really. If you've ever heard of gallstones or kidney stones, or been unfortunate enough to have experienced them directly, then you know about bio-gastroliths. I once learned more about bio-gastroliths and their effects on human health while waiting in a hospital emergency room with a fractured right radius, a consequence of a bicycle accident. Meanwhile, the patient lying next to me passed a kidney stone, which took him about an hour or so. (Let's just say that after that, I stopped feeling so sorry for myself.) Later, once the memory of his screams faded enough for me to think more objectively, I read about kidney stones. These bio-gastroliths are deposits of calcium oxalate, which form in people's kidneys as a result of calcium imbalances and insufficient fluid intake. (Fortunately for many of us, myself included, regular beer drinking is excellent preventative medicine.) Gallstones, on the other hand, are more organic than mineral and are normally composed of cholesterol, although these can sometimes have calcium mixed in as well.

Remarkably, some crustaceans, such as marine crabs and freshwater crayfish, secrete their own bio-gastroliths of calcium carbonate. The purpose of these secretions is to absorb calcium from their exoskeletons just before they molt, then put that calcium back into new exoskeletons. Hence, these bio-gastroliths are not pathological but act more like tiny biochemical ATMs, from which these crustaceans can deposit or withdraw calcium as needed. These concretions are not only in modern crustaceans but also are preserved in the geologic record, trace fossils that let us know about a former crab or crayfish presence.

In contrast, geo-gastroliths are rocks made outside of animals' bodies by normal geological processes—whether igneous, metamorphic, or sedimentary—but that later somehow make their way inside while that animal is still alive. Most of these rocks are composed of silica-rich minerals, such as quartz and feldspars, and originally were parts of much larger rocks that were broken down and perhaps rounded by streams over many years before residing in a gut or two.

Sizes of these rocks can vary considerably from sand-grain-sized to as big as baseballs, and even in the same animal. Size is roughly correlated with the body size of the host animal, meaning small gastroliths are in smaller animals and big gastroliths are in bigger critters. However, the width of a gastrolith seemingly never exceeds 3% the length of the animal. In terms of numbers, an animal with gastroliths could have as few as two or three, or more than a hundred. Many gastroliths are rounded and have smooth, polished surfaces, but some are duller and irregularly shaped, with angular outlines. In short, gastroliths have a few traits in common, but each can have its own personality, and a "population" of these in any given animal can be quite diverse.

How these rocks get in the belly of an animal is one of the most important questions surrounding them, but answers to that inquiry can be reduced to just two: accidental and purposeful. In the first instance, an animal could be going about its daily business and accidentally swallow a few rocks. For example, an alligator might use its snout to help dig a den, and a few rocks then get in its mouth and go down its throat while it is shoveling. Similarly, nest-building birds that are picking up ground debris, especially soil, also might inadvertently take in small stones. Alternatively, an alligator or bird might see some rocks in its environment, and this triggers a behavioral response along the lines of "must eat rocks." They then accordingly scoop these up in the mouth and swallow.

Birds certainly do this, intentionally selecting sand-, pebble-, or gravel-sized particles and ingesting them. Anyone who has owned or otherwise watched chickens knows about this behavior, in which these backyard or barnyard fowl grab grains of sediment with their beaks. I have watched videos of crows doing the same thing, deliberately walking along and selecting pea-sized gravel to eat. Most telling with respect to dinosaurs, ostriches and other large flightless birds regularly include rocks on their lists of items to get into their bellies.

However, where this seemingly simple behavioral division of "Oops!" or "I meant to do that" gets a bit more blurred is when, say, our same alligator purposefully eats another animal for a meal, but in its eagerness swallows a few rocks sticking to or just underneath the prey. Even more complicated would be if the alligator ate another animal with gastroliths inside of it. In either case, this means the gastroliths were ingested accidentally, regardless of whether the original gastroliths were inside or outside of another animal, or biologically or geologically formed. Powerful digestive juices inside that alligator then may reduce the body of its prey item, but the gastroliths might be left behind, a ghostly clue that another animal might have been there. Of course, we also do not know if the prey animal had accidentally or purposely eaten these rocks; all we know is that the rocks were being recycled, spending time in yet another animal's gut.

Now, it might seem unlikely that an animal would knowingly consume a rock, but think of how many pet owners fool their pets into taking unwanted pills by folding these into their food. Once a feeding frenzy starts, all sorts of items might unintentionally get included and make it into a gullet. Even herbivores could consume rocks that just happen to be part of the surrounding soil sticking to plant roots. Moreover, animals sometimes mistake rocks for food. For example, snails in a pond might look just like some of the stones lying along a pond bottom. So a vertebrate that likes to eat these snails, such as a fish, turtle, or alligator, might just indiscriminately ingest anything that looks like those snails, some of which will be rocks. For those of you who think "I would never make such a stupid mistake," recall all of the times your eyes were drawn to brightly colored beads that looked like hard candies and how tempted you were to pop them in your mouth (and perhaps did). Visual triggers can be quite powerful, particularly when accentuated by hunger.

All this gastrolith goodness is nice to know, but a nagging question about intentional rock eating should be asked: Why? As in, why would an animal eat a rock on purpose? Perhaps surprisingly, the reasons are many and quite sensible from an evolutionary perspective. Yet the best-understood one is to use these geologic materials as substitutes for teeth: letting the rocks do the "chewing" of food so that its surface area is increased and more easily digestible. This process, also known as *trituration*, is like having a food processor in the upper part of an animal's digestive system.

In birds that use gastroliths for just this function, their alimentary canal, from top to bottom, goes like this: mouth, esophagus, crop, proventriculus, ventriculus, small intestine, large intestine, and rectum, the latter also known as the cloaca. (Recall that in females, this is also where eggs exit.) Let's say a chicken has just eaten some grass seeds mixed in with delicious, nutritious insect larvae. This chicken may briefly chop its food with its beak, but more grabbing and swallowing happens than chewing. Her food, some of it still squirming, travels down her esophagus through a series of muscular contractions, or peristalsis. It may then reside briefly in her crop, which is an enlarged storage area at the end of the esophagus that acts as a holding room. Once ready for digesting, this mixture of grains and insects is passed down to her proventriculus, sometimes nicknamed "soft stomach." The proventriculus has low pH (acidic) fluids which start dissolving the food and otherwise regretfully informing any insects still holding on to their lives that resistance is futile.

Still, this chemical attack might not be enough for proper digestion, especially if the seeds had tough outer coatings and the insects had some hard parts, such as mandibles. This is where the chicken's ventriculus (more popularly known as a gizzard) comes to the rescue, aided by its super geo-friends, gastroliths. The food is then moved down to her gizzard, a strong, muscular organ hosting the gastroliths. The gizzard squeezes and releases, squeezes and releases, over and over, repetitively and redundantly. The hard stones inside the gizzard act like a mortar and pestle, mechanically breaking down these seeds and insect parts into tinier particles, and correspondingly increase their surface areas. This food then may be passed back to her proventriculus, which applies more stomach acids that react more efficiently with more food surface area available. Then it's back to the gizzard again for yet more grinding. Back and forth: a digestive tennis match, but one in which the final volley goes below the gizzard to the small intestine for further absorption.

This winning combination of muscular activity, acids, and geological materials works very well for birds that use gastroliths as digestive aids. However, this picture is not so simple as saying "grinding good, not grinding bad." Gastroliths serve other more subtle but important roles in bird digestion. One is in predatory birds, such as hawks, which may swallow a few pea-sized rocks, keep them in their digestive tracts for a while, and then regurgitate them before hunting. This process better prepares that bird's digestive system for upcoming meals. The gastroliths apparently are used to scour a bird's crop, proventriculus, and gizzard of any impurities (fat, mucus) left behind by previous meals, which, when coughed up, rid it of those nasty residues. Ornithologists and falconers—people who raise and train birds of prey—refer to these types of gastroliths as *rangle*.

Yet another use of gastroliths in birds is at the dietary opposite end of the spectrum from carnivores, which is in those birds that get lots of fiber in their diets by eating grasses, leaves, and other plant parts. Large, flightless grass-eating birds—such as emus or ostriches—regularly cut into leaves and grasses, or pull up plants with their beaks. Down their necks these plants go, and once they find their way past the crop, they begin to be digested. However, for plants like grasses, their long, hard-to-process fibers may start to bunch and bind like a tangled ball of string. If you have ever pushed or driven a lawnmower through a patch of tall grass, you may have experienced something similar to this, where the lengthy stems and leaves wrap around the blades, requiring regular stopping of the motor and pulling these out. Now imagine this happening in your stomach, in which the grasses form a big, knotted mass and get stuck there, rendering constipation most fatal. Fortunately for these birds, though, gastroliths not only help to grind these plant fibers but also separate them enough to prevent balling up and turning into impassable objects.

Additional health-related roles of gastroliths include their possible use as mineral supplements, especially if these are calcium-rich rocks, like limestone. Read the label on the bottle of nearly all over-the-counter antacids, and under "Active Ingredients" you will see calcium carbonate listed. This means the antacids are made of either calcite or aragonite, which also happen to be the primary minerals in most limestones. These minerals, once they encounter and react with low-pH acids in your stomach, increase the pH, making it less acidic and more basic. In this reaction, the calcium carbonate dissolves, liberating calcium from its bonds with carbon dioxide, which is released as a gas. (This is where former geology students fondly remember identifying limestone in lab exercises, in which they repeatedly dropped diluted acid on these rocks and satisfyingly watched them fizz.) How would we know that a bird or other animal ate limestone? We would not, unless we caught it in the act. The limestone, once inside an animal's stomach, would quickly disappear, leaving no sign it was ever there, except for maybe sudden releases of gas.

No firm evidence supports the notions that gastroliths are used for other self-medication, such as for re-introducing gut bacteria, or destroying parasites—the latter the internal equivalent of crushing a tapeworm between two rocks. There is also no reason to suppose that stones are used to either alleviate or increase appetite, thus fulfilling a recipe for "stone soup."

Nonetheless, a few birds, mammals, and even humans sometimes ingest clay or soil as a digestive aid, a practice called geophagy. This is not such a crazy practice, as some clay minerals have adsorptive properties, in which toxins latch on to the minerals and are later passed out of the body. For example, kaolinite—a white clay common in Eocene and Cretaceous sedimentary deposits of Georgia—has been long used for medicinal purposes, dating back from its traditional use in West African societies, and then continued in slave communities in North America. This remedy was used more broadly in Kaopectate™, a commercial liquid used to treat stomach ailments, in which kaolinite was an active ingredient. However, ornithologists who have studied geophagy in South American birds—such as macaws and parrots—found that these birds were also attracted to soils with high sodium content. Hence, consumption of these soils may provide trace elements and minerals in these birds' diets, but while also incorporating some gastroliths along the way.

Everything mentioned so far applies to geo-gastroliths in land vertebrates. What about aquatic ones? It turns out that gastroliths are indeed used in a variety of swimming vertebrates, such as pinnipeds—marine mammals such as sea lions, seals, and walruses—penguins, and crocodilians. For pinnipeds and penguins, rock swallowing may be therapeutic in the same way it is for raptors, in which these stones are used to clean stomachs, then regurgitated. Yet, considering how all of these animals swim, these gastroliths also might help with buoyancy control in all three groups of animals. This is not necessarily like a scuba diver wearing a weight belt, though, and the "buoyancy control" *vis-á-vis* gastroliths in aquatic vertebrates was questioned in a 2003 study on alligators, which demonstrated that gastroliths only added 1 to 2% to their body weights. In other words, these rocks had a negligible effect on alligator buoyancy. As a result, the explanation was adjusted to state that these gastroliths were used more for stabilizing bodies while swimming. Instead of weight belts, think of the tiny weights placed on car tires when these are balanced and rotated, helping the tires to wear more evenly.

Oddly enough, gastroliths are completely absent in whales and sea turtles, which in their evolutionary histories managed to find other ways to move from deep to shallow water and back again in a balanced way without the use of gastroliths. Yet these trace fossils show up as concentrated masses of rocks in the skeletons of big, non-dinosaurian Mesozoic marine reptiles, such as plesiosaurs. Paleontologists who found these gastroliths were at first perplexed by them, as digestion was presumed to have been an unlikely reason for their presence. After all, these vertebrates had seafood-only diets, much of which should not have required as much grinding as, say, plants from terrestrial environments. This supposition, combined with the negligible amount of weight these rocks added to a multi-ton marine reptile as ballast, added up to a big mystery. Why even have these rocks in their guts in the first place?

One possibility for these gastroliths would have been accidental ingestion of rocks from the seafloor. This might seem like a bizarre idea, but is actually backed up by even more ichnological evidence, which means it must be taken seriously. These spectacular trace fossils, evident as long, wide grooves ("gutters") preserved in Jurassic strata of Switzerland and Spain, are interpreted as gouge marks made by plesiosaurs as they swam along a sea bottom and scooped up seafood. Paleontologists concluded these were trace fossils based on the sizes of the grooves: some were as much as 60 cm (24 in) wide and 9 m (about 30 ft) long. Trace fossils this big could only have been made by marine vertebrates with large noses, and plesiosaurs were the only vertebrates living then with such massive snouts.

Based on this information, paleontologists interpreted them as a result of plesiosaurs diving to the bottom, inserting their muzzles into the bottom sediment, and snatching up hapless invertebrates hiding in that sediment that had no idea death would come from above. Any rocks on the sea bottom would have been accidentally consumed with the seafloor as a sort of by-catch. Nowadays, some whales and walruses make similar gutters while feeding, so this behavior is not completely unknown in modern animals, either.

Of course, the alternative to the "accidental" hypothesis is the opposite one, which is that these marine reptiles ate rocks on purpose. But if so, why? How about a synthesis of two previously competing explanations, in which gastroliths were used for both buoyancy control—but for balance more than ballast—*and* helped to digest their food? This idea got more support when two Early Cretaceous plesiosaurs described in 2005 from Queensland, Australia not only had gastroliths but also stomach contents. These contents included bits of clams, snails, crustaceans, and other invertebrate animals that lived on the ocean bottoms during the Cretaceous. This evidence came as a big surprise, because paleontologists had always assumed that these plesiosaurs only ate fish, and they never thought of them as bottom feeders.

Gastroliths in some plesiosaurs would have helped to grind down those hard-shelled critters, increasing the surface area of any calcium-carbonate shells for easier dissolution in stomach acids. So this evidence syncs well with the "scooping" hypothesis, while also explaining how feeding behaviors might have gotten those rocks into marine-reptile bellies in the first place. Other plesiosaurs, though, might have just accidentally consumed rocks as part of their wholesale ingestion of seafloor sediment, and gut contents of some specimens suggest just this.

This wide range of past and present animals that use gastroliths is by no means completely known, and more surprising examples surely await our detection. For instance, the recent discovery of gastroliths in pterosaurs—dinosaurs' flying cousins—was an unexpected one, but welcomed by paleontologists. In 2013, Laura Codorniú and several other paleontologists reported on an assortment of coarse sand to fine gravel in the body cavities of two specimens of the petite Early Cretaceous pterosaur *Pterodaustro guinazui* from Argentina. These pterosaurs were always assumed to have used their specialized beaks to strain their food from water, much like modern flamingos. Also similar to flamingos, they probably ate small crustaceans as part of their diet. The gastroliths, then, would have helped the pterosaurs to mash the harder-shelled crustaceans. This gastrolith presence also prompted the paleontologists to reconsider the front teeth of *Pterodaustro*, which they noticed were more robust than the teeth toward the back of the mouth. Accordingly, they proposed that the front teeth were adapted for grabbing sand and rocks. In other words, paleontologists gained new insights on the feeding habits and evolutionary history of this pterosaur because of the trace fossils they contained. Gastroliths: is there nothing they can't do?

How to Recognize a Dinosaur Gastrolith

Armed with this comprehensive knowledge of gastroliths in modern animals and a few ancient ones, it now should be a breeze to identify dinosaur-related ones, right? The answers to that rhetorical question are yes, no, and maybe, and not necessarily in that order. The likelihood of correctly diagnosing these enigmatic trace fossils depends on slavishly asking and addressing a few important questions:

- Are the suspected gastroliths inside a dinosaur skeleton?
- Are there just a few of these stones, dozens, or more than a hundred?
- Are these stones clustered in a compact mass, taking up a small volume compared to most of the dinosaur's body?
- Is this mass of stones in the right location in the dinosaur's skeleton, namely its abdominal cavity?

- Are the stones appropriately sized for this dinosaur?

Ideally, if all of these criteria are fulfilled, gastroliths will also outline the approximate size and shape of whatever organ was holding it in a dinosaur, such as a gizzard. As trace fossils, then, these can actually help fill in that part of a dinosaur's soft-part anatomy as a proxy.

However, if a dinosaur body is not associated with your gastrolith suspects, then confirming that these rocks formerly resided in a dinosaurian alimentary canal gets much tougher, albeit not impossible. When faced with an absence of bones but with some possible gastroliths, you must ask a second set of questions:

- Are these rocks in strata of the right age for gastrolith-bearing dinosaurs?
- Are these rocks in strata of the right environments for dinosaurs?
- Are these rocks the right composition to have survived time in an acidic environment: not limestone, but more like chert, quartz sandstone, or quartzite?
- Do they have the same size, shape, and surface texture as those of known gastroliths found inside dinosaurs?

A few paleontologists have suggested that if a gastrolith is found outside of the body cavity of a dinosaur or other animal that used gastroliths, with nary a bone in sight, then you should stop calling it a gastrolith and instead refer to it as an *exolith*, as in "exotic." Although this word has not yet caught on with dinosaur paleontologists—let alone ten-year-old dinosaur enthusiasts—it does help paleontologists and geologists make a mental shift when thinking about these as trace fossils. Unlike most other dinosaur trace fossils, such as tracks and nests, gastroliths might have been transported far away from where a dinosaur originally lived. In other words, whenever suspected gastroliths show up outside of a dinosaur skeleton, serious doubt tends to pummel any sunny optimism.

Yet another factor to consider with dinosaur gastroliths is that dinosaurs might have been recycling gastroliths before they died. Recall how certain birds of prey use rangle: they swallow a few rocks, use these to clean out the upper part of their digestive tracts, then cough them out. If some dinosaurs similarly ate and regurgitated gastroliths, then these rocks would have been left behind and perhaps later picked up by another dinosaur that thought "Hey, look—free gastrolith!" For accidentally ingested gastroliths, these might have passed through dinosaur bodies and come out the other end, deposited along with their feces. However grotesque it might seem to those of us who do not regularly gulp rocks, nor pick through feces for little undigested treasures, these stones could have been removed or eliminated and then ingested by other dinosaurs if no others were readily available.

What was the fate of gastroliths after the death of their host? Given that many dinosaurs and dinosaur parts were deposited after floating in rivers, lakes, or oceans, paleontologists have to keep in mind that gastroliths still in dinosaur body cavities went along for the ride too. Even worse, after floating for a few days, these dinosaur bodies could have burst open from decay. Given their greater density, gastroliths would have been among the first objects to leave a dinosaur body and sink to the bottom of a lake, stream, or ocean. Millions of years later, geologists might find them, scratch their heads, and wonder how these larger rocks got into such fine-grained sediments.

On land, gastroliths might have had a chance staying in or close to a dinosaur's body. However, once that body was opened, whether by predators, internal decay, scavengers, or just the ravages of time, these gastroliths could have scattered around those remains or been accidentally consumed by whatever was eating the dinosaur. The same would have been true for a floating carcass scavenged by sharks or marine reptiles. These animals could have either hastened gastrolith departures from dinosaur bodies while tearing open a dinosaur's body cavity, or inadvertently eaten some while enthusiastically noshing on dinosaur entrails. All of these considerations mean that gastroliths may have been deposited far away from a dinosaur body and gone through multiple "lives" as trace fossils.

In 2003, paleontologist Oliver Wings decided to find out under what conditions gastroliths might stay or go after the death of a dinosaur. In setting up this research, he asked himself what happened to gastroliths in modern dinosaurs—ostriches (*Struthio camelus*)—after they died and their gastrolith-holding bodies were placed in different environments. Following the death of two ostrich chicks from bacterial infections on a South African farm, he donated their bodies to science by watching their corpses decay for six days. One chick, the smaller one of the two—weighing only 2.1 kg (4.5 lbs), or about the size of a roaster hen—was left out in the open and on land, where both South African sunshine and cadaver-loving insects quickly got to work on it. Meanwhile, he placed the other, much larger chick, which weighed 11.5 kg (25 lbs), in an uncovered water-filled barrel. After three days, he turned over each body so that both sides had been exposed to air and water, respectively.

The results were surprising. In the land-based chick, its gastroliths stayed put. Although insects, dehydration, and bacterial decomposition successfully caused rapid weight loss, tissues tightened around the gastroliths in its crop, a biological shrink-wrap that kept all of these stones together and in one place. In contrast, the water-immersed chick lost very little flesh, but it opened up and lost all of its gastroliths, which fell to the bottom of the barrel. Although quite simple and limited to only two samples, this little experiment gave Wings and other dinosaur paleontologists a few insights on gastrolith residence times after death, and differences to keep in mind for a dinosaur carcass that stayed on land or ended up in a body of water.

Nevertheless, before such experiments, so many uncertainties about dinosaur gastroliths caused paleontologists to reach for the only weapons that made sense for handling such a desperate situation: lasers. As most people know, but most cats do not, lasers use amplified beams of light that maintain nearly the same diameter over long distances. Like all light, though, laser beams can be bent (refracted) or bounced back (reflected). These properties of light are then applicable to measuring the degree of polish on a surface. Very simply, less polish causes laser light to scatter more, whereas more polish, such as a mirror might have, causes it to scatter less.

How these properties of laser light could be applied to dinosaur gastroliths was thus related to one of the problems with identifying gastroliths outside of dinosaur bodies. Smooth, lustrous surfaces on rocks are not necessarily the result of being inside a dinosaur gut. Instead, such surface textures could be attributed to all sorts of natural, non-biological processes, such as wind and water erosion. Could laser reflectance detect the differences between rocks polished by wind or water versus dinosaur viscera?

So in the early 1990s, paleontologist Kim Manley took a look at the differences of laser-generated light scattering on genuine gastroliths, which were taken directly from a sauropod skeleton (*Diplodocus*), wave-rounded rocks from beaches, and river-rounded rocks. All three rocks had some degree of polish to them, but paleontologists assumed that gastroliths, having spent time grinding up food in a dinosaur's crop, proventriculus, or gizzard, and subjected to stomach acids, were the most polished.

So science struck with a blow delivered at the speed of light. It turned out that some beach rocks were just as polished as the gastroliths, although both of these were more polished than river rocks. This technique was refined in a 1994 study by other scientists who looked at gastroliths from much more recently dead avian dinosaurs, moas from New Zealand. These large, extinct flightless birds lived on both the north and south islands of New Zealand up until less than a thousand years ago, which is about when the Maori people arrived and found them far too delicious for their continued existence. The scientists, Roger Johnston and two others, again used laser reflectance to look for differences in polish between moa gastroliths and beach rocks. However, they also employed a video instrument that recorded the reflectance in three dimensions, rather than just a horizontal plane. They concluded that the moa gastroliths were more polished than beach rocks, stating that it was "fairly" successful. This faint praise came with an admission that the amount of polish on a gastrolith may have depended on the time these gastroliths spent in a moa's gizzard. This meant former beach rocks, swallowed by a moa but only put to work for a few weeks before that moa's death, would be nearly indistinguishable from unused beach rocks.

In 2000, scientists Christopher Whittle and Laura Onorato used a scanning electron microscope (SEM) to peer more intensely at gastroliths in general. In their study, they picked out a variety of gastroliths from dinosaurs, plesiosaurs, birds, and alligators, and then calculated percentages of areas that were pitted or gashed. At a magnification fifty times what we would see with unaided eyes, they found these gastroliths were mostly smooth, but also had a good number of pits and gashes on about 20% of gastrolith surfaces. In contrast, rocks that had been polished by waves, river flows, or wind were much less pitted or gashed than the gastroliths. These dinks and nicks on otherwise smooth surfaces are assumed to have come from rock-to-rock contacts more frequent than those in outside environments. These were promising results, and even though Whittle and Onorato used such expensive equipment for this study, they also pointed out that paleontologists could use normal binocular microscopes at the same magnification to look for these diagnostic features.

These outcomes also point to how gastroliths and microwear in dinosaur teeth tend to overlap in a few ways. Recall that minerals harder than apatite tend to scratch teeth, whether as quartz sand on plant surfaces or phytoliths in plant tissues. Similarly, stones of equal or different hardness mashing against one another would have imparted a sort of microwear on gastroliths, evident as pits and gashes. These marks also might have been from quartz sand or other minerals in the stomach, perhaps including those on or in plants. Both types of microwear also involved much muscular activity behind it, whether through gnashing of teeth or squeezing of gizzards. In principle, the only major difference with gastroliths is their exposure to low-pH stomach acids, which presumably would leave a more chemical overprint.

Still, despite these studies that used lasers, microscopes, and other technologically advanced tools to study suspected gastroliths, healthy skepticism about the recognition of dinosaur exoliths in Mesozoic rocks still lingered well after the 1990s. Regardless of these non-believers, though, the good number of gastroliths found inside dinosaur skeletons in the early part of the 21st century ensures that paleontologists have plenty more of these trace fossils to study, and bodes well for finding more dinosaur gastroliths in the future.

Dinosaur Gastroliths? Get Real!

So how important are these dinosaur gastroliths, considering the continued skepticism and theories that still surround them? First of all, despite previous attempts to diminish these important trace fossils via the almost-clever pun "gastromyths," they really do exist. Admittedly, our views of gastroliths in dinosaurs have changed considerably since their discovery. Yet a little history lesson about gastroliths helps to understand just how far our concepts about these dinosaur trace fossils have progressed, with novel insights about them emerging in just the past ten years or so.

The earliest description of possible gastroliths in a dinosaur was in 1838, when French paleontologist Jacques Amand Eudes-Deslongchamps noted about ten pebbles underneath the ribs of the Middle Jurassic dinosaur *Poekilopleuron*, which had been discovered in France. He concluded that these rocks were in this dinosaur's stomach, just like those found in fossil crocodiles from the same area. Much later, in the late 19th century and leading up to 1900, paleontologists began spotting gastroliths in dinosaurs and marine-reptile contemporaries of dinosaurs, such as Late Cretaceous plesiosaurs. Large stones were also associated with sauropod skeletons excavated in the 1870s, although the people digging out these dinosaurs did not call these "gastroliths," just "stones."

Most of the credit for the linking of gastroliths with some function in dinosaurs was bestowed upon paleontologist Barnum

Brown, who was doubly famous for originally naming *Tyrannosaurus rex* and wearing full-length fur coats while conducting field research. Sometime around 1900, Brown noticed a collection of rounded cobbles inside a hadrosaur skeleton that he identified as "*Claosaurus*." He imagined that these stones might be similar to those found in plesiosaur skeletons, which he discussed in a brief paper published in 1904. In 1907, Brown followed up this study with another paper on dinosaur gastroliths, which he very simply titled "Gastroliths." (To this day, no one knows if Brown's brevity was a direct affront to the verbose titles employed by the previous generation of Victorian-era scientists.)

Later, it turned out that Brown was wrong on two counts. For one, the "gastroliths" in this particular dinosaur were probably river stones that washed into the body cavity of the hadrosaur soon after it died. Second, the hadrosaur was misidentified and instead was a species of *Edmontosaurus*, mentioned in the first chapter and taken literally by a *T. rex* when taunting it by saying, "You want a piece of this?" Brown was also beaten to press on dinosaur gastroliths by a not-as-famous paleontologist, G.L. Cannon, who in 1906 published a short paper with twice as many words in its title as Brown's: "Sauropodan Gastroliths." This was the first publication to mention gastroliths in sauropods, and it was an idea that has stuck around since. Other paleontologists, such as Brown's protégé Roland T. Bird, later promoted this supposed association between sauropods and gastroliths. Consequently paleontologists began looking for, finding, and interpreting anomalous assemblages of rocks in dinosaurs, and especially sauropods.

So starting in the early 20th century, and continuing for quite a while afterwards, the conventional wisdom about gastroliths has been twofold. First of all, gastroliths were indeed present in some dinosaurs, but mostly in sauropods, and maybe a few other plant-eating dinosaurs such as hadrosaurs, and almost never in theropods, stegosaurs, ankylosaurs, or ceratopsians. Second, these gastroliths were used to help grind up hard-to-digest food in gizzard-like organs because sauropods did not have the right teeth for chewing their food. Related to these two assumptions was the tacit agreement that theropods lacked gastroliths because they were all carnivorous, with big, sharp, pointy teeth, powerful jaws, and marvelously corrosive stomach acids. In other words, only wimpy herbivorous dinosaurs needed these digestive aids.

We now know that this story about gastroliths and dinosaurs, neatly framed, displayed in a prominent place, and highlighted with artfully angled lighting, is mostly wrong. Given what we've learned about the varied uses of gastroliths in modern animals, combined with a broader knowledge about dinosaur diversity, evolution, and behavior, and all topped with healthy distrust about what constitutes a real dinosaur gastrolith, this picture, much like paleontologists' wardrobes, has changed considerably since the days of Barnum Brown.

First of all, the dinosaurs that are now the most likely to have gastroliths are *not* the plant eaters, but theropods. Yes, you read that right: some of those meat-eating, über-macho, Mesozoic killing machines of yore may have needed a little help from their geological friends. Among the theropods documented thus far with gastroliths are: the Early Jurassic *Megapnosaurus* (previously known as *Syntarsus*); the Late Jurassic *Nqwebasaurus*; the Early Cretaceous *Caudipteryx*, *Shenzhousaurus*, *Sinocalliopteryx*, and *Sinosauropteryx*; and the Late Cretaceous *Sinornithomimus*, among others.

Well, of course these theropods had to use gastroliths, you might say. These are small, effete theropods, some of which, such as *Nqwebasaurus*, had nubby teeth, and *Sinornithomimus*, which completely lacked teeth. But then tell that to more imposing theropods such as the Late Jurassic *Allosaurus* and *Lourinhanosaurus*, the Early Cretaceous *Baryonyx*, or the Late Cretaceous *Tarbosaurus*, all of which have had possible gastroliths directly associated with their skeletons, too. Granted, the supposed gastroliths found with these larger, well-toothed theropods may be evidence of accidental ingestion, whether through gulping down a prey animal with gastroliths or swallowing stones underneath the body of an animal as it was eaten. In fact, the few gastroliths in the abdominal cavity of the small theropod *Sinocalliopteryx* have been attributed to inadvertent gulping, unknowingly including them with a meal. Still, some large theropods have definite gastroliths in their bodies, thus directly disputing the previously held idea that only sauropods and plant eaters made use of gastroliths.

There is one additional and important facet of this continuing debate that has yet to be mentioned regarding theropods and gastroliths. Remember the very first description of gastroliths by Eudes-Deslongchamps in *Poekilopleuron*, in 1838? Well, *Poekilopleuron* was not only a theropod, but also a big one, estimated to have been about 9 m (30 ft) long. From a dinosaurian perspective, this matched an average-sized *Allosaurus* and was about ¾ the length of an adult *Tyrannosaurus*. Yes, that's right, gastroliths in dinosaurs as a concept actually began with them in a very large theropod, not sauropods. Unfortunately, we can't study the original bones of *Poekilopleuron*, as these were lost in bombing raids during World War II. Nor can we study the original gastroliths as they too have been lost, which from an ichnological standpoint is an even greater tragedy. Still, we are left with this satisfying piece of paleontological history that connected theropods with gastroliths, which we are now revisiting with much vim and vigor through their discovery in many other theropods of varying clades and sizes.

What's even more exciting about this renewed recognition of gastroliths in theropods is how it coincides with the evolutionarily based revelation that birds are dinosaurs, as many modern birds have gastroliths, too. Of the small theropods found with undoubted gastroliths in their abdominal cavities, some of these—such as *Caudipteryx*, *Sinocalliopteryx*, and *Sinosauropteryx*—also have feathers. Although gastroliths are trace fossils and mostly reflect behavior, which is more fluid and open to interpretation and variety than genetically determined anatomy, natural selection must have favored theropods with the visual acuity to pick out the right rocks (ones that were rich in silica) and the mental ability to think "Must eat rocks," as those rocks would then help them with digestion and overall good health.

The gastrolith connection between non-avian feathered theropods with gastroliths in early birds was bolstered by the discovery of gastroliths in the Early Cretaceous bird *Yanornis martini* of China. These gastroliths, mostly composed of sand- and gravel-sized quartz particles, were in exactly the same place where the bird's gizzard should have been. Another insight provided by this avian rock collection was how an absence of gastroliths in other specimens of *Y. martini*, but the presence of fish remains in one, implied that this species might have been switching its diet seasonally. Some modern shorebirds do the same, eating seeds and insects in the spring through fall, but chowing down on seafood during the winter. So this Early Cretaceous bird may have died before the winter while its gizzard was still full of gastroliths, which it would have needed to fully digest and process fibrous seeds and insects.

Other theropods with gastroliths include a few species of ornithomimids. In 1890, nearly a hundred years before the theropod–bird connection was firmly established, O. C. Marsh named one dinosaur *Ornithomimus velox*; bones of another species, *O. edmontonicus*, are the most common Late Cretaceous theropods in North America, and Asian ornithomimids are not exactly rare either. Because "ornithomimid" means "ostrich mimic," it seems only appropriate now that some of these dinosaurs, like modern ostriches, also had gastroliths. Ornithomimids with gastroliths include the Early Cretaceous *Sinornithomimus*, the Late Cretaceous *Shenzhousaurus*, and a dozen specimens of an unnamed species, all from China. The most recent, and one of the most exciting examples of gastroliths in an ornithomimid, was reported in 2013. More than a thousand rocks, all relatively small, were in a skeleton of the Late Cretaceous *Deinocheirus mirificus* of Mongolia. The hundreds of gastroliths from the dozen unidentified ornithomimids ranged from sand- to gravel-sized and were angular to rounded, affirming how gastroliths are the most diverse and unpredictable of dinosaur trace fossils. All of this points toward how these so-called "ostrich mimics" were 70 million years ahead of ostriches in using gastroliths, meaning these modern birds are actually the "mimics," not their non-avian predecessors.

These gastroliths, combined with their toothless condition, also imply that ornithomimids were herbivorous, lending to the still-radical concept of vegetarian theropods.

Given this knowledge both old and new about gastroliths in theropods, what about sauropods and gastroliths? The long-held assumption is that huge sauropods, many of which only had puny, pencil-like teeth, used gastroliths to grind their food. However, this idea is now seriously doubted. The biggest problem with the previous explanation is that these gastroliths, like those used for "buoyancy control" in marine reptiles, are too few to have made any real difference in digestion. For sauropods to have actual functional "gastric mills," they would have needed many more gastroliths than the ones found in sauropod skeletons so far.

For instance, in modern ostriches and other birds that employ gastroliths to help with their food, these rocks make up about 1% of their total body mass. The Early Cretaceous theropod *Caudipteryx* matches this ratio, implying that these rocks served a similar purpose in its lifestyle. However, for sauropods, the proportion between gastroliths and estimated body mass was about 10% that of birds. So if an ostrich were scaled up to 50 tons (scary thought), then it would need about 500 kg (1,100 lbs) of gastroliths to digest its food, which is about the weight of the largest Harley-Davidson motorcycle. Yet the greatest mass of gastroliths described thus far from a sauropod (*Diplodocus*) was only 15 kg (33 lbs), which is about the weight of a Huffy bicycle. Also, gastroliths are relatively rare in sauropod skeletons, and if used for something as essential as food processing, these trace fossils should be much more common. This huge disparity between extant and extinct gastrolith-using animals led paleontologists to conclude that these stones surely were not used for the same purposes, effectively pulverizing the "gastric mill" hypothesis. Alternatives may not be so exciting, but include: accidental ingestion, especially if rocks were adhered to plant roots; mineral supplements, such as for calcium or trace elements; or used as separators for keeping fibrous food from bunching in their stomachs.

Nonetheless, the disproving of one explanation for gastroliths in sauropods doesn't mean they suddenly vanished. For example, one specimen of the Late Jurassic sauropod *Diplodocus hallorum* (formerly named "*Seismosaurus hallorum*") of northern New Mexico had more than 200 gastroliths, which were carefully mapped inside and around its former body cavity. These gastroliths were all igneous and metamorphic rocks with varying degrees of polish to them, and most were about 2 to 8 cm (<1–3 in) wide. An assemblage of 115 gastroliths, totaling about 7 kg (15.4 lbs), was also packed into a small area within a skeleton of the Early Cretaceous sauropod *Cedarosaurus* from Utah. Other sauropods with gastroliths directly associated with their bones include: *Apatosaurus, Barosaurus,* and *Camarasaurus* of the western U.S., as well as *Dinheirosaurus* of Portugal (all Late Jurassic sauropods), and the Early Cretaceous *Rebbachisaurus* from Morocco. In one instance, 14 gastroliths were directly associated with the remains of a juvenile *Camarasaurus* scavenged by *Allosaurus*, the latter indicated by tracks, toothmarks, and dislodged teeth. Several specimens of a Late Triassic prosauropod from South Africa, *Massospondylus*—the same dinosaur affiliated with nests, eggs, and babies mentioned in a previous chapter—had gastroliths too. Moreover, the skeleton of another prosauropod, the Early Jurassic *Ammosaurus* of North America, had gastroliths. In short, although gastroliths are relatively rare, having been found in less than 4% of all sauropod skeletons, these trace fossils are abundant enough in them that they should not be ignored, either.

So imagine you are a 20-ton sauropod and walking along a stream bank during the Late Jurassic. At some point during your stroll, you get a serious hankering for some geo-gastroliths, and you don't know why, because your brain is smaller than that of a 0.075-ton (150 lbs) human. How do you pick the right rocks? The cerebral processing needed to discriminate between "good" rocks (solid silica-rich ones) and "bad" rocks (crumbly ones or limestone) can't be too complicated. One thing is for sure, though: it can't be based on smell, because most rocks lack good scents. It also can't be from sound, although the distinctive crunching of silica-rich metamorphic or igneous rocks underfoot might have given off tones different from those of, say, limestone or shale. How about touch? Your feet might have somehow felt the right rocks based on their size, shape, and surface texture, but that would have required an unexpected sort of sauropod sensitivity akin to the princess and the pea.

Hence you are probably left with sight and taste. Depending on your visual spectrum, brightly or darkly hued rocks of a certain size might catch your eye. If too small, though, these rocks might not be worth the effort. If too big, you risk choking to death, which would be very bad for transmitting your genes. So you needed proper search images for potential gastroliths before lowering your head to the ground, grasping a rock or two (or three) with your teeth, and gulping it down. Perhaps during the short time a rock was in your mouth, taste might have had some influence on your choice, too. Do taste buds in your mouth help say "Yes, swallow this!" and you comply? Or does the opposite happen, a spit-take of rejected rocks, striking and killing nearby small theropods like stray missiles?

Imaginative scenarios aside, sauropods, theropods, and other dinosaurs that ate rocks on purpose must have applied some sort of pattern recognition to select the ones best for them, for whatever reason. Based on the presence of gastroliths in the Early Jurassic prosauropod *Massospondylus*, this ability and its outwardly expressed behavior probably started in dinosaurs early on, such as in the Late Triassic Period. From an evolutionary view, one of the more interesting aspects of gastroliths is how prosauropods, sauropods, and theropods (birds, too) all share a common ancestor as saurischians.

In contrast, only a few ornithischians have gastroliths. These include the Late Cretaceous ankylosaur *Panoplosaurus* of Canada, the Early Cretaceous ceratopsian *Psittacosaurus* of China, the Late Cretaceous ornithopod *Gasparinisaura* of Argentina, and nearly no others. So although gastroliths are rare in the vast majority of dinosaurs, they are much more likely to be in saurischians, showing up in those dinosaurs from the Late Triassic through the Late Cretaceous, and then continuing in birds. That means this viewing, recognition, grabbing, and eating of stones has been going on in dinosaurs for nearly 200 million years, and will keep on happening as long as birds are around and also need gastroliths.

Speaking of birds, another type of dinosaur behavior that was possible, but difficult to test scientifically, is that parent dinosaurs taught their young by example, showing them how to select rocks for their own internal collections. Mammals are of course well known for passing on skills to their offspring, such as how mother grizzlies teach their cubs to fish for salmon. But in the past few years, behavioral scientists are gaining a greater appreciation for how some species of birds, such as crows and ravens, learn through experimentation and watching one another, while also imparting newly acquired knowledge to their chicks. If any dinosaurs performed similar behaviors, such as teaching their offspring how to become rudimentary geologists, then this would be a glimpse of dinosaur learning.

One way for paleontologists to test this outlandish idea—that parent dinosaurs taught juvenile dinosaurs how to find the rocks they need—would be to look for gastrolith-bearing skeletons of the same species and at different growth stages. For instance, do both smaller skeletons (juveniles) and larger skeletons (adults) of the same species bear gastroliths, or do just the adults have them? If both have gastroliths, are these the same types of rocks, implying that they might have been picked out together and at the same places? What are the proportions of gastroliths to body sizes of the juveniles and adults: do they correlate, or are the numbers small enough in one of the size groups that accidental ingestion must be considered? These questions and more can be applied if gastroliths are some day found in multiple generations of the same species of dinosaur, allowing us to wonder beyond mechanistic explanations and consider the bendability of behavior.

Gastrolith Ghosts of the Triassic

Let's say you're a paleontologist and you've been asked to study a geological formation dated from the time when dinosaurs ruled the earth, or at least when they co-ruled it with insects. All of the rock types, sedimentary structures, invertebrate trace fossils, and geochemical data in this formation show it was formed in environments that should have been inhabited by dinosaurs, such as river valleys, lakeshores, or seashores. Yet after weeks in the field spent scrutinizing these strata, you realize they not only lack dinosaur bones, but also are devoid of tracks, nests, eggs, toothmarks, coprolites, or other obvious evidence of dinosaurs.

The only possible sign of a former dinosaur presence are some annoyingly regular, rounded, polished quartz pebbles and cobbles, which are locally abundant in some sedimentary layers. With a great sigh, knowing that your colleagues will castigate you, you are forced to consider that these rocks might be gastroliths, the only clues that dinosaurs were present. With much resignation, you write "Gastroliths?" in your field notebook, draw a little sad face next to this entry, and wonder what to do next.

Along those lines, in an article published by Robert Weems, Michelle Culp, and Oliver Wings in 2007, they argued that the presence of many unusual, rounded, polished rocks in the Bull Run Formation of northeastern Virginia demonstrated that dinosaurs lived there about 200 *mya*. They came to this conclusion despite a complete lack of dinosaur bones, tracks, nests, eggs, coprolites, and other fossil evidence of dinosaurs in the Bull Run Formation. This meant these gastroliths, which were also actual examples of exoliths, constituted the only signs that dinosaurs had been there and then. It was an ichnologically wonderful study, one that demonstrated how more than a hundred years of cumulative knowledge about dinosaur gastroliths can be applied to quite reasonably state "dinosaurs were here" on the basis of a pile of rocks.

The paucity of dinosaur fossils other than gastroliths in the Bull Run Formation was frustrating, because it was the right age (Late Triassic) for preserving evidence of early dinosaurs in North America. Unfortunately for dinosaur hunters who live east of the Mississippi River, however, this depressing situation is the norm. Despite more than 200 years of paleontological study, Mesozoic rocks of the eastern U.S. are embarrassingly shy about revealing dinosaur bones, although a few places—such as in Connecticut,

Maryland, Massachusetts, and Virginia—have plenty of dinosaur tracks. This general lack of dinosaur bones is credited to poor preservational conditions—not enough dinosaur bodies getting buried quickly and in anaerobic environments—and what may have happened to bones after burial, such as dissolution by acidic groundwater. Yet many of these environments may not have been all that good for preserving dinosaur tracks, either. For example, only a few theropod tracks have been found thus far in Late Triassic rocks of northeastern Virginia, despite these having been formed in valleys with abundant river floodplains and lakes. Finally, paleontologists and geologists in the eastern U.S. face a challenge that is not so overt in the arid western U.S. states: plants. This lush vegetation not only covers up or otherwise obscures rocks, but also destroys original bedding and other primary features, such as dinosaur tracks, via root activity and soil formation, often prompting eastern U.S. geologists to jokingly praise the virtues of defoliants.

Weems and his colleagues built their case about the Bull Run gastroliths by noting a number of traits that matched those of known dinosaur gastroliths:

- These rocks stuck out as unusual concentrations of pebbles, gravel, and cobbles weathering out of the Bull Run Formation, most of which is composed of finer-grained strata such as siltstones and sandstones.
- More than two hundred of them were recovered from a small area of about 30 m^2 (323 ft^2), just slightly more than the size of a baseball diamond.
- All of the rocks were made of quartz. Most were quartzite (metamorphosed sandstone), and the rest originated from quartz veins that formed in either igneous or metamorphic rocks. No limestones, sandstones, or other sedimentary rocks are represented.
- Nearly all were 2 to 5 cm (<1–2 in) wide, and the biggest was about 10.3 cm (4 in) wide.

- Their proportions, such as ratios of lengths and widths, were almost identical to those of gastroliths described from the Early Cretaceous sauropod *Cedarosaurus*.
- Nearly all were rounded, with few angular corners.
- All had some degree of polish to them, and over their entire surfaces.
- Their colors ranged from dark yellow to yellow-orange to brown to dark red.
- Their surfaces had plenty of pits and gashes in between polished surfaces, details revealed through a closer look with an SEM.

Do any of these traits sound familiar? Yes, they should. The paleontologists further concluded that the gastroliths were likely used as gastric mills, helping to mix and wear down food in gizzard-like organs. So these stones were good candidates as dinosaur gastroliths. Still, the paleontologists had a few more questions: Which dinosaurs carried these rocks in their bellies, where did they get them, and how far did they travel from their source area?

The answer to the first question was simple: prosauropods. This was based on the size of the gastroliths—the largest of which could only have been in relatively big animals—and their resemblance to gastroliths in the sauropod *Cedarosaurus*. However, sauropods were rare in the Late Triassic and did not become more widespread until the Middle Jurassic Period. So the closest ecological analog to these in the Late Triassic would have been the largest herbivores at that time, which were prosauropods. Furthermore, the size range of the gastroliths (2–5 cm wide) compares well to that of gastroliths in the Early Jurassic prosauropod *Massospondylus*. Unfortunately, the body fossil record for Late Triassic prosauropods in North America is quite scanty, with no bones or teeth known from the eastern U.S., and only a few prosauropod tracks. These trackmakers might have been the same gastrolith-bearing dinosaurs that lived in the area of the present-day Bull Run Formation, or at least been related to them. But with so little other evidence for prosauropods, to say anything more would require some great speculative leaps. Nevertheless, these gastroliths represent a great start in better understanding where the first large-bodied dinosaurs lived and how they lived during the Late Triassic.

The second set of interrelated questions—where did these dinosaurs find the original rocks, and how far did they travel with their rocky payloads—was a tougher one to figure out, but answerable by looking for similar stones in other Late Triassic formations in the same area of Virginia–Maryland. The best match came from the Manassas Sandstone, which has conglomerates; these are sandstones that also have chunky bits, such as pebbles, gravel, and cobbles. It turned out these coarser-grained sediments—most of which were also composed of quartz and quartzite—were a good fit for the Bull Run Formation gastroliths. Assuming that the gastroliths are gravestones, marking the final resting places of prosauropods, then these rocks moved a minimum of about 20 km (12 mi) under dinosaur control, and probably much further. If true, this would be a great example of how dinosaurs had already begun changing their environments, transporting large sediments to places where rivers did not reach.

This sort of study demonstrates the great potential of gastroliths as trace fossils, filling in the gaps in lieu of dinosaur bones or tracks and other traces. At first glance, gastroliths might seem like the most boring and misunderstood of dinosaur trace fossils, holding little of the aesthetic appeal of footprints or raw ferocity of toothmarks. But when treated with respect, and through the awesome healing power of science, these humble rocks greatly extend our knowledge of dinosaur presence and behavior. Despite all we know now about gastroliths, though, I still think back to that summer of 1983 when I held that strange rock from the Morrison Formation in Wyoming and later wondered whether it spent time inside a dinosaur before resting in my hand. One thing is for sure, though: if other paleontologists or I were to go back to that same spot today and find that rock, I am confident we could better answer that question.

CHAPTER 8

The Remains of the Day: Dinosaur Vomit, Stomach Contents, Feces, and Other Gut Feelings

Dinosaur's Digest

Dinosaurs had to eat. What we would like to know, though, is just what dinosaurs ate, how they digested this food, how often they ate, and how their eating patterns affected everything else around them. These questions can be answered partly by looking at dinosaur jaws, teeth, arms, legs, and rare soft parts. After all, a theropod with a mouth full of sharp, pointed, and serrated teeth, set in jaws with attachment sites for powerful muscles, accompanied by stout recurved claws on hands and feet, would have become a mite peckish at a vegetarian restaurant. Yet there is also a point where paleontologists admit a dinosaur's anatomy cannot tell us everything about what went into or out of its body. For example, who would have known that Late Cretaceous hadrosaurs crunched wood? That Early Cretaceous ankylosaurs swallowed fruit? That

Late Cretaceous sauropods grazed on grasses? That Early Cretaceous feathered dinosaurs gulped their avian cousins?

So dinosaur trace fossils come to our rescue again, enlightening us about what passed into or exited the dark passages of dinosaur entrails. We already know how toothmarks help with interpreting what some dinosaurs ate. Microwear on teeth informs us about chewing and whether dinosaurs were low-level grazers or not. Gastroliths tell us how these stony implements assisted dinosaur digestion. But those trace fossils were just the appetizers for expanding our knowledge about dinosaur diets and the ecological consequences of their feeding. More trace fossils—with exotic-sounding names like enterolites, cololites, urolites, and coprolites—await our consumption, absorption, and appreciation, explaining more deeply what dinosaurs ate and how these traces affected other lives in dinosaur ecosystems.

A Brief Tour of Modern Excretions

Based on the malodorous, steaming pile of metaphors related to vomit, urine, and feces crossing all cultures and used throughout history, humans have had a long fascination with bodily wastes and the functions that produce them. For example, here are a few such expressions, best uttered with impassioned aplomb: "You make me want to puke!" "You're pissing me off!" "Don't give me any crap!" "You're full of shit!" "What kind of crap is this?" "I don't give a shit!" Or, at a more juvenile level, "You poopy head!" All such phrases and the rich heritage behind them hint at how all of us, regardless of age, gender, ethnicity, or socioeconomic status, must puke, pee, and poop. It's part of what makes us animals.

This intrinsic connection to all things urinal and fecal is probably related to our mammalian heritage, in which these products became not just a way to get rid of bodily wastes but also a form of communication. For example, take your dog out for a walk and you'll expect it to stop often to sniff at virtually everything. During this walk, your dog is using its nose like you use your eyes, investigating an olfactory landscape holding thousands of invisible clues, many of which are a result of secretions by dogs and other mammals in the area, such as urinations, defecations, mucus, hair, or body scents imparted by rubbing. Whenever I track canines of any species, whether these are domestic dogs, coyotes, foxes, or wolves, I sometimes refer to these messages, seen and unseen, as "doggy e-mail." In such exchanges, canines post to one another, many of which simply say "I was here," or more emphatically "This is my territory, stay out!" This is why the phrase "marking your territory" is so commonly associated with urination, while also serving as an apt allegory in business and academic practices.

Wastes as signposts can get even more extreme, such as when carnivores compete over territory and use feces to intimidate or terrorize. For instance, while I was taking a wolf-tracking course in central Idaho, the instructors told us about what happened when wolves (*Canis lupus*) entered what was formerly coyote (*Canis latrans*) territory. The coyotes picked up the scent of a wolf pack on a trampled trail, so they left scat on the trail, announcing to the wolves "Oh, yoo-hoo! I beg your pardon, but this just so happens to be our neighborhood. We would be most appreciative if you departed from it. Ta!" The wolves responded by hunting down, killing, and eating one of the coyotes. Soon afterwards, they left scat in the same spot, but this time with coyote fur in it. (I could not help but envisage the coyotes' reactions when they came back to this place, got a whiff of the scat, and realized it contained one of their own.) The tracking instructors said they had seen trace evidence of this Mafia-like behavior not once but twice, suggesting it might have been a standard wolf response to coyote defiance. I have no idea if predatory dinosaurs had similar practices in establishing territory, but this modern example certainly lends credence to such scenarios.

Since then, whenever tracking, I have paid careful attention to the identity, placement, and contents of mammal feces with relation to one another, and have often confirmed for myself other such "scat wars." These conflicts are normally manifested by one mammal dumping in a prominent spot—such as the middle of a trail—and then on top of it will be another, fresher pile of feces, left by another mammal. For some mammals, they also make sure their feces cover as much real estate as possible. For example, hippopotamuses do this by rapidly flicking their tails back and forth as they excrete, flinging poo with wild abandon. This windshield-wiper action instantly expands their territory, too, as other animals flee from this fecal assault.

Birds are different. Appropriately enough for a clade represented by about 10,000 species, these modern dinosaurs eat a wide variety of foods, and this sustenance is digested and excreted in myriad ways. As far as I know, though, their wastes are not used for staking claims over territory. This is understandable because most birds are more influenced by sight and sound rather than smells. Their digestion is also more complete than that of most mammals, and instead of the mostly solid feces or purely liquid urine of mammals, each emitted through different orifices, birds excrete uric acid out of a single orifice, the cloaca. This waste is liquid in some birds but more solid in others. Bird droppings that are more solid typically wear a little white cap of liquid uric acid on one end. Prideful car owners everywhere have experienced such calling cards from songbirds; when flocking, these birds leave an impressive amount of waste. Nonetheless, raptors and other carnivorous birds are famous for their white-liquid sprays, which sometimes squirt several meters behind them from their roosts.

However, probably the most charismatic of avian sprayers are penguins. In a paper published in 2003, two researchers, Victor Meyer-Rochow and Jozsef Gal, became intrigued with the radial patterns of feces created by two penguin species, chinstrap (*Pygoscelis antarctica*) and Adélie (*Pygoscelis adeliae*), which they formed by jet-propelled defecating around their ground nests (the penguins, that is, not the researchers). As a result, the researchers calculated the physics behind these intriguing traces, taking into account a variety of factors such as: horizontal distances covered by the spray, poo density, as well as cloacal diameter, shape, and height above the ground. After making a series of measurements, including penguin cloacal diameters (that would have been fun to watch), they figured these fecal blasts exited at 2.0 m (6.6 ft)/sec, although each volley only took about 0.4 seconds. Once they figured values for density and viscosity of the liquid, which they unappetizingly compared to olive oil, all numbers were plugged into equations, yielding estimated pressures of 10 to 60 kilopascals. For comparison, a kilopascal is 1% of normal atmospheric pressure, so 60 kilopascals is more than half an atmosphere, and much more forceful than what any known human can do. So if someone placed a defecating penguin's cloaca in front of you, the force generated by its stream would be only about one-fifth that of a garden hose, albeit with uric acid instead of water.

Before wading any more deeply into urine or feces, though, let's talk about the other end of the voiding spectrum: puking. As many individuals and human societies have learned, purging the upper part of an alimentary canal, however discomforting it might feel at the time, can be therapeutic. In our species, vomiting mainly evicts toxic substances before these are absorbed any further into our bodies, but it can also be a reaction to severe physical or psychological trauma, or induced artificially by manually triggering a "gag reflex" in the back of the throat. In other mammals, eating rotten food or drinking bad water can cause vomiting, but it also can be part of a healthy lifestyle. For instance, cats cough up hairballs, which are a result of daily grooming, or grasses they chewed earlier. Nonetheless, as many pet owners and trackers know from experience (and shoe cleaning), these excretions are often placed deliberately, such as at doorway entrances, on sidewalks, or trails used by other animals. Hence, regurgitated food can serve a dual purpose, similar to that of urine or feces: keeping digestive tracts healthy, but also sending territorial communiqués to anyone who might be "listening."

Again, birds are different. Sometimes not all food makes its way through a bird gut, but instead returns from where it came. Probably the most well known of such deposits are cough pellets, also known as gastric pellets. Owls and some other predatory birds produce these regularly, and it is perfectly normal and healthy. Owls or hawks start these pellets by: grabbing a mouse, chipmunk, or squirrel; eating it whole or in chunks, with fur, bones, muscle, and organs mostly intact; and digesting most of the protein in the proventriculus and gizzard. These birds then eject the undigested parts from the proventriculus, all packaged in a neat little cylindrical mass consisting of fur and bones. For smaller mammals swallowed whole, its entire skeleton might be extractable from this pellet.

Gastric pellets vary according to the size of the bird turning its head and coughing, as well as what it ate. For instance, I have seen candy-bar sized pellets made by wading birds, such as herons and egrets; these are filled with fish scales and bones, or sometimes crayfish parts. For seagulls, their much smaller pellets include lots of molluscan or crustacean body parts, the latter reflecting bountiful fiddler crabs that might be living in nearby coastal environments. Tracks associated with these cough pellets normally show the coughers with their feet together (side by side) and a pellet in front of the tracks.

Birds have yet another reason for purposeful vomiting, and that is out of a sense of parental responsibility. Most people know about this behavior after watching heart-warming (and foot-chilling) documentary films about penguins in the Antarctic. In these films, dutiful adult penguins go out to sea, swim around, eat lots of crustaceans, some get eaten by leopard seals, and the survivors go back on land to hurl crustacean-flavored goo into the mouths of their appreciative young ones. However disgusting this behavior might sound to us, it is a perfectly fine method for ensuring that hungry chicks get fed and fed well, particularly for those chicks that require extended parental care in or near their nests.

How about urine and feces in other vertebrates that live on land, such as some amphibians and reptiles? Frogs, toads, and salamanders certainly leave scat, but also excrete urine. I have seen toad trackways accompanied by central grooves—caused by a toad turd that stubbornly clung onto a rear end before finally detaching—as well as toad urination marks. Naturally, for those amphibians and reptiles that spend most or all of their time in lakes, streams, or other water bodies, their world is also their toilet, with urine fulfilling a "dilute and disperse" strategy, and with feces falling to the bottom and contributing to the nutrient cycling of these ecosystems.

Among those aquatic reptiles voiding in their environments are crocodilians, which are most readily compared to dinosaurs in many people's minds. Crocodilians have extremely thorough digestion, which is achieved by a combination of strong stomach acids and a long residence time for food once down the hatch. This digestion causes their feces to emerge as homogenous white or gray rounded cylinders, bearing little evidence of what might have contributed to them: fish, frogs, birds, raccoons, or cocker spaniels. Alligator feces have always impressed me in this respect, and whenever I encounter these traces, my field dissections of them reveal no visible bits of bones, feathers, or fur. These traits contrast greatly with what I have experienced with large mammalian carnivore scat, such as those from cats, canids, or bears. For instance, scat I have seen by mountain lions (*Puma concolor*) or wolves typically contains lots of hair and macerated bone. For black bears (*Ursus americanus*) that have eaten elk or other protein-rich meals, their scat is large, flattened, and composed of pinkish masses of digested flesh and bits of bone or fur, or if they ate plants, it will emit grassy or fruity scents. Scat from grizzly bear (*Ursus arctos*) is similar to that of black bear scat, only much larger, occasionally containing little bells, and smelling like pepper spray.

As mentioned before, urine is a big deal in mammals for scent marking, but not so much in birds, reptiles, and amphibians. In ratites, they "urinate" (get rid of liquid waste) first, and then defecate; meaning for them, #1 is indeed followed by #2. But remember that birds use just their cloacas for waste disposal. So even though male ratites and some other birds, such as ducks and geese, have penises—some of which, incidentally, are worthy of applause or dread—these lack an enclosed urethra. In fact, these birds must first get their lengthy penises out of the way so they can urinate and defecate out their cloacas.

Describing feces, whether from mammals, birds, crocodilians, or other vertebrates, can be enjoyable, using terms that evoke instant search images: for instance, pelletal, cord-like, tapered, or blocky. For humans, the standard descriptive method for feces is the Bristol scale, used in clinical situations. In this scale, stools are rated from 1 to 7, with 1 as separate, hard, rounded nuggets and 7 as completely fluid. In between these two extremes, most of the weight of a stool (about 75%) comes from water, and of the ~25% solid stuff, almost a third of this can be from dead bacteria, whereas the rest can be undigested food, which if recognizable also enters into a description. These proportions vary with all sorts of factors, such as what was eaten or overall health. Speaking of health, feces also can contain living or dead remains of parasites, ranging from one-celled protozoans to tapeworms, the latter reaching lengths more than twice that of a stretch limousine.

All of these fun facts about feces and other by-products of digestion would go to waste if it were not for also realizing how feces have changed the world. This is the fault of plants, which as we all know by now are organisms completely unencumbered by morals and ethics and thus are free to advance their evolutionary agendas by directing animals to carry out their wishes. This tyranny started with seed plants, which developed in the Devonian Period (about 400 million years ago), then escalated into outright slavery with the evolution of flowering plants in the Late Jurassic Period, about 145 to 150 million years ago. Flowering plants were particularly crafty, developing soft, delicious, juicy, sugary fruit that surrounded seeds. It was evolutionary bribery at its best: I give you food, you bear my children, and oh, by the way, give them a ride while you're at it. You see, while the fruits are digestible, the seeds are not. Hence, these pass through animal digestive tracts unscathed and come out the other end, covered in warm fertilizer.

What flowering plants started, then, was a combination of seed dispersal and seed planting, which non-avian and avian dinosaurs likely helped throughout the end of the Mesozoic. Today, birds and mammals have shaped entire landscapes by eating fruit, carrying seeds in their guts, and pooping. This is a topic we will revisit toward the end of this book as we consider how dinosaur traces molded our modern world and will continue to affect it.

Last Suppers and Permanent Constipation: Dinosaur Stomach and Intestine Contents as Trace Fossils

If put in the same order as an alimentary canal, trace fossils related to dinosaur feeding typically start with toothmarks and end in coprolites. So what trace fossils are between these two end members? Gastroliths certainly qualify, and as discussed before, enough of these are preserved in dinosaur abdominal cavities to inform us about dinosaur behavior. But what about the digested food itself? Does any of this get preserved? Yes, indeed. Preserved contents of stomachs, intestines, and anything else that resided in a dinosaur's gut—but did not make it out in one form or another before that dinosaur died—are also trace fossils. What about dinosaur hurling? In ichnology, this counts too; it just needs to have been preserved in the fossil record, and then ichnologists would very happily label these as trace fossils.

Before going on any further, a little more jargon is needed to know how paleontologists name such trace fossils. Most broadly, any former food item associated with the digestive tract of a dinosaur (or any other fossil animal, for that matter) is called a *bromalite.* Bromalites include *enterolites* (fossil stomach contents), *cololites* (fossil intestinal contents), gastroliths, *regurgitalites* (fossil puke), and *coprolites* (fossil poop). *Emetolite* is yet another term proposed for a regularly regurgitated deposit, such as a cough pellet.

For now, we will focus on enterolites and cololites. Paleontologists are always very excited whenever they find such trace fossils, considering the amount of information these provide about a specific dinosaur's diet. Of course, paleontologists also must take care whenever interpreting possible dinosaur last meals. After all, fossils stacked on top of one another or otherwise jumbled together can give the illusion that plant or animal remains are inside an animal's body cavity. Add compression from a few tons of overlying rock, and formerly three-dimensional and separate body fossils can get squished together into flattened masses. Thus paleontologists must make sure that the suspected food items are actually in between the ribs or pelvis of the dinosaur, rather than above or underneath it.

This situation applies to the Late Triassic theropod *Coelophysis bauri* which, based on supposed stomach contents, was once falsely accused as the Hannibal Lecter of dinosaurs. *Coelophysis* was a slender, greyhound-sized theropod and is one of the most abundantly represented dinosaurs in the fossil record, with hundreds of complete specimens in former river deposits in northern New Mexico. Among these were a few skeletons of adult *Coelophysis* that apparently contained the remains of juveniles between their ribs. The "cannibalism" hypothesis was originally interpreted and promoted by noted paleontologist Ned Colbert, who had excavated and studied *Coelophysis* since the late 1940s. "Eating children" thus became one of the most commonly told *Tales from the Crypt* for *Coelophysis*, and for decades most people accepted it because it came from Colbert. Its plausibility was further supported by how many modern carnivores eat their rivals' offspring. In terms of Mesozoic grim fairy tales, also recall those *Tyrannosaurus* and *Majungasaurus* toothmarks in bones of their own species, confirming that at least a few theropods made daily specials out of their relatives.

Great story, but too bad it was wrong. In 2002, paleontologist Robert Gay took a closer look at these *Coelophysis* specimens and found that the "juvenile *Coelophysis*" bones were misidentified. Instead, they belonged to *Hesperosuchus*, a small crocodile-like animal that lived at the same time as *Coelophysis*. Another study done by Sterling Nesbitt and others in 2006 confirmed that it was *Hesperosuchus* in the belly of the beast, not another *Coelophysis*. In yet another study in 2010, Gay pointed out that supposed "stomach contents" in one *Coelophysis* made up a larger volume than its original stomach, so this was not an enterolite, either. Still, the idea that *Coelophysis* looked to its kin for an occasional meal did not go away, as another adult specimen was found with juvenile bones next to its mouth. Interestingly, this deposit, because of its position and jumble of bones, was interpreted as a regurgitalite. If so, did this *Coelophysis* purge itself just before dying and entombing, or did this digestive rejection reflect some sort of remorse? It probably was not the latter, as hungry dinosaurs surely lacked any such taboos.

Fortunately, paleontologists did not have to fixate solely on *Coelophysis* for examples of theropods with stomach contents, as many more were uncovered from the 1990s on. For instance, in 2011, Jingmai O'Connor and two other paleontologists reported a specimen of *Microraptor*—a small feathered theropod from the Early Cretaceous of China—with bird bones between its ribs. The bones, consisting of a left wing and both feet, were undigested, without toothmarks, and the feet were closer to the front of the *Microraptor*'s body cavity. All of this implied that the bird, which was an adult, was snatched headfirst and eaten whole or in chunks. These contents comprised a wonderful find, as they confirmed that *Microraptor*—which had feathers on all four limbs—was likely a tree-dwelling dinosaur, hunting fully grown birds far above the ground. However, this did not mean *Microraptor* ate only birds, as another specimen also had a mammal bone in it.

A couple of Early Cretaceous theropods from China, *Sinosauropteryx* and *Sinocalliopteryx*, also have intriguing enterolites. *Sinosauropteryx*, a fuzzy theropod covered by a thin coat of down, is famous as the first of many feathered theropods discovered in China starting in the 1990s. As a special bonus, the first specimen also had a lizard in its gut, but farther down—closer to its hips—were two eggs. Did this dinosaur die after attempting to reenact a scene from the movie *Cool Hand Luke* (1967)? No, as any swallowed eggs would have never made it intact to the intestines. That meant this *Sinosauropteryx* was female, pregnant, probably had dual oviducts (remember *Troodon* and the paired eggs in its nest?), and had just gained some needed sustenance before giving birth, but instead died and became part of the fossil record; its loss, our gain.

Another *Sinosauropteryx* contained jaws from three different small mammals. These remains demonstrated how this dinosaur terrorized mammals more than a hundred million years before dinosaur-worshipping humans made movies depicting similar scenarios, perhaps reflecting an ancestral memory. Fortunately for Mesozoic mammals, at least one struck a blow against its dinosaurian oppressors: stomach contents of the Early Cretaceous *Repenomamus*, a badger-sized mammal that also lived in China, included the bones of a baby *Psittacosaurus*, and thus started a long-time tradition of mammals eating dinosaurs.

Based on other stomach contents, another feathered theropod, the Early Cretaceous *Sinocalliopteryx* of China, also ate birds and non-avian dinosaurs. One skeleton had two specimens of the bird *Confuciusornis* in its body cavity, as well as acid-etched bones of an unidentified ornithischian dinosaur. In this instance, two birds in the gut was worth one neat hypothesis, as the birds were nearly complete and in the same digested state, meaning they were likely caught and gulped quickly, one after the other. Yet *Sinocalliopteryx* had no apparent adaptations for climbing trees, and at more than 2 m (6.6 ft) long, it was likely too big to scale a tree trunk anyway. A different specimen of the same species also contained a "drumstick" (leg) of another theropod—probably *Sinornithosaurus*—an example of theropod-on-theropod action that did not end well for one of them. All in all, these body fossils of *Sinocalliopteryx* and their enclosed trace fossils tell us that *Sinocalliopteryx* was probably a formidable ground-hunter.

Luckily, not all Cretaceous theropods with stomach contents were magically restricted to China. Ably representing North American theropods and their meals, David Varricchio—introduced in a previous chapter—discovered a Late Cretaceous tyrannosaurid (*Daspletosaurus*) in Montana with a few anomalous bones in its abdominal region. These consisted of four tail (caudal) vertebrae and part of a lower jaw (dentary) from unidentified juvenile hadrosaurs. Notice the plural: the vertebrae belonged to a hadrosaur that was likely about 3 m (10 ft) long, whereas the dentary came from one that was about 1 m (3.3 ft) long. Juvenile hadrosaurs typically had a thin layer of outer (cortical) bone on their caudal vertebrae, but this was missing; the spongy bone underneath was exposed

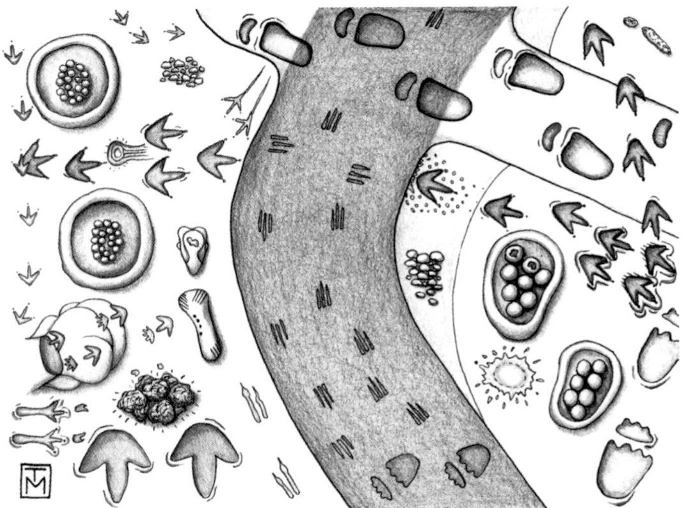

FIGURE 1. An imagined Cretaceous landscape devoid of dinosaurs, but filled with their traces, some of which later became trace fossils.

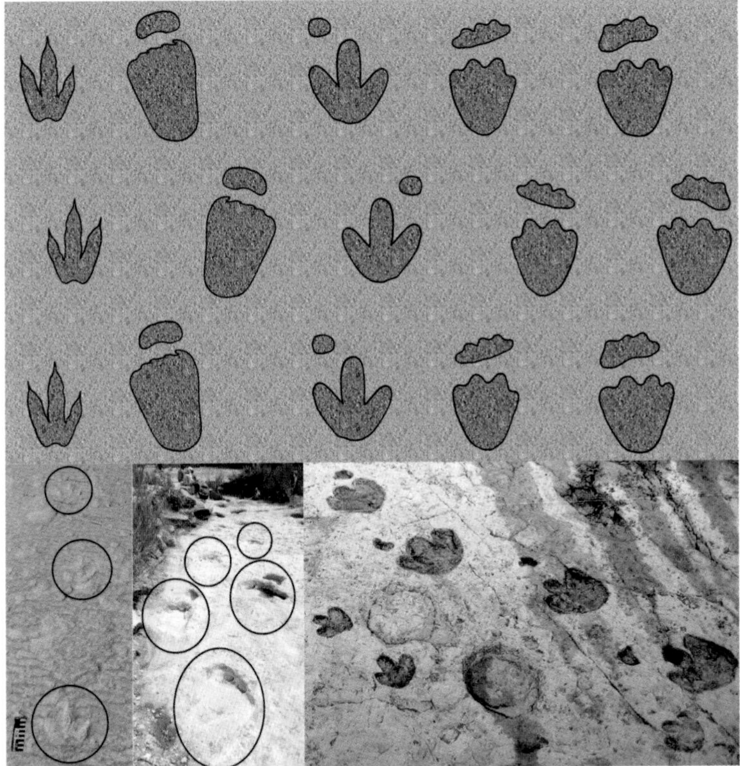

FIGURE 2. Known trackway patterns of some dinosaurs, including (top, left to right) theropods, ornithopods, sauropods, ankylosaurs, and ceratopsians; examples of trackway patterns (bottom, left to right) made by a Middle Jurassic theropod (Wyoming), an Early Cretaceous sauropod (Texas), and Late Cretaceous ornithopods (Colorado). The ornithopod tracks show both bipedal and quadrupedal movement by their makers.

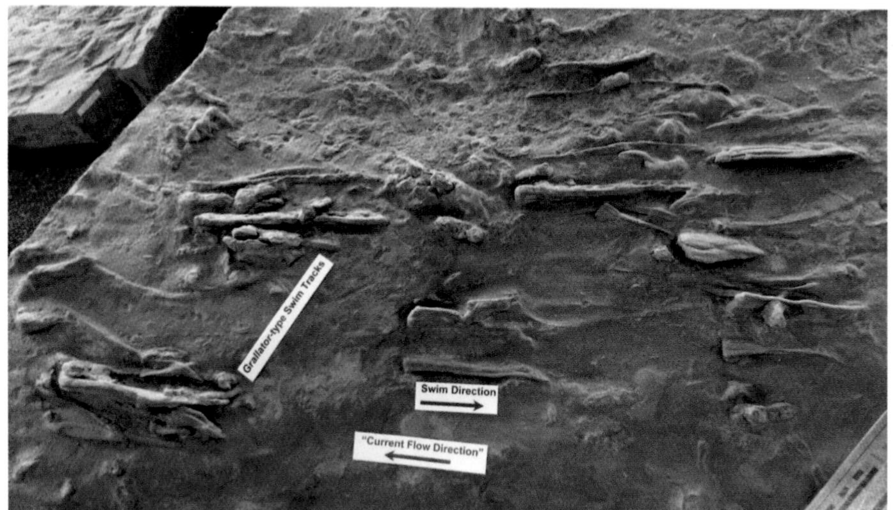

FIGURE 3. Swimming dinosaur tracks from the Early Jurassic of Utah. These were probably made by theropods that swam from left to right against the current flow, with their claws digging into the bottom. The tracks were later filled in with sand, which made natural casts of the tracks. Tracks on display at the St. George Dinosaur Discovery Site at Johnson Farm, St. George, Utah. Scale (lower right) in centimeters.

FIGURE 4. Parallel and evenly spaced Late Jurassic sauropod trackways, showing probable herding, at the Purgatoire River tracksite near La Junta, Colorado. (Notice how the people seeing these tracks can't help but walk along them, stand and stare at them, or simply sit in them.)

FIGURE 5. Panorama of the Lark Quarry tracksite in Queensland, Australia. Nearly every dent on the rock surface is a dinosaur track, a small sample of the nearly 3,000 Late Cretaceous footprints preserved there.

FIGURE 6. Close-ups of the three-toed dinosaur tracks at Lark Quarry, with the largest one (top) from a large theropod or ornithopod, and the smaller ones also either from theropods or ornithopods. Note how most of the small ones are pointing the opposite direction of the big one. Why?

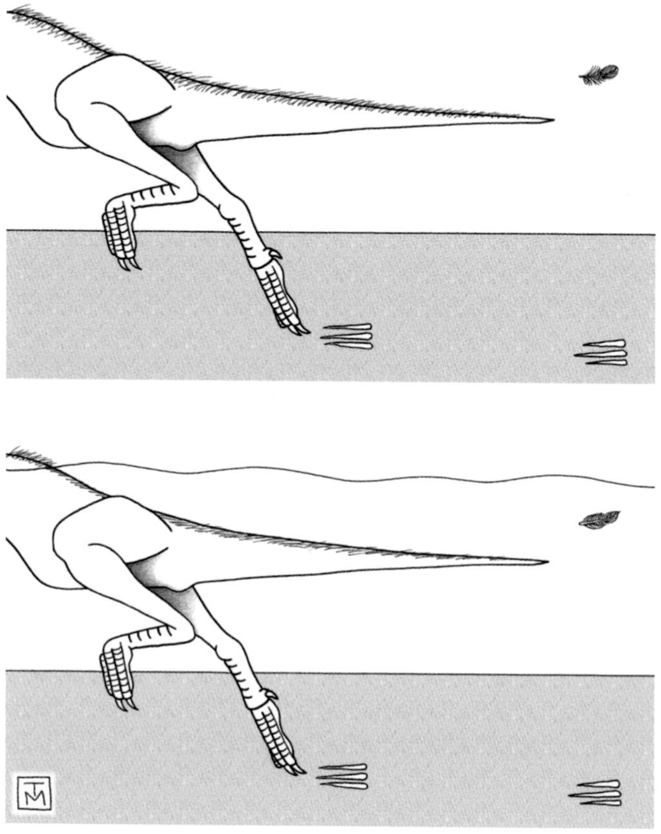

FIGURE 7. How a running theropod (top) and swimming theropod (bottom) of the same size could have made tracks looking like those in Lark Quarry.

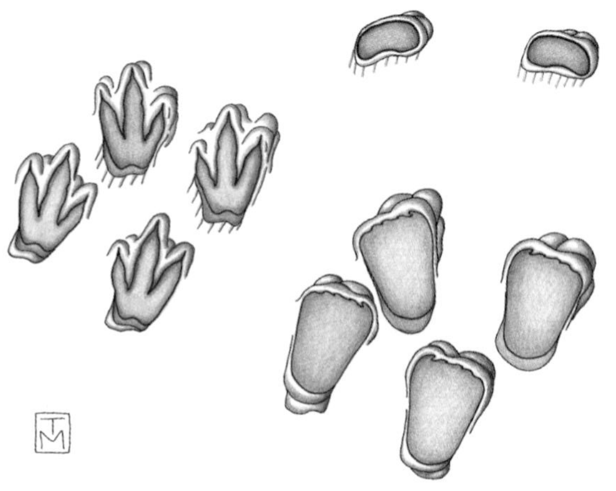

FIGURE 8. Hypothetical dinosaur trackway patterns associated with sex, such as from bipedal theropods (left) and quadrupedal sauropods (right).

FIGURE 9. Late Cretaceous insect cocoons in rock (center, lower right) that was part of a *Troodon* nest (Montana). The cocoons likely belonged to burrowing wasps that lived in the soils on and around the dinosaur nests.

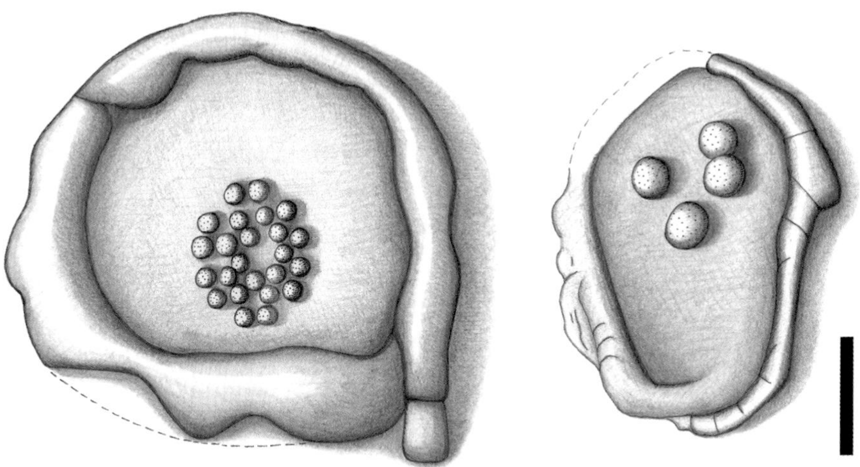

FIGURE 10. Late Cretaceous *Troodon* nest structure (Montana), artistically reconstructed and with egg clutch in center (left), and Late Cretaceous titanosaur nest structure from Argentina with only part of its original egg clutch (right). Both nests at the same scale, scale bar = 50 cm (20 in).

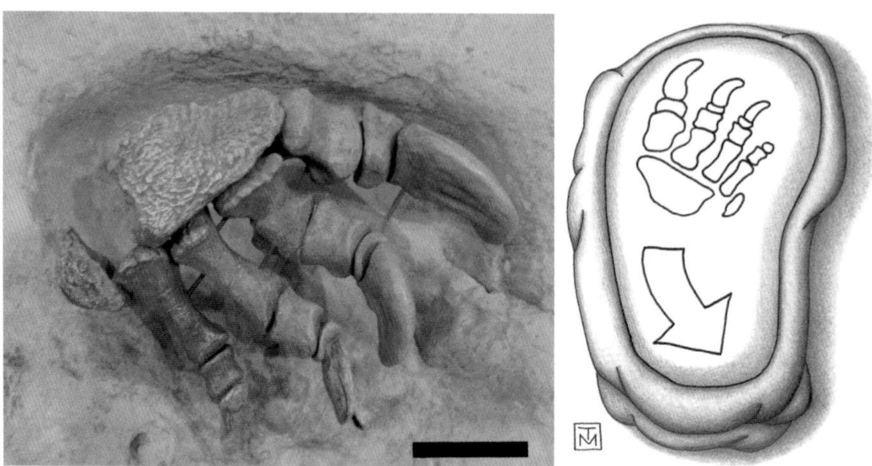

FIGURE 11. Right rear-foot anatomy of a sauropod suspended above a track, showing how its anatomy corresponds with the track below it (left); digging motion needed by rear foot to make nest (right). Foot and track part of a display at Dinosaur Valley State Park (Texas), and diagram informed by Vila *et al.* (2010) and Fowler and Lee (2011). Scale bar (left) = 10 cm (4 in).

FIGURE 12. Field photograph documenting the discovery of the first known dinosaur burrow and the ornithopod dinosaur that made it, *Oryctodromeus cubicularis*. The sandstone spiraling down from the reddish mudstone outlines the burrow, and the plaster jacket surrounding the dinosaur bones also shows the approximate location of the burrow chamber. (*Photograph by David Varricchio.*) Brush (upper left) = 2.5 cm (1 in) wide.

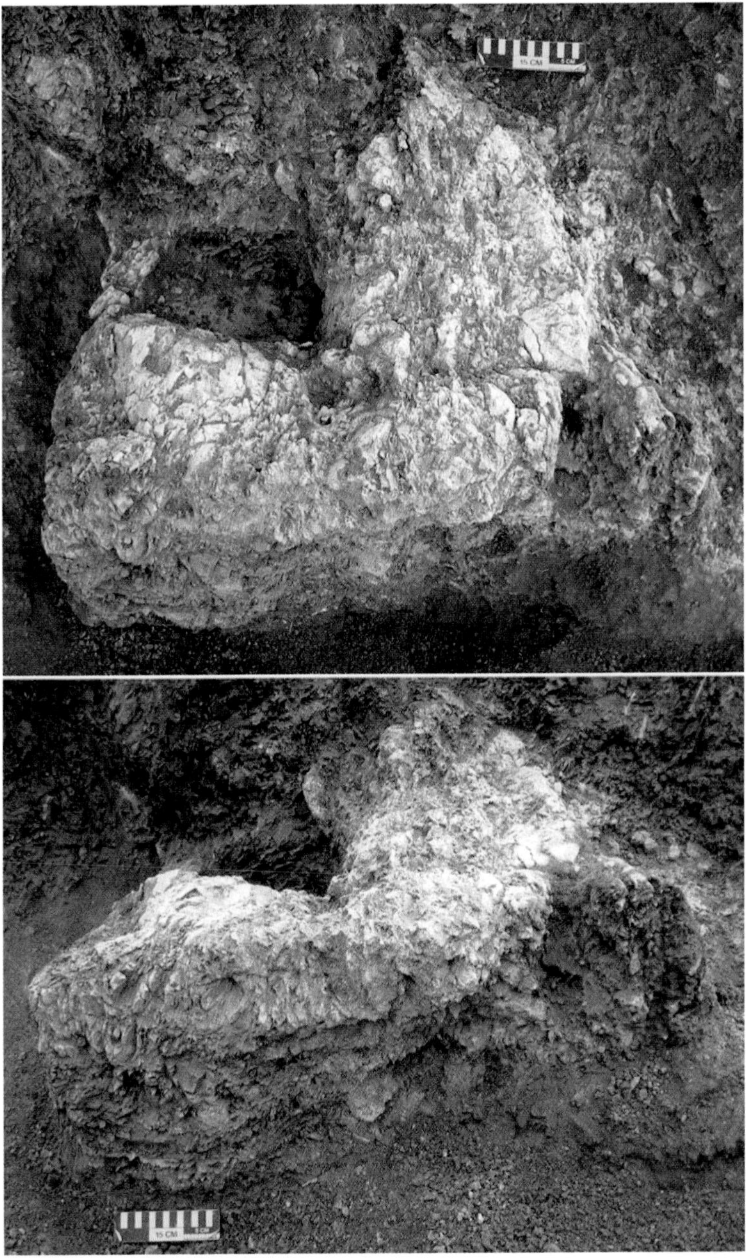

FIGURE 13. Field perspectives of excavated dinosaur burrow attributed to *Oryctodromeus*, preserved as a sandstone cast of the original burrow: overhead view (top) and side view (bottom). The burrow chamber was cut off from the main "tunnel" when the plaster-jacketed bones were recovered from the site. Notice also the small burrows coming off the right-hand corner and left-hand corners of the main burrow, interpreted as coming from commensal animals. Scale = 15 cm (6 in).

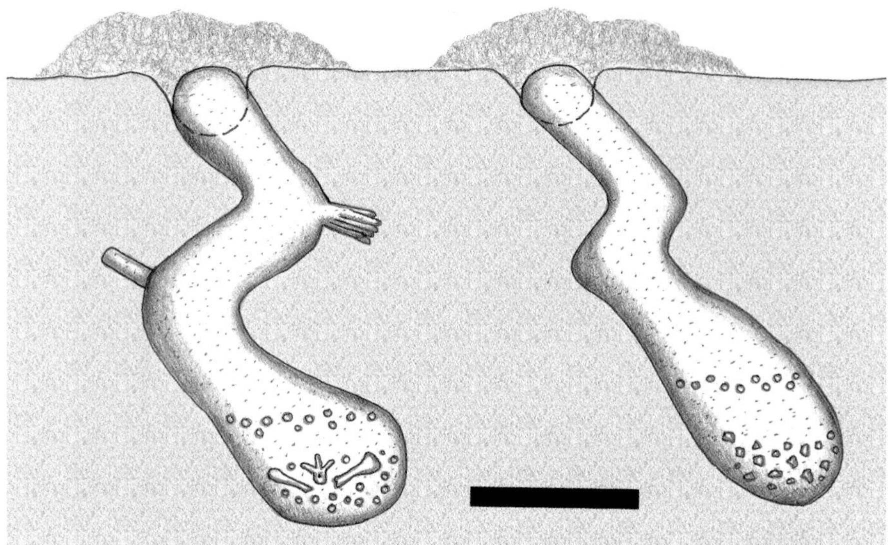

Figure 14. Comparison between Late Cretaceous dinosaur burrow from Montana (left) and purported Early Cretaceous dinosaur burrow from Victoria, Australia (right). Each had a coarser sediment fill (pebbles), although only the Montana one contained bones and had commensal burrows. Both burrows at the same scale, scale bar = 50 cm (20 in).

Figure 15. Borings in Late Jurassic dinosaur bones, attributed to bone-boring termites. Obviously, these trace fossils were made after the dinosaur was already dead. Specimen is about 15 cm (6 in) wide, and is on display at Dinosaur National Monument (Utah).

FIGURE 16. Toothmarks in an *Apatosaurus* ischium, which, based on their sizes and spacings, were probably made by the large theropod *Allosaurus*. Specimen is on display at Dinosaur Journey Museum (Museum of Western Colorado), Fruita, Colorado; scale in centimeters.

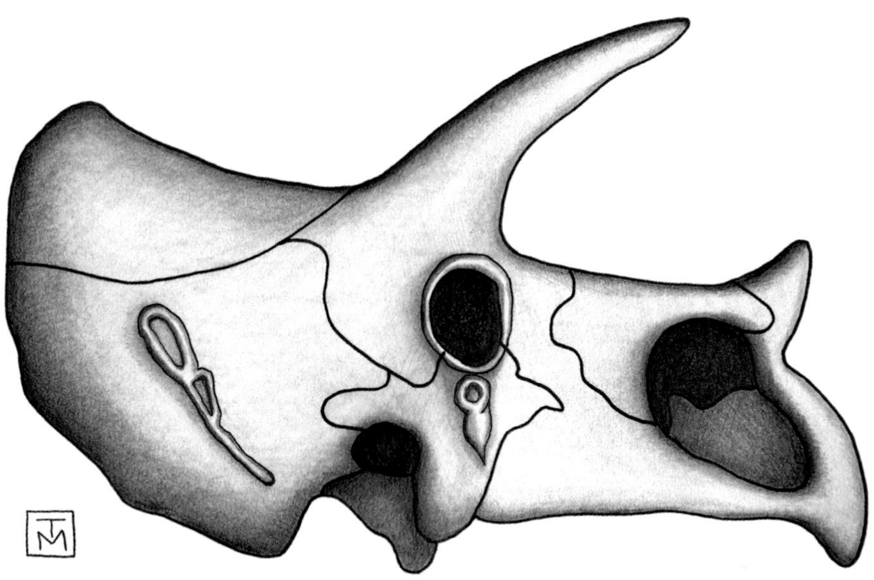

FIGURE 17. Right side of a *Triceratops* skull with wounds on its squamosal (left) and jugal (center) bones, a trace fossil of where another *Triceratops* collided with it. Sketch informed by figures and descriptions by Farke *et al.* (2009).

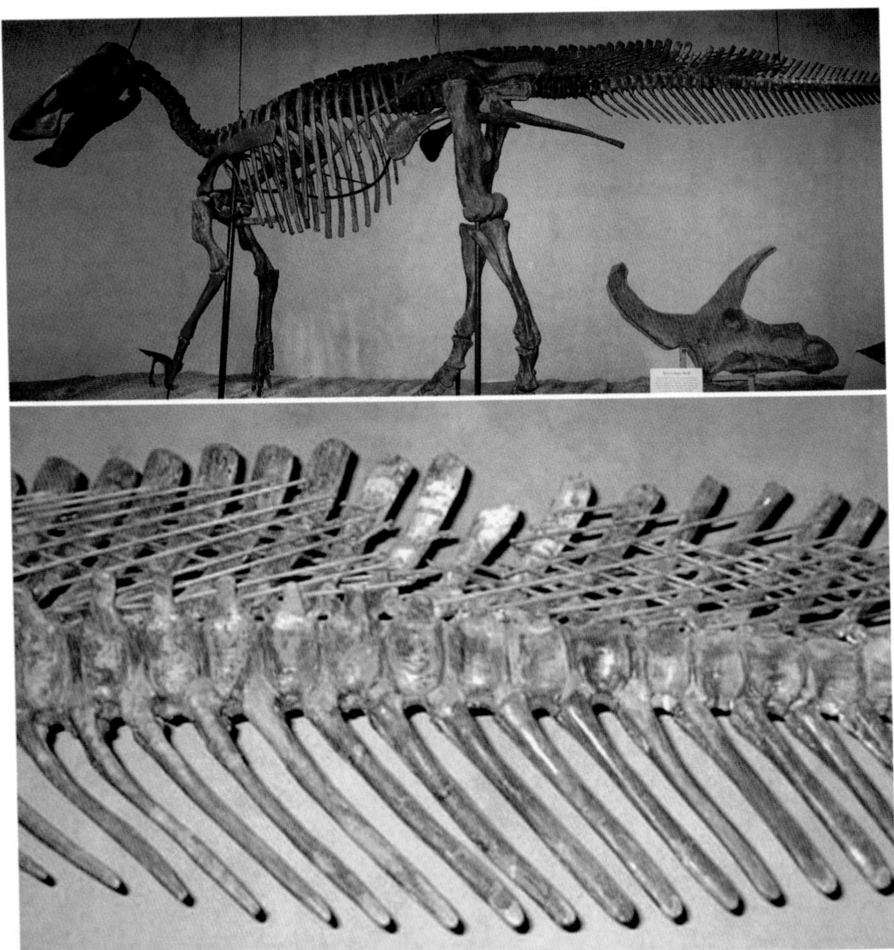

FIGURE 18. The Late Cretaceous hadrosaur *Edmontosaurus* (top), with a *Triceratops* skull for scale, and a close-up of its tail bones (bottom) with a healed bite mark. The latter is a trace fossil of a theropod attack that the hadrosaur escaped (lucky for it, unlucky for the theropod).

FIGURE 19. Possible dinosaur gastroliths, or just rocks? A dilemma posed by smooth, rounded, and polished stones in many Mesozoic sedimentary formations made while dinosaurs were alive. Specimens are on display at the College of Eastern Utah Prehistoric Museum in Price, Utah.

FIGURE 20. Gastroliths that were formerly in the abdominal cavity of an Early Cretaceous marine reptile, with a partial rib included. Specimen at Kronosaurus Korner in Richmond, Queensland, Australia.

FIGURE 21. A mass of small gastroliths in the abdominal cavity of the Early Cretaceous theropod *Caudipteryx* from China. Specimen is a replica (cast) of the original and is on display in the Carnegie Museum of Natural History, Pittsburgh, Pennsylvania.

FIGURE 22. Modern carnivore feces (left) left as a territorial marker by a coyote (*Canis latrans*) on the middle of a well-used trail in Georgia, with the author happily pointing to it and approving of its presence; a cough pellet (right) filled with fiddler-crab parts, deposited by a laughing gull (*Leucophaeus altricilla*) on an intertidal sandflat, Georgia. Scale bar (lower photo) = 5 cm (2 in).

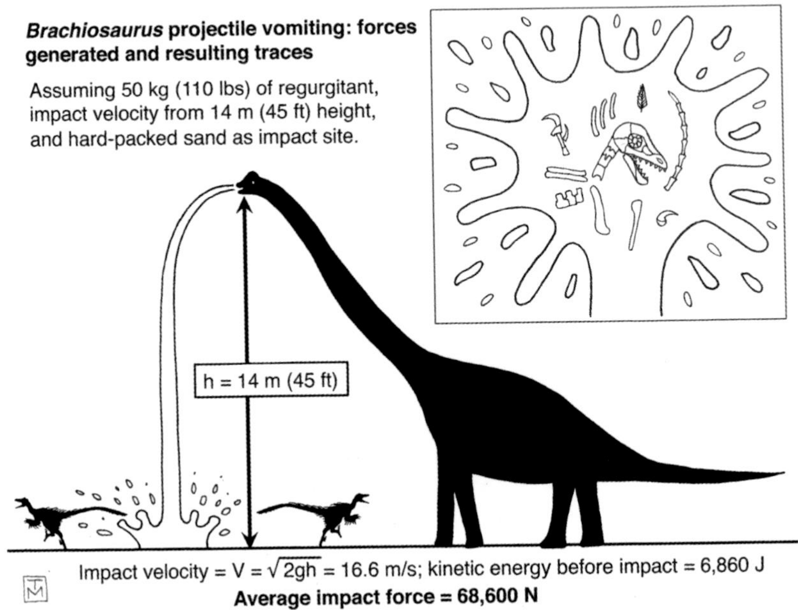

FIGURE 23. *Brachiosaurus* projectile vomiting and physics associated with such tracemaking activity (left), and hypothetical trace resulting from this behavior, with main impact crater, associated stream, and small-theropod victims (right).

FIGURE 24. Odd depression in a Late Jurassic limestone, Colorado (left), identified as a sauropod urination structure (urolite), and an undergraduate student providing scale while helpfully demonstrating its possible origin; artistic recreation of dinosaur urolite from the Early Cretaceous of Brazil (right), based on a photograph by Fernandes *et al.* (2004). Scale bar (right) = 10 cm (4 in).

FIGURE 25. Coprolites attributed to the Late Cretaceous hadrosaur *Maiasaura* of Montana and evidence that these were attracting some invertebrate admirers. Coprolite (top) containing dung-beetle burrows (located left center and upper right); coprolite (bottom) in the field with fossil snail in it (located near its center). Scale in upper photo in centimeters; specimen part of traveling display from Museum of the Rockies, Bozeman, Montana.

FIGURE 26. The clearest example of a dinosaur track—probably from a small ornithopod—from Victoria, Australia; when found in 1980, it was the only one from southern Australia. Specimen is in Museum Victoria, Melbourne, Australia, and is about 10 cm (4 in) wide.

FIGURE 27. The most abundant concentration of dinosaur tracks discovered thus far in southern Australia, discovered at Milanesia Beach, Victoria. Sandstone block (top) hosting the tracks, which were interpreted as small theropod tracks. Map (bottom) of where the tracks are located on the rock surface; scale bar = 15 cm (6 in).

FIGURE 28. Close-up of one of the best-preserved theropod tracks in the first-discovered rock slab at Milanesia Beach (top), compared to the track of a sandhill crane (*Grus canadensis*) from Georgia (bottom). The tracks are about the same size, scale in upper photo in millimeters.

FIGURE 29. Modern and fossil comparisons between traces of theropod dinosaurs, avian and non-avian, respectively: trackway of a Southern cassowary (*Casuarius casuarius*) on beach sand (left) at Mission Beach, Queensland, Australia; Early Jurassic theropod trackway (right) in sandstone, southern Utah. Scale on left = 20 cm (8 in) and on right = 15 cm (6 in).

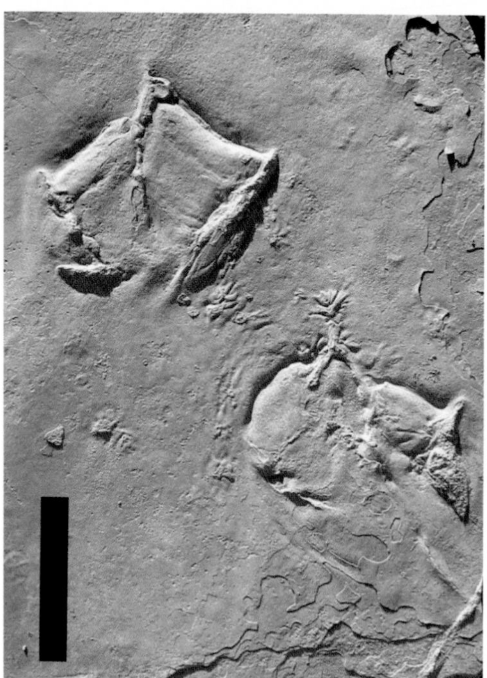

FIGURE 30. Fossil bird tracks from the Early Cretaceous of Victoria, Australia (top) and the Eocene of Utah (bottom). The Cretaceous track is from the right foot of a heron-like bird that came in for a landing from flight on a sandy river floodplain; scale in centimeters. The Eocene tracks are from a web-footed bird similar to a duck or flamingo, made on a muddy lakeshore; scale = 10 cm (4 in).

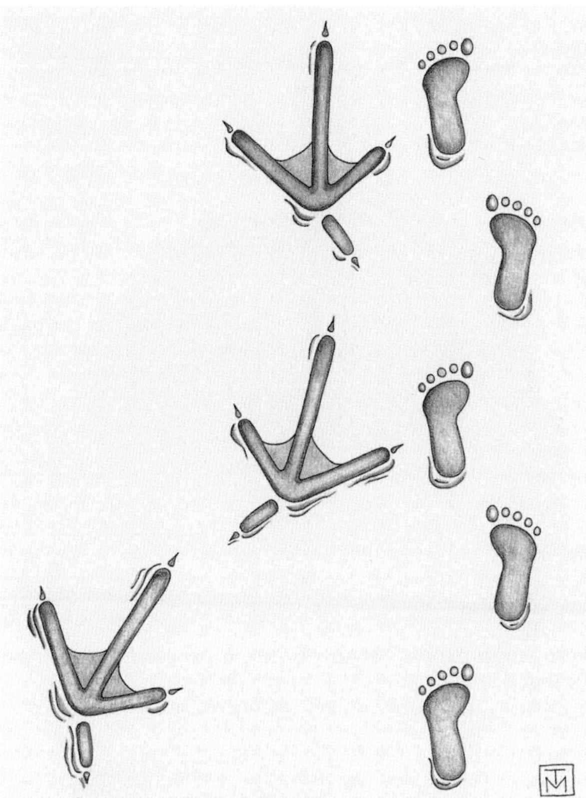

Figure 31. Speculative trackway of the giant Marabou stork (*Leptoptilos robustus*) that decided to go "hobbit hunting," turning to stalk a small human (*Homo floresiensis*). Both species shared the island of Flores (Indonesia) during the Pleistocene Epoch.

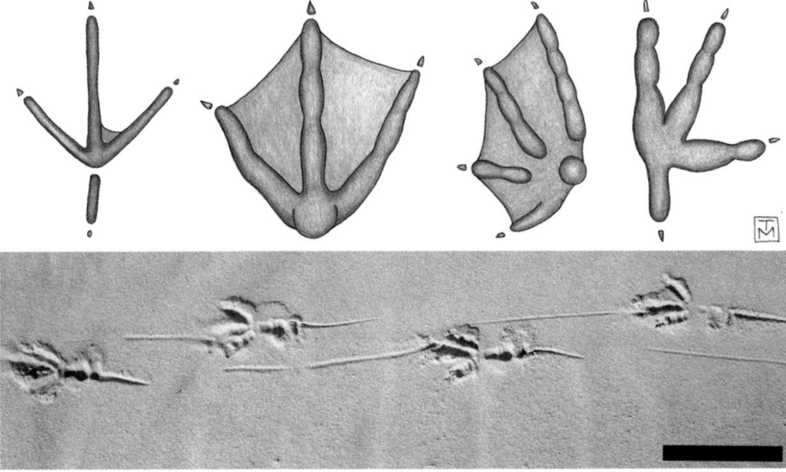

Figure 32. Main forms of bird tracks, from left to right: anisodactyl, palmate, totipalmate, and zygodactyl; a typical bird walking-trackway pattern, made by a boat-tailed grackle (*Quiscalus major*) on Sapelo Island, Georgia. Scale bar in bottom photo = 10 cm (4 in).

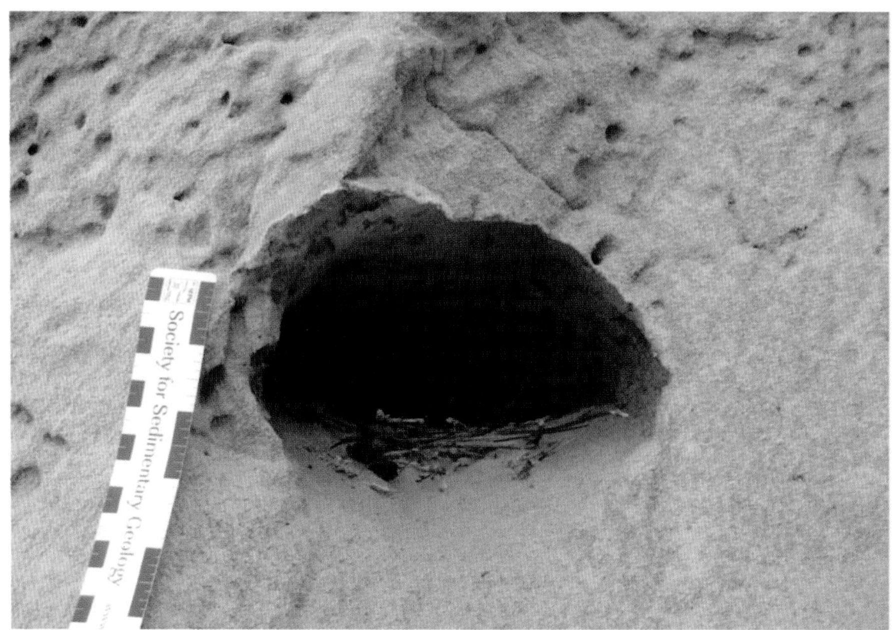

Figure 33. Modern bird burrow, probably from a northern rough-winged swallow (*Stelgidopteryx serripennis*), which it dug into a soft Pleistocene sandstone on Cumberland Island, Georgia. Scale in both centimeters and inches.

Figure 34. Big sauropod tracks in Early Cretaceous sandstones, exposed at low tide near Broome, Western Australia. Spousal unit (Ruth Schowalter) for scale.

FIGURE 35. Nesting ground of Australasian gannets (*Morus serrator*) with regularly and densely spaced nest mounds on an elevated marine platform: Muriwai Beach (North Island), New Zealand.

FIGURE 36. Theropod track forms—avian and non-avian—that were likely seen by Charles Darwin in South America (Argentina) and Edward Hitchcock in North America (Massachusetts) at about the same time during the early 19th century. Track (top) of greater rhea (*Rhea americana*) in Patagonia, Argentina; scale in centimeters. Theropod dinosaur track (right) from the Connecticut River Valley similar to those described by Edward Hitchcock. Scale = 5 cm (2 in) in both photographs; *photo of dinosaur track by Patrick Getty.*

and pitted. Low-pH stomach acids in the theropod's proventriculus probably dissolved cortical bone on the caudal vertebrae, and the dentary was likewise in bad shape. So based on their identity as juvenile hadrosaurs, appearance, and location inside a *Daspleto-saurus* skeleton, Varricchio concluded that both hadrosaurs used to be food. Although this was not enough evidence to say whether this *Daspletosaurus* actively preyed on these two differently aged hadrosaurs or munched on already-dead ones, the close proximity of these bones as gut contents implies that the theropod must have eaten one immediately after the other.

The Early Cretaceous theropod *Baryonyx walkeri* of England is yet another large theropod with stomach contents, but surprising ones. The first specimen of this dinosaur was discovered in a clay pit mine in 1983; paleontologists Alan Charig and Angela Milner named it several years later, in 1986. It died before growing to full adult size, but was still an impressively large theropod at about 10 m (33 ft) long, or half the length of a healthy tapeworm. Yet its most noteworthy features were: a snout shaped more like that of a crocodile; a kink in its jaw also like that of a crocodile; nearly a hundred teeth in its jaws; robust arms; huge claws on its "thumbs"; and leg proportions that suggested it could walk on all fours. This was a big predatory dinosaur but built unlike most known at that time, thus provoking a good question: With adaptations like these, what did it eat?

In a more detailed study of this specimen published in 1997, Charig and Milner concluded that all of these traits made *Baryonyx* well suited for grabbing, handling, chomping on, and eating fish. In other words, it acted like a grizzly bear, albeit a massively up-scaled one. Preposterous? Not when the same specimen also had acid-etched fish bones and scales in the area of its former stomach. These bits belonged to a bony fish identified as *Scheenstia*, a common fossil in Early Cretaceous rocks in England, France, and Germany. Consequently, Charig and Milner speculated that *Baryonyx* was comfortable wading into and swimming in lakes and streams to find food. So thanks to this combination of body and trace fossil evidence, paleontologists began to think more often of some theropods as *piscivorous* (fish-eating), whether as a main part of their diet or whenever they felt like having fish. (As learned previously, paleontologists were further encouraged to adopt this formerly strange idea of fish-eating theropods when they found thousands of theropod swim tracks in Early Jurassic rocks of Utah.)

This same specimen of *Baryonyx* included lots of other ichnological extras, such as the remains of a juvenile ornithopod (identified as *Iguanodon*), gastroliths, and its own broken bones. The *Iguanodon* remains included neck, back, and tail vertebrae as well as arm, leg, finger, and toe bones. Thus the "grizzly bear" analogy holds up well for *Baryonyx* in this respect, as these modern carnivores are not just restricted to eating fish but also deer, elk, caribou, moose, and other animals. The *Iguanodon* bones were eroded, probably by stomach acids, just like the hadrosaur bones in *Daspletosaurus*.

Charig and Milner also mentioned gastroliths as being present and associated with the skeleton; sadly, they gave no other description of these trace fossils. Oddly enough, they mention "gastroliths" in their 1986 paper, but only "an apparent gastrolith" in the 1997 one. Did the others get misplaced during that eleven-year gap? However, if *Baryonyx* only had one or a few gastroliths, such rocks could have been accidentally grasped and swallowed as it scooped up a slippery fish. Lastly, a few broken bones in *Baryonyx* were explained as the trace fossils of other dinosaurs that irreverently stepped on its body after it had already died and been buried.

Nearly all of the trace fossils mentioned thus far were enterolites, food items that only made it down as far as the proventriculus, gizzard, or whatever else functioned as a stomach in a dinosaur. What about cololites, in which this food was more digested, passing through the intestines for further nutritive absorption, and almost ready to explode from a dinosaur's other end but never quite made it?

Only a few such trace fossils are known in dinosaurs, but one of the best is also in the most complete dinosaur from an entire continent, *Minmi paravertebrata*. This Early Cretaceous dinosaur lived about 115 *mya* in Queensland, Australia, and was an ankylosaur;

recall that ankylosaurs were armored dinosaurs that lived throughout much of the Jurassic and all of the Cretaceous periods. I have seen a cast of this specimen displayed at a small museum in Richmond, Queensland, and it is incredible. Although flattened so that it looks like Cretaceous road kill and about the size of an adult sheep, nearly every hard part of its body is intact and articulated.

Fortunately, this exceptional preservation included its gut contents, which consisted of a compacted mixture of leaves, seeds, small fruits, and other parts from a variety of plants inside the area of its hip. All of these plant remains were the size of coarse coffee grounds, measuring only a few millimeters across. "Then it must have had gastroliths!" you shout excitedly. In response, I smile, nod, and thank you for outwardly expressing such enthusiasm for those trace fossils, but then gently inform you that this ankylosaur had no gastroliths. This means that it more likely chewed its food thoroughly in its mouth before swallowing, rather than relying on its stomach to reduce food to a more digestible texture.

Okay, time to get skeptical. Dinosaurs, like other vertebrates, probably spilled their guts after dying, whether from decomposition, scavengers, or a combination of the two. Once opened, a dead dinosaur's body cavity would have invited the outside world to intrude on its inner spaces. This means river currents could have carried in the finely ground and varied plant material and deposited it in the open body cavity of this *Minmi* after it died. Therefore, these plants might not represent its final meal after all.

Good thinking, but in my preceding description of *Minmi* and its cololite, I omitted a key piece of information, one that crushes that hypothesis like ankylosaur jaws would a Cretaceous plant. It turns out this specimen of *Minmi* was deposited in a Cretaceous sea, and no river was anywhere near its body when it was laid to rest in a shallow marine grave. Central Queensland abounds in fossils from this Cretaceous seaway that divided Australia then. In the same museum where I stared at the cast of *Minmi*, hundreds of fossils of ichthyosaurs, plesiosaurs, ammonites, squid, clams, and other sea life surrounded it, all from the same rocks.

No serious paleontologist suggests that the heavily armored ankylosaurs paddled out to sea, just as medieval knights were not likely keen about going for a swim in full armor. So this *Minmi* must have been a "bloat-and-float" dinosaur. It died on land with a belly full of terrestrial plants and was filled with enough gas from decomposition to buoy it into the ocean. Once out there, it somehow escaped scavenging by marine carnivores; no toothmarks or other signs of nibbling were on its body. Once it sank, it also must have made quite an impression on any sea life living on the bottom before burial. Most amazing of all, this is an example of a dinosaur that carried its trace fossil with it after death and on a long journey to sea.

Dinosaur Puke

So how about food that a dinosaur's gastrointestinal tract rejected instead of passing through, or more quiescently as gastric pellets like those emitted by predatory birds today? Tragically, examples of dinosaur regurgitalites are either extremely rare or unrecognized from the fossil record. One regurgitalite inferred from Late Triassic rocks contains pterosaur remains, but is thought to have come from a large fish rather than a land-dwelling animal. I also mentioned one example interpreted in 2009, which was apparently fossilized beside the mouth of a *Coelophysis*. As much as I want to believe this is Triassic puke, the coincidence of vomiter and vomit seems a little too fortuitous. For instance, how did this former meal stay put while river currents buried its supposed producer?

Nonetheless, I am hopeful that dinosaur regurgitalites are more common than previously supposed, and paleontologists just need a combination of training and imagination to find them. So here are some criteria for finding dino-barf. For one, because the contents of regurgitalites only spent a short amount of time in an animal's gut, they will be noticeably less digested than anything coming out its other end. Still, any solid items, such as bones, may have some acid etching, while also being more broken, poorly sorted, and chaotically arranged than if they had not been chewed, swallowed, and partially digested by a dinosaur.

Unless preserved as a pellet, like a tightly packed modern-day owl pellet, vomit should have contained enough liquids to cause it to spread, especially if delivered from a great height. For a moment, just imagine the Late Jurassic sauropod *Brachiosaurus* puking its guts out, with its mouth as much as 14 m (46 ft) above the ground. This likely would have caused immediate panic amongst any animals in the vicinity, which would have tried to escape such a harrowing aerial bombardment, or at least been spooked by the unusual noises accompanying a purging sauropod. Trace fossils would thus show the following:

- Large sauropod footprints at a standing position, and behind a crater with splatter marks radiating several meters around its center;
- A concentration of bigger, more solid items in the middle;
- Run-off marks from included fluids, especially if influenced by slopes on the former ground surface;
- Trackways of other animals with ratios between footprint lengths and stride lengths showing they were moving at high speeds and fleeing in directions away from the crater;
- The remains of one or two small unlucky theropods knocked out and drowned by the initial shower of debris;
- Thousands of trackways of the "cleanup" crew, such as insects specialized at eating dinosaur throw-up. Just as there are insects that specialize in eating dinosaur poop, there may have been some that enjoyed a pre-digested (but totally vegan) meal.

Only one instance of fossilized dinosaur vomit, containing a mix of dinosaur and turtle bones in Early Cretaceous rocks of Mongolia, is inferred to have come from an actual dinosaur. Another possible dinosaurian up-chuck is embodied in a mass of four birds from the Early Cretaceous of Spain. However, the paleontologists reporting it thought it also may have been from a pterosaur. Otherwise, most Mesozoic regurgitalites are credited to marine-reptile contemporaries of dinosaurs: ichthyosaurs. These deposits, found in Late Jurassic rocks of the U.K., consist of concentrated collections of belemnite shells. Belemnites, which were squid-like animals that shared the seas with ichthyosaurs, also show up as stomach contents. However, these masses of cigar-shaped shells were found by themselves and were acid-etched. This suggested they spent time bathed in low-pH stomach acids, but because they were not inside an ichthyosaur skeleton, they were either regurgitalites or coprolites.

Coprolites were quickly eliminated as an explanation, though, because belemnite shells are pointy on one end. Now imagine being an ichthyosaur that swallowed several dozen pointy-ended shells and passed those out its other end: no, thanks. The risk of internal injury also would have been too high to pass such shells all the way through intestines too sensitive to handle such a payload. Hence these were the marine version of gastric pellets, in which ichthyosaurs snapped up lots of belemnites, kept them down long enough to digest their useful nutrients from the soft parts, and then coughed up their shells.

Other than gastric pellets or illness, what would have been other reasons for a dinosaur to barf? One would have been from the desire to feed their children warm meals. And if it were regurgitated as meals for hungry youngsters and puked right into their waiting little mouths, a trace of this in the fossil record would be very rare indeed. As mentioned earlier, though, bird parents commonly seek out food, ingest it, partially digest it, and then egest it into their chicks' mouths. Keep this in mind for later, and how other trace fossils opened our eyes to this particular nurturing behavior as a possibility in dinosaurs.

Streaming on Demand: Dinosaur Urolites

We know that dinosaurs defecated and probably threw up, but what about urination? This is a tougher question to answer than one might think. Considering how dinosaurs were evolutionarily "in between" crocodilians and birds, we could examine the liquid-waste excretion of these two groups and see which best fits dinosaurs. But the first place to start thinking about dinosaur urination is with modern birds, which then can be compared to their non-avian theropod ancestors more directly than with crocodiles. Given an appreciation for the super-soaker potential of penguin excretion, we can also think about the sorts of structures that might have been produced by dinosaurs' liquid wastes, especially if delivered from cloacas on high.

Fossilized structures formed by urination, called *urolites*, are either rare or rarely recognized, with only a few thus far attributed to dinosaurs. For anyone who has seen or made their own modern examples, these structures are best defined and most recognizable when a forceful (high-velocity, low-diameter) stream of urine hits and erodes soft sand. An idealized urination structure made under such conditions should have a central impact crater, closely associated splash marks, and, if a slope is present, linear rill marks caused by excess fluid running down that slope. Although such traces have low fossilization potential, they feasibly could be preserved if made in sand dunes with the right conditions for fossilization: for instance, if dry wind-blown sand stuck to the wetted sand and filled these structures, making natural casts of them.

So far, only two dinosaur urolite discoveries have been reported. The first of these was in the Late Jurassic Morrison Formation near La Junta, Colorado. This urolite was mentioned in a poster presentation at a paleontology meeting in 2002 and understandably garnered much media attention, especially for how the paleontologists reporting these—Katherine McCarville and Gale Bishop—tried to replicate this structure on a Georgia beach by using a combination of water buckets, funnels, hosing, and ladders. The suspected Morrison urolite, which is preserved in a former lakeshore limestone, is a shallow, oblong, trench-like depression, which McCarville and Bishop described as "bathtub-shaped," a descriptor that does not sit well when imagining it occupied by liquid waste. Because this depression cuts across bedding, scouring of some sort formed it, but in a specific spot on an otherwise nearly flat surface.

Had paleontologists encountered this feature just by itself, most might have shrugged, written "weird bathtub-shaped depression" in their field notebooks, and moved on. However, additional trace fossil evidence led McCarville and Bishop to consider sauropod pee as a possibility. The same limestone bed with the suspected urolite also holds more than eighty dinosaur trackways, divided almost evenly between sauropods and theropods. Known as the Purgatoire site (after the Purgatoire River, which runs through the area), it is one of the most spectacular dinosaur tracksites in the western U.S., including some of the best-known examples of herding-sauropod trackways. This circumstantial evidence meant that large potential peers were at the site, and the size of the possible urination structure pointed toward a sauropod as the culprit.

Based on its present dimensions, it would have held about a cubic meter (265 gallons) of liquid, which would easily overflow a small wading pool. Even the largest of Late Jurassic theropods living in this area, such as *Allosaurus*, would not have had enough juice to produce such a huge structure, no matter how much they drank and how long they held it. Also, an adult sauropod cloaca—especially those of *Apatosaurus* or *Diplodocus*—would have been much higher off the ground (3–6 m, or 10–20 ft) than those of contemporary theropods (1–2 m, or 3.3–6.6 ft). Hence, their liquid wastes would have generated more force and imparted greater erosion.

How likely is it that this enigmatic feature is a urolite? I rate it a definite "maybe." Fortunately, I have seen it in person and did not have to rely just on these paleontologists' descriptions of it or their critics' guffaws. In 2002, my colleague Steve Henderson, a bunch of our eager undergraduate students, and I visited the Purgatoire tracksite. Steve and I were co-teaching a dinosaur field course and swung by La Junta to see the incredible dinosaur trackways south of town. Our guide—U.S. Forest Service paleontologist Bruce Schumacher—happily showed us the hundreds of dinosaur tracks there, wowing students and instructors alike. He then detoured our group to the suspected urolite to talk about it as an example of "science in progress."

Although I'm still skeptical of its proposed identity, it nonetheless provided an enjoyable experience for our students, in which we asked basic scientific questions such as "What evidence supports the hypothesis?" "How would you test this hypothesis?" and more specifically "How much pee would be needed and from what height would it have been delivered to make something like this?" Urolite or not, this enigmatic structure presented us with a fine educational opportunity, and a memorable one.

Given the controversy over the Morrison urolite, I was much relieved to later find out that paleontologists Marcelo Fernandes, Luciana Fernandes, and Paulo Souto had also interpreted dinosaur urolites in the Late Jurassic–Early Cretaceous Botucatu Formation of southern Brazil. Although they found only two such trace fossils, both were beautiful textbook examples of what urolites should look like, bearing all of the marks of steady but focused streams of liquid hitting dry sand. Preserved in sandstones, these trace fossils are teardrop-shaped craters connected directly to streamlines. The structures were likely made in sand dunes, with the craters having formed by erosion of upper sand layers—caused by urinary impact—and the streamlines from liquids trickling downslope on a dune surface.

The craters and streamlines of the two specimens were nearly identical in size and form, measuring about 2 cm (<1 in) deep, 16 to 19 cm (6.3–7.4 in) long, and 11 to 13 cm (4.3–5.1 in) wide, about the size of a gravy boat. The craters would have held about 300 to 400 cc of liquid (a cup and a half), but some of this would have soaked into the underlying sand with first wetting, so the original volume was probably more like 500 to 1,000 cc. Because these trace fossils were in the same strata as ornithopod and theropod tracks, and those dinosaurs were the only animals large enough to have produced such vigorous bursts of fluid, the paleontologists concluded they were the most likely to have left such marks.

Nevertheless, Fernandes and his colleagues, just like McCarville and Bishop, further tested their results by trying to make their own structures in sand. Fortunately, this did not require any self-experimentation, such as chugging liquids and running to a nearby sand dune. Instead, they took two liters of water and poured it from 80 cm (2.6 ft) above and onto a loose sand surface with a slope having the same angle (about 30°) as the fossil ones. This procedure successfully created a structure strikingly similar to the interpreted urolites, with an elongated central crater and streamlines caused by water dripping downslope. However, their triumph did not satisfy completely, so they then "cheated" by watching a modern dinosaur urinate: an ostrich, that is. Like many countries, Brazil has ostrich farms, so these researchers simply went to one and watched these big birds take a leak. Sure enough, what they observed matched their interpretations.

Because ostriches and other ratites eliminate liquid wastes first, then solid wastes, this also would explain how ornithopods and theropods could have made both urolites and coprolites. That is if they had plumbing comparable to ostriches, versus most other birds that only produce a mixture of the two. Also keep in mind, though, that if dinosaurs peed more like birds and less like mammals, their liquid waste would have been evacuated behind them, not in front; this is how ostriches tinkle. Oh, and one more thing: Remember that modern male birds do *not* use their tools to urinate. So even if a male dinosaur had a penis, you still would not be able to tell whether a male or female dinosaur made a urolite, which is definitely not the case with any mammal urination structures I have seen. Regardless, if paleontologists are lucky enough to find urolites directly associated with dinosaur tracks, they must be careful in defining "pre-pee" and "post-pee" footprints in that trackway.

Given that this is all we know so far about dinosaur urolites, it doesn't take a whiz to figure out that more studies are needed to better recognize these trace fossils in the geologic record, which should be abundant. After all, these dinosaurs had to go sometime, so the traces must be out there. So I will suggest a fun follow-up to this research, which would be to try duplicating what was done with penguin-poo physics. In other words, figure out minimum cloacal heights and diameters of these peeing dinosaurs, velocity of flow, liquid viscosity, and other factors, and then model the resulting structures. These could then serve as search images for similar trace fossils. For example, minimum cloacal heights for a urinating dinosaur would have been about the same as their hip heights; as we learned earlier, these can be calculated from dinosaur tracks. Paleontologists who do such research could be assured of making a big splash with it, while also going against the flow of others' prejudices. Afterwards, they will be flushed with success, and their colleagues pissed off.

The Straight Scoop on Dinosaur Poop

Assume that every dinosaur pooped. If so, not all of these end products of dinosaur digestion were preserved in the fossil record. But you will have a load taken off your mind when you know that those found thus far have not gone to waste, nor remained the butt of jokes.

So let's say you found what might be a dinosaur coprolite. After all, it looks like something your dog, your neighbor's dog, or your neighbor left in the yard, except it's a rock, and quite large. In your excitement, you dash to the nearest natural history museum or university, find a paleontologist, show it to her, and announce with much fanfare and dramatic flourish, "Behold, a dinosaur coprolite!" Before doing that, though, you really need to be a good little skeptic and go through a checklist that asks the following questions:

- Was it from rocks of the same age as dinosaurs (Late Triassic through Late Cretaceous)?
- Did it come from rocks formed in a continental environment, such as a former soil, river, or lake?
- Were other dinosaur body and trace fossils in the same rocks?
- Does it fall in the right size range for known dinosaur coprolites?
- Does it contain any body fossils of what might have been digested, such as plant or bone fragments?

Only when you have answered this checklist with "yes" for every item should you take your rock to a professional scientist. Otherwise, she will tell you wearily that wrongly identified "coprolites" are the bane of her existence, rivaled only by wrongly identified "meteorites" and "gold." In this respect, the most basic questions—dealing with age, environment, co-occurrence with other dinosaur fossils—are very important. For instance, if you found this rock in Cincinnati, Ohio, I would instantly tell you it is not a dinosaur coprolite. Cincinnati's a great city with a lot going for it, but it has the wrong age rocks (Ordovician Period, 450 million years old) and wrong rocks (shallow marine limestones and shales), and hence no dinosaur fossils. As a result, its civic boosters should never add "dinosaur coprolites" to its list of local natural wonders, which would remain true even if you crossed the Ohio River and went into Kentucky.

Of these criteria, by far the most important one is that it contains body fossils of whatever the dinosaur ate. A suspected coprolite may look like fossil crap, feel like fossil crap, and taste like fossil crap, but does not qualify as fossil crap unless it holds fossil food. There had better be plant tissues, spores, seeds or pollen, bits of bone, insect or crustacean parts, or other bodily remains for a lumpy chunk of rock to qualify as a genuine, bona-fide coprolite.

The least important criterion applied to coprolites is size, and that is because of its variability. Dinosaur dung may have been as big as footballs (American or Australian rules), or it could have been as small as chocolate-covered raisins. Coprolite size would have depended on: the age and size of the defecating dinosaur; its health; time of the year; or what happened to the dung after it emerged. Even the mere act of dinosaur-cloacal pinching would have affected the size of each exiting nugget, meaning the total volume of feces might have been quite high but composed of many pieces snipped by a well-honed sphincter. Of course, falling and landing on the ground would have altered the size and shape of such deposits depending on mass, anal altitude, and relative solidity. Some feces would have gone "thud" and flattened slightly on impact. Others would have gone "splat" and spread out over a sizeable area. Those left by swimming dinosaurs might have been floaters, making no sound at all. Smell would have entered this equation, too, as dung-loving insects or other animals in the area would have picked up on any distinctive odors and hurried to indulge.

Nonetheless, by far the most common question about dinosaur coprolites I've heard is not "How do we know that this is dinosaur dung?" Instead, it is "How did it fossilize?" This inquiry is understandable, considering how the closest encounters most urban dwellers have with dung either involves a brief experience in the morning, picking up or stepping on dog doo, or cleaning a kitty-litter box. In contrast, people from rural communities, and especially those who live on farms, cannot avoid feces, as livestock and all other animals leave "land mines" often and everywhere.

First of all, preservation was really helped if these feces already had minerals in them. This means carnivorous dinosaur scat had a better chance of preserving than that coming from insectivores or herbivores because meat eaters were more likely to ingest bone, and bone is composed of apatite. Second, anaerobic bacteria in the feces could have assisted in preserving it, in which their metabolic processes caused chemical reactions that made more minerals precipitate, and do so rapidly (geologically speaking). In a few instances, this bacterially mediated precipitation replaced undigested muscles and other soft tissues, leaving ghostly mimics of these body parts in a dinosaur coprolite. And third, rapid burial, such as from a nearby river flood, would have prevented fresh droppings from getting eaten, poked, prodded, sniffed, trampled, washed away, or otherwise damaged. Under the right geochemical conditions, mineralization also could have accelerated once the feces were buried, as more anaerobic bacteria would have joined the mineralization party.

However, a dinosaur coprolite turned to stone is not necessarily the end of its journey, in which it waits patiently for a well-trained paleontologist to recognize it for its true fecal nature. Coprolites are among the few dinosaur trace fossils—such as toothmarks in bones, and gastroliths—that could have been transported before and after burial, or reburied. Before burial, a dinosaur turd could have fallen down a hill slope, had part of it rolled by a dung beetle, or carried some distance by flowing water. After burial, it could have been exhumed by wind, streams, tides, or waves, and moved to a new place before getting buried again, or not—in which case it might have weathered, eroded, and vanished from the fossil record millions of years before human consciousness. This also means a dinosaur coprolite, like dinosaur bones, feasibly could be reburied in geologically younger sediments. In short, where you find a dinosaur coprolite in the field should never be assumed as representing the same place or time where that dinosaur took a dump.

Given all of these special conditions for preserving dinosaur coprolites, they understandably are among the most precious of dinosaur trace fossils. They are also among the most difficult to attribute to a specific dinosaur. Once a coprolite is identified, paleontologists often limit themselves to saying it belongs to a carnivorous or herbivorous dinosaur based on its contents (bone or plant fragments, respectively). This is where bigger is better, in that large coprolites are easier to connect to dinosaurs big enough to have made them, whereas smaller ones could have been made by a wide range of small dinosaurs. Dinosaur body and trace fossils also help, in which paleontologists can play the much-cherished game of "match the defecator."

Of dinosaur coprolites identified thus far, the best understood ones are attributed to the Late Cretaceous hadrosaur *Maiasaura* of Montana. Some of these coprolites are quite large; although most are broken into smaller pieces, some suggest original volumes of about 7 liters (1.8 gallons). (For perspective, a regulation U.S. basketball is about 8.5 liters.) Other large coprolites are credited to Late Cretaceous tyrannosaurids, such as *Tyrannosaurus rex*; one of these is more than twice the length of a 12-inch sub sandwich. Moreover, coprolites in Late Cretaceous rocks of India have been connected to sauropods.

Nonetheless, the vast majority of probable dinosaur coprolites fall into the aforementioned nebulous categories of "carnivore" or "herbivore." For example, Early Cretaceous coprolites from Belgium have bone fragments in them. Hence, these are allied with carnivorous theropods, but nothing more can be said about them. This vagueness is especially apparent once dinosaur ichnologists admit, with much embarrassment, that we have no idea how to distinguish whether therizinosaurs, ornithomimids, ankylosaurs, nodosaurs, stegosaurs, ceratopsians, pachycephalosaurs, or lots of other dinosaurs made some dinosaur coprolites. We also have not yet discovered a dinosaur coprolite showing any evidence of insect eating, nor of clear omnivory in which, say, a dinosaur had a salad with its steak.

Of these, coprolites showing that some dinosaurs ate insects as a regular part of their diet would qualify as fantastic finds. This seemingly un-dinosaur-like behavior, which is extremely common in modern birds, was proposed for the bizarre theropod *Mononykus* from the Late Cretaceous of Mongolia and a few other theropods. So if insect-bearing coprolites of the right size were found in strata of the right age, environment, and place as *Mononykus* bones, this would be one way to confirm an idea that is now mostly speculative.

Dinosaur Dung, Conifers, Insects, Bacteria, and Snails: A Love Story

Every day, I give thanks to dung beetles. My thanks is offered for what these insects do to keep our planet clean, because otherwise we would be up to our waists in waste. The tight relationship between large herbivores and their diligent insect cleanup crews is easy to witness today: wherever elephants, cows, horses, or other plant-eating mammals loosen their bowels, dung beetles are not far behind. Why are these beetles and other insects, such as dung flies, attracted to feces? Because it is irresistible as baby food. Some beetles roll balls of this nutritious stuff as take-out, which they push into burrows, lay eggs on them, and seal off the burrow. Other beetles burrow below these patties, or into a patty itself, and lay eggs there so that the beetle larvae are surrounded by food when they hatch, which they can then devour. (It's dinner *and* a nursery!)

Amazingly, dung beetles have been performing this essential ecological service virtually unchanged since at least the Late Cretaceous

Period, and we know this because of dinosaur coprolites. Thanks to the careful and insightful collaboration of paleontologist Karen Chin and entomologist Bruce Gill, they convincingly showed how the Late Cretaceous hadrosaur *Maiasaura*, dung beetles, and conifer trees interacted with one another as part of a food web about 75 *mya*.

As is often the case in ichnology, body parts had little to do with this discovery, which Chin and Gill documented in 1996. Plenty of dinosaur bones and eggs were in the same area, as this was the same place near Choteau, Montana where *Maiasaura* and *Troodon* nested, preserved in the Two Medicine Formation. Yet as of this writing, not one body fossil of a dung beetle—legs, wings, abdomens, antennae, or anything else—has been recovered from the rocks of that area. The only body fossils involved in this research came from the conifers, which were represented as blackened bits and pieces in calcite-cemented coprolites.

As mentioned before, some of these coprolites were big, spanning about 34 cm (13 in) wide, although others were mere cobbles. They were also plentiful, showing up in sixteen spots within a square kilometer. Such a coprolitic concentration might imply that the area was a Cretaceous latrine. However, a more likely scenario is that dinosaur feces were nearly everywhere on dry land then, but a few places—like river floodplains—were better at burying them rapidly, which aided in their fossilization.

To have such fine fossil feces in the same field area as dinosaur eggs, babies, adults, and nests was a very lucky find and thrilling enough in itself. Yet when Chin and others also realized these coprolites had burrows in them, their paleontological importance skyrocketed: a dreamy ichnological two-for-one deal. Some of the burrows were open, but in others the insects had actively filled them, having packed a mixture of sediment and dung behind them and leaving distinctly visual "plugs." The burrows also varied considerably in size, from about a millimeter to 3 cm (1.2 in) wide, all of which were insect-sized and with circular outlines. This variation implied that more than one species of insect made them.

Which insects made these burrows? First of all, they had to be ones that loved tunneling into dinosaur manure. This narrowed down likely candidates to two major groups, dung flies and dung beetles. Dung flies are relatively small and normally just lay their eggs on feces; their larvae then hatch on this food supply and start chowing down, which they continue doing until they pupate. In their life cycles, dung flies do not dig wide and lengthy burrows into the dung, let alone backfill them. But dung beetles do. Accordingly, Chin and Gill focused on these insects as the most likely suspects for these trace fossils.

Direct observations of many modern dung-beetle species and their traces served as guides for figuring out how these Cretaceous beetles' lives depended on dinosaur waste. Dung beetles today employ three different strategies in handling feces: tunneling, dwelling, or rolling. Tunnelers burrow into and below a patty, making and storing a brooding chamber with dung and eggs. Dwellers make themselves at home in the patty itself, digging out brooding chambers so that the larvae emerge in the dung-beetle equivalent of a candy store. Rollers scrape dung off the surface of a pile, shape it into a big ball, and leave the neighborhood with their prizes, evoking Sisyphus as they roll dung balls larger than themselves. They later stuff these dung balls in burrows dug elsewhere. Given the three choices, Chin and Gill figured these Cretaceous burrows were from tunnelers, which had burrowed into the dinosaur feces while it was still gooey, gathered some of this organic goodness, and placed it into nearby burrows.

The most exciting conclusion drawn from this discovery was how *Maiasaura* interacted with and affected plants and insects in its surroundings, which in turn provided a sketch of how a Late Cretaceous ecosystem might have functioned, and with a dinosaur as a possible keystone species. As a large herbivore, *Maiasaura* may have had an impact comparable to elephants in savannah ecosystems today, in which dung beetles played an important role in the flux and flow of elements consumed by such herbivores.

However, another mystery about the Two Medicine coprolites was how the pieces of conifer wood had become so blackened. The answer came from within, as in fossil bacteria that originally lived in *Maiasaura* guts. In a paper published in 2001, geochemist Thomas Hollocher, Karen Chin, and two other colleagues detected both abundant body fossils and chemical signatures of anaerobic bacteria in the coprolites. These bacteria invaded vascular tissues in the wood and left distinctive black organic residue called *kerogen*, the same mix of organic compounds in oil shales. The simplest explanation for how these bacteria got into the plant tissues is that they were in the dinosaurs' intestinal tracts. This made sense, as any modern herbivores likewise have gut microflora that aid in breaking down cellulose and other compounds in consumed plants.

This discovery of bacteria that lived inside a dinosaur was important enough. But the bacteria also did paleontologists a 75-million-year-old favor by helping to fossilize the coprolites. Once these researchers examined thin sections of the coprolites under microscopes, they realized that the calcite in the coprolites was probably precipitated in two stages: inside the vascular tissues of the fragments, then in the areas between the fragments, including the dung-beetle burrows. They proposed that bacteria could have initiated this precipitation, starting with live bacterial colonies in the original feces hardening these droppings. Once these proto-coprolites were buried, calcification would have continued, turning what was originally dark, mushy, and smelly into just dark and rocky.

Yet the story of these coprolites does not end with these two studies. As often happens in paleontology and other sciences, this research raised more questions. For instance, as Chin looked more closely at thin sections of the fossilized wood, she realized that something was rotten in the Cretaceous. The hadrosaurs had not been masochistically masticating hard, fresh, living conifers. Instead, they went for long-dead and already-decayed wood. The fossilized wood lacked lignin, a connective tissue that holds wood fibers together. Without this "glue," wood falls apart. In modern forests, fungi aid in this disintegration, which rots the wood throughout. Then wood-boring insects—such as termites, beetles, and ants—break it down further. Although the fossil wood was too ground-up to tell whether insects had bored into it, Chin found some evidence of fungal damage in the wood fragments.

So Chin asked the best question of all: Why? As in, why would a dinosaur eat decayed wood? The nutritional value of the wood fiber itself would have been negligible, hardly worth the effort involved in chewing and digesting it. So there had to be something more behind this behavior than just exercising jaw muscles. The fungi on and in the wood must have provided some sustenance, but probably not enough to keep a hadrosaur going. Also recall that *Maiasaura* is the "good mother" dinosaur, with a scientifically earned reputation for its child-raising skills. Think about the disappointment (not to mention hunger) baby dinosaurs would have felt if their parents simply brought them degraded wood to eat.

This is when Chin thought about both woodpeckers and vomiting. The first part of her reasoning—woodpeckers—was prompted by how these birds feed. As most people know, woodpeckers drill into dead wood not to eat it but to gain access to yummy and protein-rich larvae of wood-boring insects living just under wood surfaces. Woodpecker parents also do this for their chicks, flying back to nest cavities with insect treats for their offspring before drumming up their own meals. So perhaps these hadrosaurs were also breaking up wood to get insects, and carrying these back to their nests.

However, this same strategy would not have worked very well for a 6-to-7-ton *Maiasaura* trying to feed more than a dozen ravenous hatchlings, especially as their appetites grew with them. The image of a hadrosaur mother or father wearily carrying one beetle grub at a time in its mouth to a nest, dropping it into one of many competing maws, then going back for another, is too absurd to consider. It also did not explain why their coprolites contained wood, meaning the hadrosaurs didn't just break wood, but swallowed it. So one solution that neatly explained such wholesale consumption of woody tissues is that these dinosaurs were eating large volumes of this decomposing wood for the insects (and perhaps some fungi). These caring dinosaur parents then transported these MREs (meals-ready-to-eat) in their stomachs back to their nests, where they obligingly regurgitated them into the waiting mouths of their hatchlings. Admittedly, this is a difficult hypothesis to test more directly, unless paleontologists some day find juvenile *Maiasaura* skeletons with enterolites matching the content of the adult's coprolites.

What more could be learned from these coprolites, the gifts that kept on giving? It turned out that Cretaceous dung beetles were not the only ones taking advantage of bountiful supplies of nutritious dinosaur dung: snails got into the act, too. I remember these gastropods surprising me in 2009 while on a field trip with about thirty other paleontologists, led by Dave Varricchio, Frankie Jackson, and Jack Horner. Most participants would have admitted that the *Maiasaura* and *Troodon* nest sites were the main draw of the field trip for them. Yet I was equally excited to know that we would also be strolling through the same area with the hadrosaur coprolites I had seen nine years before when studying fossil insect cocoons and burrows there. Would I notice anything different about the coprolites this time?

Well, yes. The coprolites, just like before, were easy to spot in the field area, sticking out on the eroded surface of the Two Medicine Formation as dark, baseball- to soccer-ball-sized lumps. Although some of us had seen them before, this did not dim our enthusiasm and we all stopped to marvel at how these formerly squishy vestiges had survived long enough for us to witness them, at that moment, now hard as rock. Sure enough, the finely broken and blackened wood chips of Cretaceous conifers were held together by whitish calcite cement. A few even had circular outlines or lengthwise sections of burrows, the right size to have belonged to dung beetles. We were in awe. Still, as is typical with coprolites, bathroom humor quickly overtook lofty ideals and higher cognitive functions. Within minutes, we regressed to photographing one another squatting above these 75-million-year-old trace fossils and giggling at the resulting images in our digital camera viewscreens.

Suddenly, a moment of seriousness intruded when I picked up one coprolite—probably to pose in another picture with it—and noticed a snail's beautifully coiled shell embedded in its matrix. Puzzled, I turned the sample all around and scanned the surface for more such fossils, automatically practicing a basic tenet of science: testing whether or not the initial observations could be repeated. In this instance they were, as I found at least three snails in this coprolite. Excited, I mentioned this to whoever was closest to me, who fortunately shared my interest in seeing something new, too. With an independent observer who had been cued on the same search image, we quickly found more coprolites containing fossil snails sprinkled between the wood and burrows.

What explained this seemingly odd association with snails and dinosaur poop? My first thought, said jokingly in the field with others, was that the snails were happily grazing on fungi growing on a decaying log. But then a hadrosaur came along, took a big bite of the wood, chewed, and swallowed it, snails and all.

This idea, however amusing, was wrong. We did not know it at the time, but Karen Chin had already finished studying these snails and surmised how they got into feces in the first place. First of all, none of these gastropods could have survived a trip down a dinosaur's alimentary canal. If any had been gulped down as inadvertent escargot, their thin calcareous shells, once bathed in the low-pH stomach acids of a hadrosaur proventriculus, would have dissolved instantly. As a result, Chin concluded these mollusks, which were in 40% (6 of 15) coprolites she studied, were either eating feces or whatever might have been growing on the feces. In her paper, published later in 2009, she identified at least seven species of snails, four terrestrial and three aquatic. These ecologically distinct snails painted a complicated post-pooping history for the coprolites before they were buried and fossilized. Some no doubt had been dropped on dry land, where terrestrial snails happily grazed on them. However, some of these land nuggets might have been covered later by nearby floodwaters, or did high dives and made big splashes when exiting their former hosts. Aquatic snails then would have been attracted to them before their final burial and fossilization.

These Two Medicine Formation coprolites thus demonstrated the potential for such trace fossils as a powerful means for understanding how one species of dinosaur—*Maiasaura peeblesorum*—could relate to its inner and outer ecosystems. More than most other fossils from their time, they expanded our consciousness of a Late Cretaceous web of life: gut bacteria, conifers, beetles, and snails, all connected through a dinosaur.

Should I Stay or Should I Go: Dinosaur Gut Parasites

Speaking of ecosystems, individual dinosaurs harbored their own diverse microflora and microfauna, using their bodies as places to live, eat, and reproduce in what is sometimes called a *microbiome.* As most people know, everyone, humans and animals alike, carry—on them or inside them—trillions of bacteria, fungi, skin mites, and occasional harmful (pathogenic) microbial or animal parasites. Animals may consist of lice, ticks, fleas, worms, leeches, or whatever else feels like getting a free meal without that person or other animal host noticing it. Not all of this flora and fauna are necessarily harmful to the host, though; for example, some gut bacteria help to produce vitamins K and B12, and about 70% of all bacteria are harmless to humans. Still, parasites seem to get more attention than beneficial members of this microbiome because of the genuine harm they cause, whether through disease transmission, psychological effects, or impurifying our precious bodily fluids.

Parasites fall into two broad categories: *endoparasites* (living inside), such as tapeworms, and *ectoparasites* (living outside), such as ticks. Despite all we know about parasites and how they cycle through multiple animals today, we didn't know for sure whether dinosaurs had these as parts of their microbiomes. Instead, parasites were just assumed for dinosaurs, especially massive sauropods, which must have represented "The Promised Land" for many parasitic Mesozoic species. Indeed, the main premise of *Jurassic Park* (the book and movies) was how mosquitoes and other blood-sucking insects (which are ectoparasites) had ingested dinosaur DNA in their meals.

Although some insects or other Mesozoic arthropods could have parasitized dinosaurs, we had no actual evidence of their having hosted endoparasites. So it was time to look at modern analogs for guidance. Knowing that most modern endoparasites live at least part of their life cycles in animals' guts, we also know that evidence of these parasites—live or dead bodies, eggs, and dormant stages (*cysts*)—is often included in host animals' feces. For that reason, dinosaur coprolites might likewise hold such signs.

Indeed, some do. In 2006, two paleontologists, George Poinar and Art Boucot, documented three types of endoparasites in a dinosaur coprolite: protozoans, which are one-celled amoeba-like organisms; trematodes, a group of flatworms that includes flukes; and nematodes, which consist of "round worms" and horsehair worms. The coprolites were from an Early Cretaceous deposit in Belgium that also contained many skeletons of the ornithopod dinosaur *Iguanodon*, which were discovered in the 1870s. This did not mean, however, that the coprolites were dropped by *Iguanodon*, as some had bone fragments, pointing toward meat-eating theropods as the culprits. The coprolites, first identified in 1903, ranged from 2 to 5 cm (1–2 in) wide and 11 to 13 cm (4.3–5 in) long, or human-sized. Poinar and Boucot took one of them, broke it up, subjected it to strong acids, and centrifuged it: the most action its contents had received in about 125 million years. What survived this preparatory process were protozoan cysts, a trematode egg, and nematode eggs. Incredibly, one of the nematode eggs even held the coiled body of a baby nematode, which had just missed hatching after exiting a dinosaur's body in its feces.

Although all of these fossils were new to science, they closely resembled modern pathogenic species seen in amphibians, reptiles, and birds. So thanks to this dinosaur coprolite and its 125-million-year-old secrets, biologists interested in the evolution of endoparasites have a minimum time for when such perfidious behaviors began: at least in the Early Cretaceous, but more likely extending back into the Jurassic.

Carnivorous Movements

Everything we know about nearly all large theropods points toward their carnivory: teeth, jaws, claws, toothmarks in bones, bones in abdominal cavities, and with some of those bones etched by stomach acids, to name just a few items. Still, nothing tells us about how much meat could go through a theropod's system on a given day like a gigantic tyrannosaurid turd. Fortunately, paleontologists have not just one but two such massive coprolites. Both are credited to tyrannosaurs because of the following: co-occurrence with known tyrannosaur bones; age (Late Cretaceous); environment (river floodplains, which tyrannosaurs may have preferred for hunting); size (extra-long baguettes, anyone?); broken pieces of bone; and, most unexpected of all, fossilized muscle tissue in one coprolite. Both were found in Canada, one in Saskatchewan and the other in Alberta, each of which have body fossils of tyrannosaurids, such as *Albertosaurus* and *Tyrannosaurus*.

The first known coprolite attributed to a tyrannosaurid came out of fluvial (river) deposits in the Frenchman Formation of Saskatchewan. This formation dates from the latest part of the Cretaceous Period, about 66 *mya*, just before an extraterrestrial object delivered some bad news to all non-avian dinosaurs. The best way to describe this coprolite requires applying an overused word, but justified in this instance: awesome. When found in the field, it was partly eroded, but after having been recovered and put back together, it measured 44 cm (17 in) long, 13 to 16 cm (5–6 in) wide, and had a minimum volume of 2.4 liters (0.6 gallons). Almost half of it was composed of finely ground bones which were from a young ornithischian dinosaur; this bone meal was held together by apatite.

The paleontologists who were lucky enough to study it—Karen Chin and three others—had little doubt it was a dinosaur trace fossil and related to carnivory. They considered the possibility that it might be a regurgitalite, an enormous cough pellet that became fossilized. However, it was held together so well by apatite and included such tiny bone fragments (some of which were sand-sized), a fecal origin made much more sense. (A cough pellet, in contrast, would have held more complete bones.) Considering its great size and how both its geologic age and location overlapped with that of *Tyrannosaurus rex*, this coprolite was most likely made by that massive theropod. In their 1998 paper reporting their find, the paleontologists crowned it as a "king-sized coprolite."

Once this coprolite was linked with *T. rex*, it revealed much about the carnivorous behavior of the world's most famous dinosaur—or at least one individual of that species—and some of it was surprising. For one, the bone was broken into tiny pieces, which implied that this tyrannosaur either chewed its food thoroughly or it nipped bone while feeding. Yet tyrannosaur teeth and jaws were more suited for slicing and crushing, not grinding for hours or nibbling delicately. Furthermore, toothmarks in *Triceratops* and other dinosaur bones show that *T. rex* punctured bone. Thus one explanation for so many small bone fragments is that it did come from bone chipped with each puncture, which went along for the ride with any consumed flesh. Another possibility is that these bone bits came from this tyrannosaur scraping meat and sinew off long bones with a sideways motion of its head. In this speculative scenario, numerous denticles on its serrated teeth would have taken off minute pieces of bone.

Another unexpected point raised by all this bone was how it was still there. Remember how big modern reptilian predators—such as crocodilians—digest bones so well that almost none of it shows up in their scat? They do this by retaining food in their guts for a long time. In contrast, undigested bone suggests that the meal raced through the dinosaur and was not given enough time to be completely absorbed. Did this tyrannosaur have the runs, caused by endoparasites? Probably not, but a short residence time for food also could have made room more quickly for seconds, thirds, fourths, and dessert.

Despite the impressive size of this *T. rex* coprolite, it was surpassed by one from the Late Cretaceous Dinosaur Park Formation of Alberta, reported in 2003. Studied again by Karen Chin and five other paleontologists, it was 64 cm (25 in) long and as much as 17 cm (7 in) wide. Like the coprolite from Saskatchewan, it contained many pieces of bone and was cemented with apatite. Nevertheless, it differed in two important ways. First, *Tyrannosaurus rex* could not have made it because this dinosaur was not alive until about 10 million years after this fecal mass peeked out of a cloaca. The Dinosaur Park Formation did, however, have the tyrannosaurids *Aublysodon*, *Daspletosaurus*, and *Gorgosaurus* living there, so one of these theropods could have dropped it.

The second way it differed was even more extraordinary. In this coprolite was recognizable muscle tissue, some of it preserved in three dimensions. In their analysis, the paleontologists were astonished to see: striated cell-like structures revealed in thin sections; bundles closely resembling muscle cells and connective tissues in SEM photographs; and high concentrations of carbon in and around these structures. All of these observations meant the paleontologists were looking at fossilized meat, in which minerals had faithfully mimicked the original muscle cells.

Never before had a tyrannosaurid meal been preserved in such intimate detail. This 3-D snapshot of its former excrement required very unusual conditions, such as brief gut-residence time (food rushing through, resulting in incomplete digestion), dumping it as a cohesive mass on a river floodplain just before it flooded, rapid burial, and anaerobic bacteria in the feces that helped to precipitate minerals in and around the muscle cells. Based on the distinctive broken bits of bone throughout the coprolite, these paleontologists figured these all came from the same animal, such as a pachycephalosaur.

Just for some biological trivia, these two tyrannosaurid coprolites also tell us something about the soft-part anatomy of their makers, namely their cloacas. Based on widths of the coprolites, their minimum cloacal diameters must have stretched to 13 to 17 cm (5–7 in) to allow these fecal masses to bid adieu to their gastrointestinal tracts. However, the actual diameter was probably greater, considering that the original feces lost volume with dehydration before burial. Too much information? No, it's more like the more you know, the more you wonder.

Just how long has such carnivory been a part of dinosaurs? According to coprolites, as long as there were dinosaurs. At the opposite end of the geologic-age spectrum from tyrannosaurid coprolites are the oldest probable dinosaur coprolites. These are about 228 *mya*, coming from Late Triassic rocks of Argentina and the same strata bearing bones of two of the oldest dinosaurs in the fossil record, *Eoraptor* and *Herrerasaurus*. Reported in 2005 by paleontologist Kurt Hollocher and three colleagues, these coprolites, like the tyrannosaurid ones, had bone fragments and were cemented with apatite. Out of the ten found, two were studied in great detail. Both of these were similarly sized, measuring about 2 cm (0.7 in) wide and 3.1 to 3.7 cm (1.2–1.4 in) long, about the average size of what might be encountered in a dog park.

Like the *Maiasaura* coprolites, these had invertebrate burrows in them, although only a few millimeters wide. Hence they were probably not from dung beetles but perhaps fly larvae or worms. Because *Herrerasaurus* was a carnivorous theropod, and its body parts were the most common of any carnivore in those strata, plus it was about the right size for these coprolites, Hollocher and his cohorts proposed that they came from that dinosaur. Still, other archosaurs may have been responsible for them, too.

These theropod coprolites, from the Late Triassic through the Late Cretaceous, demonstrated how carnivory was a dietary mainstay in this lineage for at least 160 million years. Furthermore, this evolutionary tradition continues today with predatory birds, although fortunately for us and many other animals, none of these theropods are producing feces on the scale of a tyrannosaur.

Passing Grass: The First Grasses and the Dinosaurs That Ate Them

Today we take grasses for granted. They occupy most of the soils in our lawns, parks, pastures, athletic fields, and university quadrangles, but are also ubiquitous in our daily diets. Grasses we eat include corn, wheat, or rice, and others we drink in a fermented form, like barley, oats, or hops. With regard to the latter, paleontology, geology, and many other sciences could not be done without these grasses and their by-products, or at least these sciences would be a lot less fun.

The conventional wisdom about grasses is that they first evolved from non-grassy flowering plants during the Cenozoic Era—in just the past 65 million years—and especially took off in just the past 30 million years or so, first in South America, then in North America. Grazing mammals, including the earliest horses, were always thought to have facilitated the spread of amber waves of grain. Indeed, changes in horse dentition and limbs were likely linked to changes in grassland habitats over time. Dinosaurs, in contrast, had absolutely no place whatsoever in this comforting story of co-evolution between grasses and mammals. Having all vanished soon after a big rock arrived 65 *mya*, non-avian dinosaurs only contributed their recycled elements to the soils feeding grasses, with their bones ground to dust under the hooves of grass-grazing ungulates.

Fortunately, thanks to some Late Cretaceous sauropod coprolites from India, we now know this story needs updating. Two studies, done in 2003 and 2005, showed that sauropods—specifically, titanosaurs—were eating different plants than expected, and the latter study revealed that these plants included grasses. These results surprised nearly everyone for two reasons: few people suspected that grasses had evolved during the Mesozoic, and almost no one expected to find fossils of them in dinosaur coprolites.

The first detailed report on these coprolites was by Prosenjit Ghosh and five of his colleagues in 2003. How did they know these coprolites came from sauropods? First of all, these grayish masses in the Lameta Formation of central India were actually first identified as dinosaur coprolites in 1939. Their makers were then deduced on their occurring in the same strata as titanosaur bones. Moreover, although they were not very large by titanosaur standards—the biggest were only about 10 cm (4 in) wide—they were big enough not to have come from any other herbivorous animal in those strata. The researchers also noted how the coprolites had dried out and cracked slightly before burial and fossilization, meaning they were originally deposited on dry land.

Ghosh and his colleagues detected vascular plant remains in the coprolites, as well as fungal spores, algae, and plenty of bacteria. However, their most significant finds came from a chemical analysis of the coprolites, which helped them to narrow down which types of plants the titanosaurs ate. To figure this out, these scientists calculated a couple of stable-isotope ratios in the coprolites—for carbon and nitrogen—and compared these ratios to those in the feces of modern animals, such as deer, camels, buffalo, and big cats (leopards and tigers).

Just to back up with some basic definitions: isotopes are variations of the same element, but with different atomic weights. For example, the isotopes of carbon (C) are ^{12}C, ^{13}C, and ^{14}C. Of those isotopes, ^{12}C and ^{13}C are stable, but ^{14}C is not, as it undergoes radioactive decay and changes to another element. Nitrogen (N) isotopes are ^{14}N and ^{15}N, and both of these are stable. Then what are stable-isotope ratios? In this instance, these scientists calculated $^{12}C/^{13}C$ and $^{14}N/^{15}N$. Plants, through different means of photosynthesis, take in carbon and nitrogen isotopes in distinct ways, which is then reflected by their stable-isotope ratios. For instance, C_3 and C_4 plants—so called because of the number of certain carbon compounds they form—have dissimilar ratios, because C_4 plants absorb ^{13}C more easily than C_3 plants. In short, these ratios are chemical signatures that, under ideal conditions, persist in fossil plants, even if they went through the gut of a dinosaur and were buried for about 70 million years.

As it turned out, the carbon-isotope ratios showed that the titanosaurs ate C_3 plants, nearly matching a value for birds, and they were much closer to ratios of modern herbivores—goats, camels, and buffalo—than carnivores. Additionally, C_3 plants make up nearly 90% of all modern vegetation, and include grasses. However, this did not mean these titanosaurs were eating grasses specifically; plant pieces in the coprolites matched those of conifers and other non-flowering plants. But these dinosaurs were certainly consuming plants with similar modes for photosynthesizing, and the isotopic signatures in their coprolites were close to those of modern plant-eating mammals. Furthermore, the nitrogen stable-isotope ratio matched that of animals that do not use fermentation in their guts (usually aided by bacteria) to help digest their food. So again, this was more like birds and less like mammals that use bacteria in their foreguts (small intestines) or hindguts (large intestines) to ferment their food.

These geochemical clues gleaned from sauropod coprolites certainly gave paleontologists better insights on what these dinosaurs ate and how they digested their food. Yet, as much as ichnologists hate to admit it, they sometimes need a few body fossils in their trace fossils to better interpret the latter. So in 2005, Vandana Prasad and four of his colleagues hit the jackpot, finding grass phytoliths in the same Late Cretaceous coprolites from India studied by Ghosh and others.

Remember phytoliths? These are microscopic bits of silica precipitated by plants and residing in their tissues, which also left microwear on dinosaur teeth when they ate these plants. In many plants, phytoliths can be "fingerprints" for identifying plant clades, and sometimes a specific species. Fortunately for dinosaur ichnologists, phytoliths are also resistant to acids, so these can pass relatively unscathed through an animal's gastrointestinal tract. This evolutionary innovation by plants thus helped earmark a minimum time for when grasses showed up in Mesozoic ecosystems (65–70 *mya*) as well as when dinosaurs started grazing or browsing on them. Not mammals, but dinosaurs.

This discovery of grass phytoliths in sauropod coprolites thus fulfilled the maxim of "You never know until you look," especially when amended by saying "You never know until you look in dinosaur feces," or other trace fossils associated with dinosaur digestion. From enterolites, to regurgitalites, to cololites, to coprolites, which can be connected to gut bacteria, plants, snails, insects, mammals, birds, and other dinosaurs, these trace fossils tell us inside stories and intimate details of dinosaur lives.

CHAPTER 9

The Great Cretaceous Walk

Because of the sparse and uneven record of dinosaurs in Australia, their fossil footprints are more valuable here than anywhere else on Earth.

—Thomas H. Rich and Patricia Vickers-Rich, *A Century of Australian Dinosaurs* (2003)

Looking for Traces in All the Wrong Places

Dinosaur tracks are hard to find. This humbling realization struck me during the third week of a month-long excursion in May–June 2010, while doing field work along the craggy coast of Victoria, Australia. Just the year before, paleontologist Tom Rich of Museum Victoria invited me to look for trace fossils made by dinosaurs and other Cretaceous animals that might be preserved in the rocks of Victoria. Yet as was often the case with looking for fossils of any kind, there was no guarantee of success. During our time in the field, he and I had already searched more than a hundred kilometers of coastal cliffs and marine platforms east of our present location, with only a few fossil finds in all of its vastness. Now we were working our way through sites to the west, and so far nothing else had been found worth writing home about.

So a bit of stubbornness underpinned our visit to Milanesia Beach, located in southwestern Victoria, Australia. Milanesia Beach is about a three-hour drive from the big city of Melbourne, but like many places in Australia it feels isolated, far away in space and time from a world where people sip lattes, use mobile devices, or drive cars, sometimes all simultaneously. A testament to its relative inaccessibility is that, despite its having a beautiful beach framed by dramatic sea cliffs, the people who normally see it are not swimmers, surfers, or sunbathers, but hikers. Even then, it is only a brief waypoint for those people as they otherwise enjoy the gorgeous scenery of one of the most famous walking routes in Australia, The Great Ocean Walk. In fact, this trail inspired me to dub, tongue-in-cheek, my excursion along the Victoria coast as "The Great Cretaceous Walk."

On Monday, June 14, 2010, we were not visiting Milanesia Beach to hike or enjoy the bucolic countryside. Instead, Tom Rich, local guide Greg Denney from the nearby town of Apollo Bay, and I were there to look for fossils from the Early Cretaceous Period, at about 105 million years ago. The landscape was certainly very different back then. It was a time when Australia was close to the South Pole, and dinosaurs presumably walked across broad floodplains of rivers that coursed through its circumpolar valleys. Since then, Australia had drifted north and now was just below the equator. Of course, many animals and plants native to the place had gone extinct, whereas some lineages evolved into the distinctive life of Australia today, such as its rich diversity of marsupials found nowhere else in the world. More than a hundred million years can really change a place.

Milanesia Beach, though, was a new place for me, and it might as well have been new for Tom, as he had not visited it in more than twenty years. His main reason for looking at its rocks was for fossil bones, especially those of dinosaurs or small mammals. As an ichnologist, I was there to look for trace fossils. I knew most of these vestiges of life would consist of burrows and trails made by invertebrate animals like insects, crustaceans, or worms. But if we were really lucky, these rocks might also reveal trace fossils of vertebrates, such as the burrows or tracks of mammals, dinosaurs, or other backboned animals.

Unfortunately, during the preceding three weeks of field work I had only found a few invertebrate trace fossils (burrows) and no vertebrate trace fossils: no tracks, nests, burrows, toothmarks, gastroliths, coprolites, or anything else that might tell of a former vertebrate presence. Similarly, Tom had not yet found a single scrap of bone. We seemed more than due for a big break.

More Bones than Traces

Throughout this book, I've lauded the great advantages of dinosaur trace fossils over bones for all of the insights they give us about dinosaur behavior. Other than telling us about behavior, and especially how dinosaurs interacted with one another and their environments, one of the best benefits of dinosaur trace fossils is their overall great abundance compared to bones. For the many reasons explained before, in Mesozoic rocks of most places in the world you are much more likely to find a dinosaur trace fossil than a dinosaur bone.

Yet there are a few regions that have Mesozoic rocks of the right ages and environments for dinosaurs where dinosaur bones are more common than their trace fossils. One such place is Victoria, Australia, where I accidentally began some research projects in 2006. Up until then and there, I had only dabbled with dinosaur trace fossils in the U.S., mostly through having seen many dinosaur tracks and other trace fossils in the western U.S. As mentioned previously, I was also helping several colleagues with the description of a burrow made by the Cretaceous ornithopod *Oryctodromeus*. In terms of writing about dinosaur trace fossils, I had done a chapter about them in two editions of a dinosaur textbook. Other than this, I was largely ignorant of dinosaur trace fossils. Most of my training, research, and teaching dealt with other traces, both modern and fossil, made by a wide variety of animals, invertebrate and vertebrate. Dinosaur trace fossils, such as their tracks, nests, gastroliths, toothmarks, and coprolites, were fascinating but did not occupy my every waking thought.

Ironically, then, my interest in dinosaur trace fossils was kindled in an area of the world where they are scarcer than dinosaur teeth, starting with the first day I laid eyes on Cretaceous rocks of Victoria. Although I knew about Lark Quarry, the so-called "dinosaur stampede" (or "dinosaur swim meet") site far to the north in Queensland, and a few other dinosaur tracksites in the northern and western parts of Australia, I knew next to nothing about any dinosaur tracks in the southern part of the continent. Later I found out this couldn't just be attributed to laziness or disinterest on the part of other paleontologists; as of 2006, there really were very few dinosaur tracks or other trace fossils known from there.

Just for comparison, let's take a look at other continents and their dinosaur trace fossils. In North America, the U.S., Canada, and Mexico have tens of thousands of dinosaur tracks and plenty of other trace fossils, such as nests, gastroliths, toothmarks, and coprolites. South America? The same. Africa, Europe, Asia? Ditto. But as of my writing this, no dinosaur nests have been reported from Australia. Not even body fossils associated with dinosaur nests—such as eggshell fragments and embryonic bones—are known from there either, let alone egg clutches. No research has been done on Australian dinosaur gastroliths. In all of Australia, not one study has been done on dinosaur toothmarks. Not a single dinosaur coprolite has been interpreted. It's almost as if the dinosaurs in Australia didn't give a crap.

So what caused Australia to become so poor in dinosaur trace fossils, ranking only above Antarctica in quality and quantity? Dinosaurs certainly lived there, as evidenced by theropod and ornithopod bones and teeth recovered from Early Cretaceous strata (120–105 *mya*) in Victoria, and theropod, ornithopod, and sauropod bones in central Queensland. The latter area is now bursting with Early Cretaceous dinosaur skeletal material, reminding paleontologists of the beginning of the "Great Dinosaur Bone Rush" during the late 19th century in the western U.S. Still, why are their tracks, nests, gastroliths, toothmarks, and coprolites so rare? Or is it just an apparent scarcity, a combination of wrong conditions for preserving these trace fossils in most of the Mesozoic rocks there as well as not knowing what to look for?

The answer is probably complicated. Regardless, the best way to reach for it was to get out, walk around, and look for trace fossils in the rocks there.

Back to the Cretaceous

It was a fine day on the Victoria coast, started by crisp morning temperatures, a mild breeze, and overcast conditions, but with no signs of the antipodal winter thunderstorms—accompanied by rain, gusting winds, and powerful waves—that had kept us off the coastal outcrops for much of the previous week. Earlier that morning, Tom Rich and I drove from where we were staying in Apollo Bay, picked up Greg Denney at his home, and parked our vehicle near a trailhead, about two kilometers (1.2 miles) uphill from the beach. The walk down to the outcrops, punctuated by muddy, slippery patches, promised a vertically challenging slog later in the day, just when we would be most spent from our explorations below.

Greg, who joined us to scout rocks of the Eumeralla Formation composing the dramatic cliffs near Apollo Bay, had a longstanding relationship with Tom as a field assistant and friend. He also had the good fortune of growing up next to one of the most famous dinosaur sites in Australia: Dinosaur Cove. In the 1980s, Greg and his father, David Denney, assisted Tom, Pat Vickers-Rich, and a crew of volunteers with some of the most technically difficult conditions any dinosaur dig site should ever have to endure, detailed by Tom and Pat in their book *Dinosaurs of Darkness*, published in 2000. In our more recent ventures, Greg had quickly proved a valuable asset in our field endeavors, suggesting roads and parking spots for our field vehicle and advising on safe access points to outcrops.

Greg had also become my ichnological apprentice during our previous week together in the field and quickly became quite good at spotting small fossil invertebrate burrows in Cretaceous outcrops. I would have liked to credit his rapid success to my extraordinary teaching abilities, but instead chalked it up to his spending much of his life outdoors. After all, he had already trained his eyes to pick up small details in his natural surroundings, such as wallaby tracks, echidna dig marks, and kangaroo feces. These skills had not been sullied by the constant distractions of "big-city life," a challenge I face every day when not in the field and living at home in the metropolitan area of Atlanta, Georgia. I envied him these opportunities, available to him every day, and in such gorgeous places.

The thirty-minute walk to Milanesia Beach promised by the trailhead sign next to where we parked was surprisingly accurate, considering how carefully we placed our feet while walking down the steep, winding track. Toward the bottom of the trail, we also had to cross a small stream teeming with freshwater leeches, just like those I had encountered at Knowledge Creek a few years before. At the end of the trail, we were greeted by an upside-down sign bearing the usual admonitions about all of the potential forms of mayhem that awaited us if we proceeded. These signs seem to be everywhere in Australia and are meant not just for tourists, but also for anyone who might perform acts of foolishness while celebrating nature.

Once we were on the more level ground of Milanesia Beach, two choices faced us in our fossil explorations: either go to the outcrop on our immediate right—with only beach sand in front of it—or to a more modest outcrop on our left, with our path complicated by numerous blocks of rock that had fallen from cliff faces above. We looked briefly at the closest part of the outcrop to the right, but these rocks seemed too coarse-grained to have many discernible trace fossils. Fine-grained sandstones or siltstones are much better for preserving identifiable invertebrate burrows, vertebrate tracks, and other trace fossils, as opposed to conglomerates. So we chose to go left, a decision also encouraged by high waves already lapping against the outcrop on our right.

While walking parallel to the shoreline, we soon went from a sandy beach to a rocky shore. Some of the blocks of rock we passed were much smaller and more rounded than others, providing indirect clues of their relative time on the beach, in which the surf shaped the smaller and more rounded rocks far longer than bigger ones. In contrast, the larger blocks retained angular corners from their more recent breakage off nearby cliff faces. Normally in a talus field like this, I would just stroll along and not spend much energy looking at each stone for its paleontological value. Nonetheless, I did glance at them, albeit more out of a sense of self-preservation. I wanted to make sure I stepped in all the right places and didn't slip on any slimy algal films and thereby become too physically intimate with these rocks.

While ambling, we stopped occasionally to scan the rocks in the vertical outcrops, as well as larger angular blocks scattered across the upper part of the shore. The several-meter-high vertical exposures of layered shales, sandstones, and conglomerates were at the top of the beach, marking where the sea had eroded these strata. The surf crashed behind us, giving us no choice but to shout at one another as we pointed out anything of geological interest. We also warily watched the sea for any rogue waves that might catch us by surprise. Field work along the Victoria coast is treacherous enough to encourage a healthy caution in its practitioners—supported by the inverted sign at the end of the trail down—and that day was no different.

In retrospect, we were fortunate to have the winter solstice approaching, which meant the sun would begin to set close to 5:00 p.m., a constraint that urged us to use our time judiciously. Sure enough, within less than ten minutes of our arriving, Greg and I started finding trace fossils—invertebrate burrows—in fine-grained sandstones and siltstones exposed in the outcrop. One type of burrow was a stubby vertical cylinder, some of which were U-shaped. Another was a thin, reddish J- or U-shaped structure, also oriented vertically. Each burrow form was abundant in the thin strata.

These little trace fossils invoked an unprecedented excitement within me, as they provided clues to the ancient ecosystems of the area. Both types of burrows were formerly open tubes, filled with sand very soon after they were made. Furthermore, invertebrate burrows often act as sensitive indicators about the former ecology of an area, and these were typical of what you might see in a modern river floodplain. For instance, some aquatic insect larvae dig burrows in sediments under very shallow water or on the surfaces of emergent sand bars, whereas other insects—such as ants, bees, and wasps—can only make nests above water. These trace fossils looked more like aquatic insect burrows to me, probably used for combined feeding and dwelling.

The presence of these burrows alone was scientifically important, and when put in the context of having been formed in a polar environment, they were doubly significant. Insects and other invertebrates in polar environments cannot burrow into frozen sediment. Rather, they wait until late spring or summer to make their domiciles or brooding burrows, after the uppermost layers of sediments have thawed out. Or they wait until new, soft surfaces have been formed by sediment deposited by spring run-off of melt waters. Moreover, the physical sedimentary structures associated with the trace fossils—ripple marks and cross-bedding—also indicated a healthy flow of water. These structures would have more likely formed during a polar spring or summer following snow melts.

Along those lines, I had published a paper in 2009 about physical sedimentary structures—such as ripple marks, mudcracks, and so on—and traces—such as invertebrate burrows and vertebrate tracks—next to the Colville River on the North Slope of Alaska. Hence, the rocks in front of us, when combined with what I had learned from that Alaskan riverbank a few years beforehand, almost acted like a time machine. The Cretaceous rocks of Australia and the modern sediments of Alaska could be compared as polar ecosystems, despite being separated by more than 100 million years and thousands of kilometers.

Yet another justification for my growing elation was that these burrows closely resembled trace fossils I had seen in rocks at another place: Knowledge Creek, just a few kilometers east of us. Knowledge Creek is the place where the only well-documented dinosaur track from the Eumeralla Formation was discovered. In 1980, a little more than thirty years before Tom, Greg, and I stepped foot on Milanesia Beach together, Tom and Pat Vickers-Rich found this track. It was probably made by a small ornithopod dinosaur, which was not surprising to them, seeing that nearly all of the skeletal remains of dinosaurs in Victoria belonged to such dinosaurs.

I had visited Knowledge Creek three times in recent years—2006, 2007, and 2009—but did not find any other definite dinosaur tracks there, only a few vague outlines. Still, I was lucky enough to have discovered possible dinosaur burrows and invertebrate trace fossils there, the latter nearly identical to the ones we were seeing that morning at Milanesia Beach. Similar sedimentary rocks and trace fossils at Knowledge Creek and Milanesia Beach implied that similar environments had produced these rocks. So perhaps the conditions at both places were conducive to dinosaurs walking across floodplain surfaces, allowing their tracks to get preserved well enough that they would be identified some day.

After photographing and noting the locations of these trace fossils, Greg and I continued down the eastern extent of the beach. Tom, on the other hand, had already gone well ahead of us, looking for bones. I think he was glad that Greg was helping with my ichnological investigations, which allowed him to concentrate more on finding dinosaur bones or those of other vertebrates. In his previous scouting of Milanesia Beach with others of his body-fossil-hunting ilk more than twenty years before, they had not found any bones or teeth. As a result, they had since written it off as a place to look for such fossils. But he also knew that during that elapsed time, rock falls and weathered surfaces might have revealed previously hidden fossil bones, new to human eyes.

Happy about the invertebrate burrows and their host rocks, I turned to Greg as we sauntered down the beach and said, casually,

"Now all we have to do is find some of those other things we've been looking for. But I won't say what, because I don't want to jinx it!" Greg grinned and said "Righto, say no more!" Somehow he knew I was talking about dinosaur tracks. In our previous forays, I had mentioned these trace fossils as something we should be looking for at every turn.

Tracing through Recent Time

What was the reason for such vigilance about dinosaur tracks? Well, up until that day, only four dinosaur tracks had been found in all of the Cretaceous rocks of Victoria, a part of Australia just a bit smaller than California. The first track—the one discovered by Tom and Pat at Knowledge Creek—was only about 10 cm (4 in) wide and long. They were doubly fortunate that day: once for finding the track, and twice for having the right tools to collect it. Using a hammer and chisel, they carved it out of the rock, put it in a backpack, and hiked out of the site. Upon their return to Melbourne, they immediately placed the specimen in the Museum Victoria fossil collection. This track later became the template for thousands of reproductions used for science education in Victoria, and photographs of it appeared in many popular articles and books. In other words, it became iconic: For many people, and for more than twenty-five years, this was *the* dinosaur track from the Cretaceous of Victoria.

In 1989, another isolated dinosaur track was found during a fossil hunt by a group led by Tom Rich. They discovered it at a place called Skenes Creek, which was more than 30 km (19 mi) east of Knowledge Creek, but with rocks the same age as those at Knowledge Creek, Dinosaur Cove, and Milanesia Beach. This specimen was also a small ornithopod track and nearly the same as the one from Knowledge Creek. Fortunately, the field crew had a rock saw with them, so they neatly cut the surrounding rock into an easily transportable square slab, and likewise took it to Museum Victoria. Once deposited, it received a catalog number and sat in the museum for the next 21 years. I saw it in its museum drawer in May 2010—only three weeks before our sojourn to Milanesia

Beach—and verified it as the second dinosaur track reported from the Eumeralla Formation. However, it still has not been formally described in the scientific literature, so it remains known to only a few paleontologists and does not get any of the public notice showered on its almost-identical twin from Knowledge Creek.

The other two dinosaur tracks from Victoria were both made by large theropods and are east of Melbourne, at the Dinosaur Dreaming dig site in the geologically older rocks (115–120 million years old) of the Strzelecki Group. I found one in 2006 during my first visit to Dinosaur Dreaming. Sadly, Tom Rich—who was with me at the time—had not yet accepted my ichnological abilities and disregarded it as a dinosaur track. A year later, in 2007, a dig site volunteer, then-student Tyler Lamb, found a nicely defined track of about the same size only about 5 m (16 ft) from the main dig site. Yes, I felt vindicated, but also very happy for Tyler that he'd discovered such an important specimen. Later in 2007, I reported on both tracks in a poster presented at the Society of Vertebrate Paleontology meeting in Austin, Texas. This poster got a fair amount of media attention because it was touted as evidence of a "polar predator" from the Cretaceous, a mixing of theropods and polar dinosaurs that somehow inspired the public imagination.

So there you go: four definite dinosaur tracks from all of the Cretaceous of Victoria after more than a hundred years of paleontological research in those rocks. The two tracks from 105-million-year-old rocks west of Melbourne—Knowledge Creek and Skenes Creek—were attributable to small ornithopod dinosaurs, like *Leaellynasaura*. Large-sized theropods, about the size of a smaller version of *Allosaurus*, likely made the two 115-million-year-old tracks in rocks east of Melbourne. None of these tracks were preceded or succeeded by another track made by the same dinosaur, which meant, unlike many other places in the world, including in Australia, Victoria had no known dinosaur trackways like ones at Lark Quarry in Queensland or Western Australia.

This is where someone in the know about such things might ask, "What about other dinosaur trace fossils?" As covered before, there were supposed dinosaur burrows from Knowledge Creek, and that was it. No dinosaur nests, gastroliths, stomach contents, or coprolites have been interpreted from the Cretaceous rocks of southern Australia. I had heard from a few paleontologists that some of the dinosaur bones from this area had theropod toothmarks on them, but these were apparently rare too. No one had even done a study on tooth microwear. Although there seemed to be a good number of invertebrate trace fossils, I was traveling in uncharted territory for dinosaur trace fossils, an ichnological analog to "The Ghastly Blank," an endearing term applied to the desert interior of Australia.

In contrast, dinosaur bones—most of which had been found by many people working with Tom and Pat during the past thirty years in this part of Australia—indicated a more diverse fauna was there than this apparent lack of trace fossils might suggest. Moreover, these dinosaurs should have at least been leaving tracks on soft sands and muds during the springs and summers in between harsh polar winters. Yet these tracks seemed to be hiding from paleontologists. We deal with gaps in the fossil record all of the time, but this was a big one for the fossil record of polar dinosaurs in Australia.

Back to the Cretaceous Again

All of this paleontological history was running through my mind and was why I tried to temper any anticipation during our excursion along Milanesia Beach. Yet seeing the physical sedimentary structures and small burrows in that outcrop told me we were looking at the former deposits of river floodplains. These environments would have been perfect for preserving dinosaur tracks made during a polar spring or summer. Regardless, I reminded myself to stop entertaining such thoughts and just be a cold, clear-headed, objective scientist: you know, a dismal pessimist.

Just to dissipate these inward distractions, I decided to look intently at more of the little burrows in the outcrop. At this point Greg abandoned me, and I didn't blame him. He walked ahead to join Tom, who was already several hundred meters east of us. Meanwhile, I took more photographs and measurements of the trace fossils we had found twenty minutes before. Once that was done, I moseyed along, scanning both the outcrop and large boulders strewn across the beach. While walking, I still carefully picked where my feet landed, avoiding those nearly invisible slippery algal surfaces on the rocks. I had already fallen a few times during more than a hundred kilometers of walking along the Victoria coast and did not want to add another bruise, bump, or scrape to my three-week-old collection.

To this day I don't know why, but one large rectangular block of rock among dozens along the shoreline compelled me to stop and take a moment to have a second look at it. All I can imagine now is that this sensation stemmed from more than ten years of tracking animals in the sands and muds of the Georgia barrier islands and other places in the world, a collective experience that led to an intuitive recognition of something worth noticing on the periphery of my vision.

I looked to my left, then down at the top surface of the boulder. There was a small three-fold impression, looking vaguely like the middle three fingers of a human hand. It was close enough to touch, so I did. My own three middle digits molded to the indentations, confirming what my eyes had seen but not quite believed.

It was a small dinosaur track.

After a quick inhalation, almost trembling, I dared to look at the rest of the rock, scanning from left to right. More patterns of three came into focus, one after another, each identified faster than the previous one. Within about ten seconds, I realized the bumpy surface was loaded with dinosaur tracks.

One of these footprints—a chicken-sized one only about 7 cm (2.8 in) long—was close to the edge of the slab. With my heart beating faster, I then did something I almost never do with modern tracks, which was backtracking. I shifted my focus behind the track nearest me to see if any similar ones preceded it. Sure enough, there was another of the same size, aligned with the previous one. Hesitating in disbelief, I backtracked one more time. Another track was exactly where it should be, at a distance nearly identical to the space between the other two, although slightly off the line of travel. One, two, three steps in sequence, with a slight rightward turn; it was a preserved motion from more than 100 million years ago, made by a small theropod dinosaur on a river floodplain during a polar summer. It was the first known dinosaur trackway in all of southern Australia, and the first polar-dinosaur trackway from the Southern Hemisphere.

There was no time to celebrate; I needed to get to work. I looked more closely, feeling the rock surface. Then I began sketching what was there and marking locations of the tracks, using graph paper in my geological field notebook to make a scaled drawing that served as a sort of "track map." Unlike taking photographs, drawing forced me to look at the rock and its fossil tracks repeatedly, carefully, and critically, a time-honored observational technique I teach to my students.

This method soon paid off, for within about ten minutes I found a few more tracks, subtle ones that either consisted of very faint toe impressions or were missing parts. This partial preservation of fossil tracks is typical. Most fossil tracks are registered on surfaces below where the animal actually walked as undertracks, like those made by a pen or pencil on an underlying sheet of paper. What then happens, erosion wipes away the "true tracks" made on the uppermost surface, taking away those footprints that may have shown skin impressions, toe pads, claw marks, and minute movements of individual toes. This seemingly unfair erasure of information means that the underlying impressions of a track have a better chance of making it into the fossil record than ones that resulted from direct contact with a foot. The penalty paid by such preservation, though, was a loss of details. For instance, two of the tracks before me were only apparent as paired marks made by sharp claws from two toes, when they should have had three fully formed digits.

A quick initial count of the rock surface yielded about fourteen tracks, all showing three or fewer toes. The surface itself only had an area of about 0.7 m^2 (7.5 ft^2), or the size of a typical dining room table for a family of four, so it held a lot of information in a small space. This was a busy little piece of real estate during the Early Cretaceous.

I used a handheld GPS unit to document the location of the rock, and it quickly determined the latitude–longitude coordinates of where I was standing. These were saved as a waypoint in the unit, but I also wrote them in my field notebook just in case the unit somehow ended up in seawater, was smashed by a falling rock, or both. Strong waves crashed behind me, acting as a reminder for me to back up this record of my position using old-fashioned analog methods.

This was about when Greg noticed I had become rooted to the same spot for nearly 45 minutes and writing intently in a notebook. Curious, he left Tom and came back down the beach to see what had been holding my attention. He was all smiles as he walked up to where I sat on another boulder in front of the track-bearing slab.

"What'd you find?" he asked cheerfully.

I grinned back, gestured toward the slab surface, and said, with a mixture of pride and awe, "Dinosaur tracks."

Greg's jaw dropped, and he briefly looked like a stunned mullet as his eyes took in what was there. In silence, I enjoyed watching him re-discover each dinosaur footprint, a wonderful moment to share with a field compatriot. Once he regained his voice, he exclaimed "Wow, this is fantastic!" Yes, it was.

I was curious to learn what Greg—a non-paleontologist but skilled observer—would see. I asked him to point to everything he thought was a dinosaur track. Within a few minutes of studying the surface, he quickly identified nearly every one I had detected, with only a few misses. "Nice job!" I told him, and we then went over his test results, just like I would with any eager and talented student of mine. We were getting very happy indeed.

Tom soon joined us. Like Greg, he also wondered what had held our interest so raptly, and he asked the same question as Greg: "What'd you find?"

My reply to him was slightly different, though, as I got a little professorial and answered his question with a question: "What do you think is there?"

Tom looked coolly at the rock surface, bringing to bear more than 40 years of paleontological experience, with more than 30 of those years spent studying the Cretaceous rocks and fossils of Victoria. A moment passed, then he pointed to the best-preserved dinosaur track and said, matter-of-factly, "Looks like a dinosaur track."

I nodded. "Yes, it is. See any others?"

One by one, he pointed to each track, and like Greg he found nearly every one I had identified. This was another great example of a little scientific principle called repeatability. That is, a scientist should be able to have her or his results repeated independently by other scientists.

I wasn't through with Tom, though, and asked him to take a close look at the little chicken-sized track nearest me. "Anything special about this one?"

That stumped Tom. "Help me out. What am I supposed to see?"

"Look behind it. See anything like it?"

He quickly put his finger on the small track preceding the one nearest me.

"Good. Anything behind that one?" I said.

"That one," he said, putting his finger on another identical small-sized track.

I beamed again. "It's a trackway."

When that little bit of information sank in, Tom allowed himself the indulgence of a very small Mona Lisa-like smile. Of all people in the entire continent of Australia, there were maybe five who would appreciate the significance of what was in front of us now, and Tom was one of them.

So with this affirmation from my field partners spurring me on, I resumed measuring track dimensions and writing notes, then took photographs. Greg helped with the data collection, acting as a scribe with my field notebook as I measured lengths, widths, and depths of the footprints with digital calipers.

As we did this, I could tell Tom's mind had gone somewhere other than gathering data about the dinosaur tracks. At some point he asked to borrow my tape measure, and he immediately went about measuring the length, width, and thickness of the slab. He sat down, wrote in his notebook, and then revealed what he was thinking.

"Greg, do you think we could get a front-end loader down here to take this up?"

I was surprised by Tom's question but shouldn't have been. As an ichnologist, I've never been much of a collector. My work normally consists of describing, sketching, measuring, and photographing trace fossils, recording their locality information, then saying goodbye, sometimes never seeing them again. After all, I did not have either a museum with storage space or research labs at my disposal, let alone collection managers and the means for picking up large heavy rocks. On the other hand, Tom has all of these amenities available to him, along with a collecting permit. In Great Otway National Park—which is where we were—no fossils could be taken without the written permission of the Australian government, but Tom and Pat had that.

So Tom wasn't planning to let this big hunk of rock stay here on Milanesia Beach, but instead was already plotting how to put it in Museum Victoria. His scribblings consisted of back-of-the-envelope calculations of the approximate length, width, and height measurements of the rock. This then yielded its approximate volume, which he then multiplied by the probable density of the rock. Because it had alternating layers of sandstone and siltstone, which consisted mostly of quartz and other silica-bearing minerals, it was about 2.7 grams/cubic centimeter, or more than 2.5 times the density of water. This yielded an estimated weight of about 700 kg (1,500 lbs) for the slab in front of us.

Nearly a year later, in early June 2011, Tom—with the help of Pat Vickers-Rich, their daughter Leaellyn (yes, the namesake for the dinosaur *Leaellynasaura*), Greg Denney, David Pickering from Museum Victoria, and personnel from Parks Victoria—succeeded in carrying out his quixotic goal. With the help of a front-end loader and heaps of cooperation, this block of rock with dinosaur tracks was transported safely off the beach. It and the other slab of rock with dinosaur tracks next to it was then put on the back of a flatbed truck and deposited in Museum Victoria in Melbourne, where both are now stored for future reference and further study.

Oh yeah, there was another rock, also with dinosaur tracks. Greg discovered that one, handily providing yet another reason why academic paleontologists are so dependent on non-academic folks to make significant contributions to our science. Greg's discovery was also a lesson in hubris for me. I was so giddy (and more than a little fulsome) about finding the first block, I hadn't even bothered to look around to see if any more like it were nearby.

Fortunately, he did. So while I was writing field notes about the tracks on the first block and imagining future fame—perhaps an appearance on Comedy Central, which represents the pinnacle of public acclaim for scientists in the U.S.—he was looking at every nearby boulder. Suddenly, at some point he started acting like a larrikin, running around the beach and looking for a piece of driftwood. In less than a minute, he found a 2″ × 4″ board (you'd be surprised at what washes up on these beaches) and quickly placed one end under a boulder only a few feet away from where I sat. Clearly channeling the spirit of Archimedes, he began flipping the rock, using the board as a lever.

"Greg, what are you doing?" I asked, as Tom and I watched him.

"This rock is the same as that one!" he shouted.

Tom and I swiveled our heads back and forth between the slab with the dinosaur tracks and the one Greg was attempting to turn over. He was right. They matched perfectly, although the one he was flipping was upside-down, its uppermost surface hidden from view. We ran over to help, and soon we were looking at its other side.

We gaped accordingly, as on the top surface of the rock were more dinosaur tracks. I pulled a small brush from out of my field vest and cleaned off the beach sand from the rock surface, and we stared at the lithified sand from a Cretaceous river floodplain, some of which had molded into the three-toed shapes of dinosaur feet. Just three hours beforehand, when we first descended onto Milanesia Beach, only two confirmed dinosaur tracks had been found in the Eumeralla Formation of western Victoria, and four in all of southern Australia. We now had twenty-four in front of us. You could say we were having a good day.

The most brilliant methods are often the simplest, and the one Greg used for finding the tracks handily fulfilled that dictum. He had noticed the thickness of the first block, as well as its thinly interlayered sandstones and siltstones. Within these alternating layers—a little more than halfway down its thickness—was a gray siltstone, looking much like a different flavor in an otherwise monochrome layer cake. He then glanced around to see if any other nearby boulders shared those traits, and matched one with the one I had discovered. Seeing that the gray siltstone bed was less than halfway down its thickness, he correctly surmised that it was wrong side up, and intuited that the dinosaur tracks could be on its top surface. He was right.

However, there was little time to revel and otherwise pat each other on our backs, as it was now mid-afternoon on a near-winter day. This meant we were losing our sunlight, and we still had a long uphill hike ahead of us back to the car park. Tom quickly measured the second slab's dimensions and figured out its weight. He reckoned it was about 400 kg (880 lbs), smaller than the first one but still too massive for any (or all) of us to haul up the trail.

I hurriedly photographed the overall surface and started sketching the forms and locations of the tracks on the surface. I also noted any similarities or differences between these newly found tracks and the ones found a few hours earlier. One of the last items on the agenda was to record a few digital video clips of me talking about the dinosaur tracks, which I knew later could be edited into an informative video to accompany a press release. Looking at those clips later was amusing, because you could see that I was trying very hard to be informative and objective but also couldn't help but show how thrilled I was with what had just happened.

Greg again assisted me in recording measurements as I used the digital calipers to gather data on the dinosaur tracks from this second block: number of toes, length, width, depth, and so on. He wrote down the numbers in my field notebook as I read them aloud. Only then did I take close-up photographs of individual tracks with a photo scale next to each, documenting their dimensions and other details. Much later, back in the comfort of my office at Emory

University in Atlanta, Georgia, I re-measured the tracks from photographs to double-check these results, just in case I had misspoken or Greg had miswritten the data that afternoon. Because I did not know whether I would see these tracks again—notwithstanding Tom's relentless plotting to acquire them for his museum—I was being extra careful in ensuring that the results could be repeated, checked by others, and otherwise pass peer review in the future.

As the sunlight faded, we decided we had had enough and vowed to come back the next week to better assess the geological context of the tracks, study and photograph them more, and figure out how to recover them from this high-energy, wave-filled place. Milanesia Beach had been generous to us that day, and as we all trudged up the steep trail, we were filled with laughter and excitement about these gifts offered by the dinosaurs, the Cretaceous floodplains, and, much more recently, coastal erosion.

The three of us came back the next week, along with Monash University geologist Mike Hall and my wife Ruth, who had just arrived in Australia to join me. Once we relocated the tracks, I recorded more measurements and descriptions while the others conducted a reconnaissance of the nearby cliff face and looked for the probable source of the dinosaur-track-bearing bed. They had a merry adventure doing this, climbing over hill and dale, while I, doing my best Marlin Perkins *Wild Kingdom* impression, stayed safely on the beach, waving to them once in a while as I looked up and saw their tiny figures crawling along the outcrop. Ultimately, they were able to narrow down the likely zone from where the tracks came, but it was inaccessible, on a sheer vertical face that kept saying "You'll all die!" if anyone tried to approach it. (This one needed no warning signs.) So gravity had been our friend, and I expect it will continue to be, as we can simply wait for it to provide us with more dinosaur tracks on Milanesia Beach in the future.

Walking Forward, One Step at a Time

We knew that lots more science had to be done before we could share our finds with the rest of the world, but we were confident that

that day would come. Sure enough, exactly one year later, on June 14, 2011, my coauthors and I—Tom Rich, Pat Vickers-Rich, Mike Hall, and Gonzalo-Vazquez Prokopec—received the good news that our scientific article had been accepted for publication in the Australian paleontology journal *Alcheringa*. Our peers confirmed what we knew that day one year before: we had discovered the best assemblage of polar dinosaur tracks in the Southern Hemisphere, the first dinosaur trackway from the southern part of Australia, and extended the geographic range of dinosaur finds in Victoria farther to the west. It was a most pleasing outcome from an otherwise frustrating field season.

The less technical picture that emerged from this discovery was this: Imagine small theropod dinosaurs—probably adorned with colorful feathers, ranging in size from chickens to cranes—making tracks while walking around on a river floodplain after the spring floods accompanied by a polar thaw. The tracks fell into three sizes, almost like they could be fitted for shoes, one of which was quite small (those would be the "chickens") and two others close to one another, but with one slightly larger than the other (men and women "cranes"). Although I couldn't prove it for sure, I suspected these three sizes were from a dinosaur family, consisting of still-growing juveniles accompanied by their mother and father.

These tracks inspired a compelling vision of parents and young performing their own version of a "Great Cretaceous Walk," leaving traces of their late spring or early summer foray on a wet, sandy floodplain next to a polar river, perhaps like the one I had studied in Alaska. Were they foraging, stopping occasionally to snack on invertebrates burrowing into the sand below their feet? Were they looking for others of their species, seeking companionship after a long dark polar winter? Or were they just reveling in the sunshine of a longer day, enjoying a warm breeze that fluffed their feathers as it wound through the river valley? That is the beauty of dinosaur tracks, as records of life as it happened while also encouraging fantasies of what once was, back then, and very close to the place where human eyes first recognized them for what they were.

So even though these tracks were only a drop in the proverbial bucket when it comes to paleontological discoveries, and still did not completely erase the reputation of Australia as a place forgotten by dinosaur trace fossils, they helped to affirm the most important point we paleontologists like to make about the fossil record: it gets better every day. Knowing this, we are encouraged to continue looking, and sometimes with the help of our friends we find what we're looking for, dinosaur trace fossils or otherwise.

CHAPTER 10

Tracking the Dinosaurs Among Us

My Mesozoic Moment

The large theropod tracks were fresh, and so was its scat. I looked around me, but the tropical rainforest was too thick to see very far, effectively hiding anything as big as what I knew had to be out there. Given such limited vision, I took advantage of the eyes and ears in the trees above me, listening to the songbirds for alarm calls, scolding, abrupt silencing, or other signs of unease. Satisfied that nothing was out of the ordinary, I squatted to take a closer look at the tracks and scat and began talking about both of these with my students, who gathered around to learn.

We were together on a short loop trail in a patch of rainforest in northern Queensland, Australia, and the theropod that made the traces in front of us was the southern cassowary (*Casuarius casuarius*). The students, all Americans and from my university, had enrolled in a study-abroad program to Queensland in June–July, 2007. Part of the program was a week-long field trip through the Wet Tropics region of northern Queensland, and on that day we were near the coastal town of Mission Beach. Before taking the students onto the trail that day, my colleague Chris Beck and I told them that the rainforest was an ideal habitat for cassowaries, which meant we might see one. However, we also warned them that we needed to be very cautious if we did, considering that they were, as my Australian friends might say, bloody dangerous.

When compared to modern concepts of theropod dinosaurs, cassowaries come straight out of the Cretaceous Period. They can be 2 m (6.6 ft) tall and weigh as much as 80 kg (175 lbs), making them among the heaviest birds alive. Unlike most mammals, female cassowaries are larger than the males, and would look down on most human supermodels. An adult cassowary's torso is covered with coarse black feathers, but its most eye-catching features are on its head. The head is topped with a casque, a tall, bladed crest that from the front looks as if it can saw through flesh, but also is broad when viewed from either side. The neat visual trick this accomplishes is that it makes for a more imposing profile while also increasing its height, like a Mohawk or Roman helmet. A stout but pointed beak pokes out from its face, behind which are yellow eyes dotted with black pupils, giving it an unnerving gaze. The side of its head and the skin just under its face reflects several shades of bright blue, contrasting smartly with long red-maroon flaps of skin (wattles) hanging down from the lower part of its neck.

Below, its scaly reptilian legs connect to three thick toes, all terminated by pointy claws. The inner one—digit II—really stands out, though. It is about 12 cm (5 in) long and is often described lovingly as "disemboweling," a descriptor bestowed in recognition of how this bird defends itself by leaping claws first into an attacker. In terms of Newtonian physics, this is a fine example of mass times acceleration translating into force which, when applied to a surface, transmits stress, and I mean that in the worst way. Given their weights, and that they are capable of running 45 km/hr (28 mph) and leaping more than a meter into the air, cassowaries can concentrate that forward momentum into the tiny areas of those claws, thus better enabling the splitting of an antagonist's soft body cavity.

Why are cassowaries so aggressive? Mostly because they feel a need to defend their territories, which could also include protecting nest sites and offspring: good reasons to give them plenty of room to roam. Incidentally, if you are chased by a cassowary and try to escape it by jumping in a nearby water body, be aware that they are also excellent swimmers and are known to venture into the sea. In fact, they are so comfortable in water that, much like human hot-tub or pool parties, their mating rituals sometimes happen while immersed. The only way this bird could be more terrifying is if it also flew. Fortunately, it and its ratite relatives—emus, ostriches, rheas, and others—are ground-dwellers, which make them even more comparable to medium-sized flightless theropods from the Mesozoic Era.

Cassowaries are omnivorous, eating fungi, forest invertebrates (insects and snails), and small mammals, but they are primarily known as terrific fruit eaters. So far, they have been documented eating fruits from more than two hundred species of plants, including a variety of figs. This cosmopolitan dining was partly evident in scat on the trail, which consisted of a dried, grayish-white, nitrogen-rich mass mixed with a variety of seeds. I was astonished at the size of their piles, which rivaled those of large mammals I've tracked in North America—more like black bears than turkeys. The largest pile we encountered was 15 to 20 cm (6–8 in) wide and contained seeds as big as 2.5 cm (1 in), nearly the size of ping-pong balls. Once I identified the first of such fecal masses, the students quickly recognized more along the trail in various states of disaggregation, their completeness depending on how long ago their cassowary producers had dropped them. Nonetheless, these droppings provided an ecological lesson for the students on the importance of cassowary poop as a means of fertilizing and dispersing seeds, and one communicated much more effectively than if I had droned on about it in an indoor classroom.

Although I had only been in cassowary territory once before then, my paleontological and ichnological training suited me well for detecting their tracks and other signs. Of course, their large three-toed tracks, measuring 20 to 25 cm (8–10 in) long and tipped with big clawmarks, were unmistakable in this ecosystem. In that place and time, the tracks could only be confused with those of emus, which lived in drier savannahs far west of there. Like most birds, their walking trackways consisted of a narrow, linear series of tracks, as if they were on a tightrope. This pattern only deviated where they stopped to forage or took a moment to assess their surroundings for potential threats or violations of their territory.

Although the tracks and scat showed that cassowaries in the area were frequent users of the human trail we were using, I figured they had their own worn pathways in the forest. However, I had no intention of testing this supposition by going for a stroll deeper into the forest and perhaps accidentally stumbling onto a nest site, which would have been very bad for both me and the cassowaries.

Like other ratites, cassowaries are ground nesters, in which the males scratch out a meter-wide shallow depression in the forest floor that they rim with vegetation. In this nest, they lay clutches of 2 to 5 eggs, which are light blue, oblong, and 10 to 15 cm (4–6 in) wide and long. Similar to emus, mother cassowaries leave the male to sit above and otherwise guard the clutch for about fifty days. Female cassowaries are also polyandrous, mating with however many males they feel like, and thus they may lay egg clutches from more than one father during breeding season, which is about May through October.

Once the eggs hatch, a male cassowary's parental responsibilities do not end. Instead, they spend at least seven months tending to the cassowary chicks, which are much cuter (or at least less imposing) than their parents, adorned with dark-brown stripes on a light-brown background of feathers. The chicks do not sit passively in their nest during this time, but are born to run. During their development, they follow their fathers, who tutor them in finding food, water, and otherwise establishing living spaces in the rainforest. However, these good times with Dad come to an end sometime between seven months and two years. Still, track assemblages made by such a grouping of cassowaries should show full-sized adult tracks accompanied by 2 to 5 sets of little three-toed tracks, along with scat heaps big and small, and browsing signs spread over a broad area.

The plants in this rainforest also imbued it with a Mesozoic-like ambiance. Botanists who first took a close look at the plant communities in northern Queensland were astonished to find that much of the flora consisted of evolutionary relics. First of all, many of the larger plants consist of tree ferns, cycads, and conifers, all of which had ancestors living and thriving during the Jurassic Period. Moreover, about fifteen families of flowering plants there are directly descended from plants that were alive when Australia was still attached to Antarctica, South America, and other Southern Hemisphere continents during the Cretaceous Period, and when non-avian dinosaurs were still part of the landscape.

For example, *Austrobaileya scandens*, a perennial flowering vine that composes part of the shady undergrowth in the rainforest, is the only species representing its clade worldwide. To see any of its relatives, you would have to look at the fossil record; its pollen closely resembles that of the oldest known flowering-plant pollen in Australia, dating from the Early Cretaceous at about 120 million years old. Unlike many flowering plants, flies—not butterflies or bees—pollinate its flowers, which is better understood when you sniff its broad, light-green flower and recoil from the rotten-meat smell. From these flowers, it produces plum-sized fruits with about a dozen large seeds in its middle covered by only a thin layer of fleshy goodness. Nonetheless, this fruit is on the extensive menu of cassowaries, hence this plant probably depends on them to carry its seeds farther away from where they might just fall off the vines. No doubt other local birds—including bowerbirds, fairy-wrens, honeyeaters, and riflebirds—also help with seed dispersal in the forest, but none had the voluminous guts of cassowaries. Even on a dare, I wouldn't swallow some of these fruits whole, yet cassowaries were doing it every day, transporting the seeds far and wide through the forest, while also helpfully depositing them with warm nitrogen- and phosphorus-rich fertilizer on top.

Earlier in the day, Chris and I explained to the students some of this basic information about the ecological importance of cassowaries, but also how these huge birds are endangered owing to a lethal combination of habitat degradation and premature deaths. With regard to the latter, while driving into the area that morning we saw many large yellow cautionary road signs saying "Cassowary Crossing" that also graphically depicted a car striking a cassowary. Young cassowaries are also vulnerable to predation by dogs, not being big enough to take on an aggressive canine, let alone a ferocious pack.

We were thus becoming filled with this academic knowledge that was being reinforced by reality, a most satisfying way to learn. So after walking another kilometer down the trail past more tracks and scat, everyone had relaxed, including me. Patches of sunlight reaching the forest floor surely contributed to our sense of ease. A cyclone that hit this area just a little more than a year before had ripped open parts of the canopy, illuminating the way. Otherwise, deep shade in the middle of the afternoon might have subdued our collective mood. Confident that our loud, boisterous group—with American accents, no less—had probably frightened off all wildlife for the day, I stopped paying attention to the songbirds in the rainforest canopy above and around us.

This was a mistake. Had I been eavesdropping on these avian conversations, they first would have alerted me to two people walking on the trail about a hundred meters behind us. These songbird scolds would have been followed by a brief moment of quietude, as a second large disturbance—which slipped into the forest, hid, and waited for our group and the couple to pass by—emerged and walked on the trail behind them. Instead, I only saw the couple come out of the woods and walk on a low, narrow footbridge over a stream our group had just crossed. I returned to idle chatter with a few of the students, all of us snapping photos of the gorgeous rainforest around us. We were clueless.

The low-frequency boom was subtle enough that at first I dismissed it. This was a lesson in observational blind spots, in which a lack of training in perceiving something new, but very real, can be quickly unnoticed or rejected. The second boom registered more firmly, though, and I realized it came from something alive, in the forest, and getting closer. A behavioral relic of my Cenozoic primate heritage was handily expressed as the hairs on my neck sprang up in direct response to a shot of adrenaline. Something was wrong, but I didn't know what.

Nearly every dinosaur documentary that attempts to recreate dinosaur noises uses mammalian or avian voices, or a blend of animal calls that somehow manage to reach some familiar place in our expectations of what dinosaurs should sound like. Even in fiction, such as during the unforgettable scene of the first *Tyrannosaurus* attack in the movie *Jurassic Park*, this theropod's petrifying roar actually was an amplified mix of sounds from an alligator, tiger, and elephant. Shaped by cinematic conditioning, we expect large theropods to roar or snarl, smaller theropods to hiss or growl, and herbivorous dinosaurs to moan or bay. They are almost never depicted emitting low-frequency booms just within the hearing range of humans. Yet that was exactly the voice used by the theropod that emerged from the rainforest that morning onto the trail and only about twenty meters behind me.

She was a beauty. Tall and resplendent with bright red and blue patches around her head and a mass of black feathers covering her torso, she moved with firm intention, exuding confidence. This was her rainforest, or at least her part of it. She let out another boom, which connected the sound to its producer and confirmed my previous feelings of trepidation.

It was then that I realized one of my students—Heather—was standing on the footbridge, blissfully oblivious that a dinosaur was behind her. I had already raised my digital camera to take a photo of the cassowary and the instantly replayed viewscreen included Heather in the foreground, wearing a big grin as she marveled at her surroundings, but with the cassowary in the background. Feeling a sense of responsibility, but while also wondering how many other professors have to worry about a theropod walking into their classrooms and confronting their students, I used an urgent stage-voice whisper to warn her.

"Heather, there's a cassowary behind you!"

She smiled and replied, oh-so-cheerfully, "Are you joking?" (Her question reflected how, during the previous weeks of the program, I had earned a reputation among the students as a prankster. I have no idea how this happened.)

"No!" I said with what I hoped was enough gravitas to change her mood.

My tone must have worked, because Heather turned to look over her left shoulder and saw that, yes, indeed, there was a large adult cassowary walking toward her. Quite sensibly, she froze in place, and her expression changed according to this newly perceived situation, which I dutifully recorded by taking another photo. You have to admit, however perilous the moment might have felt at the time, "Before Cassowary Awareness" and "After Cassowary Awareness" photos might have some good comic potential later in my teaching, and I was not going to waste this opportunity for the future educational benefit of others.

Fortunately for both of us, the cassowary eschewed taking the footbridge, stepped to its left, and began wading into the stream. Like Heather, I stood still, but continued to take pictures, most of which were blurred because of my shaking hands. Ambling at an even pace through the shallow water, her head turning jerkily to glance at each of us, the cassowary kept parallel to the bridge. Her crossing took only half a minute, although it felt more like five. She strode smoothly out of the water and onto the bank, only about three meters away, and circled behind me. I rotated in place to watch her, because nothing in the world would make me turn my back on an adult cassowary while I was standing on her home turf.

By now, other students farther up the trail had seen her and excitedly spread the word, resulting in their dashing directly toward Heather and me, and with the cassowary in between us. This was not good. Despite my admonishments to approach more quietly, some ran squealing with cameras in front of them, unknowingly risking more than most paparazzi might while approaching Russell Crowe, Lindsay Lohan, or Russell Crowe *with* Lindsay Lohan. Fortunately, instead of charging, the cassowary continued to walk around me, decided she'd had enough of these noisy humans, and re-entered the stream on the other side of the footbridge. Awestruck, we all stopped to watch her go. She paused briefly in the stream to step around a downed tree on the opposite bank, and then slipped into the rainforest. The whole incident, from first detected boom to her disappearance, had lasted just a little more than three minutes.

Later, on the other side of the world and safe at home in Georgia, I looked at my photo collection from that memorable day. Amidst the out-of-focus photos was one shot that came out absolutely clear. It showed the cassowary in full body and color, her legs pulling through the water, swirling the filtered sunlight, with greenish rainforest reflections on the stream surface. I often come back to this photo and stare at it as a reminder that I do not need a time machine in order to track a living dinosaur. These dinosaurs are everywhere around us, here and now, and making traces.

We call them birds.

A Brief Introduction to Modern Dinosaurs, Their Evolution, and Their Traces

No matter where you live in the world, it is nearly impossible to avoid seeing newly made theropod traces. Given about 10,000 species of modern birds, then billions of individual birds, multiplied by the thousands of traces a single bird can make in a normal lifetime, it all adds up to a whole heap of traces. Go to the most inhospitable places on the earth's surface, look long enough, and you are likely to find some sort of sign that a bird was there. Even urban environments are rich in bird traces, especially those where birds such as pigeons (*Columba livia*) and American crows (*Corvus brachyrhynchos*) have adapted and thrived. Bird traces thus add yet another, deeper dimension to our appreciation of their makers.

Bird traces include their tracks, cough pellets, beak marks—probe marks in sand, piercings in fruit, and drill holes in wood—as well as talon marks, nests, burrows, gastroliths, droppings, food caches, dust baths, and broken seashells, to name a few. Very broadly, most bird traces are related in some way to feeding and reproducing, but these and other behaviors are so incredibly varied and nuanced, it is well worth taking a second look at their traces for whatever insights they might supply.

The ubiquitous presence of birds and their traces throughout the history of humanity also led to their infusion into our lives as stories, idioms, and metaphors, even featuring prominently in world religions. For example, a Cherokee story tells how the Appalachian Mountains were made from the beating of a vulture's wings against the earth; most Native American traditions also feature birds as important parts of their spirituality. The Ngarrindjeri people of Australia have a Dreamtime story about a rivalry between the emu (*Dromaius novaehollandiae*) and brush-turkey (*Alectura lathami*) that resulted in the emu becoming flightless and the brush-turkey having smaller egg clutches in its nests.

Ancient Egyptians revered a heron-like god, Bennu, as the creator of the universe, as well as the sun god Ra, who had a human body but the head of a hawk. Incans viewed Andean condors (*Vultur gryphus*) as representing a higher realm of being in a three-level universe. Both Hinduism and Buddhism mention Garuda ("eagle" in Sanskrit), a bird-like god that was particularly skilled at dispatching snakes. In Judaism, King Solomon understood bird languages and used this knowledge to help his people. The Quran mentions the Ababil, flocks of birds that protected Mecca against an invading army. Christianity features doves in many of its teachings, including its symbolism as the Holy Spirit.

In the secular world, millions of people worship birds every day by looking for, identifying, and marveling at them. Birdwatching is among the most popular of all outdoor activities in the U.S., enjoyed by about 60 million people. In terms of an unconscious homage to birds, nearly every rock concert in the southeastern U.S. since 1973 has witnessed someone at some point yelling the words "Free Bird!" In short, these dinosaurs of the everyday have become powerful and omnipresent symbols for much of humanity, interwoven with the biological and cultural development of our species.

As nearly every precocious five- to ten-year-old can tell us nowadays, birds first evolved from non-avian theropods in the Jurassic Period. Also, for nearly as long as we have formally studied dinosaurs, we have known about fossil birds near the time of that evolutionary transition. First discovered in 1861, *Archaeopteryx lithographica*, from Late Jurassic (150 *mya*) rocks of Germany, was long regarded as the oldest known bird. However, some paleontologists now regard *Archaeopteryx* as more of a non-avian dinosaur, and two Chinese fossils, *Anchiornis huxleyi* and *Aurornis xui* from earlier in the Jurassic (circa 160 *mya*), may be closer to a hypothesized "first bird."

As far as bird trace fossils are concerned, the oldest suspected bird tracks come from Late Jurassic–Early Cretaceous rocks in various places from about 130 to 140 million years ago. Based on these and geologically younger tracks from the remainder of the Mesozoic Era—as well as the preservation of these tracks in rocks from river floodplains, lakeshores, and seashores—birds spread very quickly throughout the world and adapted to a wide variety of environments. For example, bird tracks are preserved in Cretaceous rocks formed in previously polar environments of Alaska as well as a few I recently co-discovered with colleagues in Victoria, Australia. These trace fossils and a few body fossils show that birds, within only 50 million years of evolving from non-flighted theropods, already had a worldwide distribution.

Explaining the evolutionary history of birds is a massive undertaking, and other people have written excellent, lengthy books on this topic, such as Thor Hanson's *Feathers: The Evolution of a Natural Miracle* (2011) and *Living Dinosaurs: The Evolutionary History of Modern Birds* (2011), a volume with many authors and edited by Gareth Dyke and Gary Kaiser. Accordingly, I will not attempt to duplicate their fine efforts here. Instead, I will point out that soon after the extinction of the non-avian dinosaurs, avian dinosaurs filled ecological niches left behind by their ancestors, while also carving out new ones. Birds of the past also had their own awe-inspiring qualities worth knowing, some of which made traces we know of, whereas others we may discover some day.

For example, if you took a trip back in time to the Eocene of North America, you might meet up with *Gastornis* (formerly known as *Diatryma*), a flightless bird as tall as the largest known cassowaries. Mostly because of its size, but also because of its massive hooked beak, this bird was originally interpreted as carnivorous. It also lived when the ancestors of modern horses were the size of domestic dogs. Hence, paleontologists painted lurid scenarios of these birds eating horses, which artists then literally painted. Regrettably for people who enjoy the concept of a bird saying the Eocene equivalent of "I'm so hungry I could eat a horse," no trace fossils, such as sliced or crunched little horse bones with beak marks, have confirmed this reputation for *Gastornis*. Owing to this lack of evidence, as well as recent research on this bird's skull and its shearing capabilities, this rapacious image of *Gastornis* has softened over the years. Paleontologists now suggest its huge beak was better suited for large fruits and nuts or for scavenging. Still, an encounter with one of these birds surely would have inspired at least as much fear in a small mammal as running into a medium-sized theropod during the Jurassic.

Other birds of the past also echo feelings normally evoked by dinosaurs, or at least a few of their Mesozoic contemporaries such as pterosaurs. For an example of the latter, the largest flying bird known from the fossil record is the giant teratorn (*Argentavis magnificens*) from the Miocene Epoch (about 15 million years ago) of South America. This bird had an estimated weight of 75 kg (165 lbs) and a wingspan of 7 m (23 ft). If transplanted into the Mesozoic Era, it would have been bigger than most pterosaurs, including *Pteranodon*. On land, terror birds—known by paleontologists as phorusrhacids—lived in South America and the southern part of North America from about 60 to only 2 million years ago. These flightless birds ranged from less than a meter to about 3 m (10 ft) tall, the bigger ones fulfilling their nicknames as top-niche predators in their ecosystems.

Among these terror birds was the Miocene *Kelenken guillermoi* of Argentina, which possessed the largest skull of any known bird: 71 cm (28 in) long, just smaller than that of an *Allosaurus*. Again, think of what it was like to be a little mammal during the Miocene in Argentina, with *Argentavis* in the sky and *Kelenken* on the land: not so different than living in the Cretaceous, but with a few big mammalian predators thrown in for good measure, too.

Other huge birds that preceded humans during the Cenozoic included *Dromornis* and its relatives in Australia. Like the South American *Kelenken* that also lived during the Miocene, *Dromornis* stood about 3 m (10 ft) tall but was big-boned enough to have supported about 500 kg (1,100 lbs) on its frame. Its relatives, dromornithids, thrived in Australia until only about 30,000 years ago, including *Genyornis*, a bird almost as large as *Dromornis*. This means that people who first colonized Australia about 50,000 years ago probably saw, interacted with, and ate these birds. Scientists still do not agree on whether the extinction of dromornithids was linked to human hunting, alterations of dromornithid habitats, or climate change. But die out they did.

Other colossal flightless birds managed to evolve on islands much smaller than Australia and well into the time of humans. For example, the elephant birds of Madagascar, which consisted of species of *Aepyornis* and *Mullerornis*, were as much as 3 m (10 ft) tall and weighed about 400 kg (nearly 900 lbs). These herbivorous birds evolved to great sizes while isolated from the rest of Africa, and until about a thousand years ago the only primates they had encountered were lemurs. Sadly, all elephant birds were extinct by the end of the 18th century, likely victims of a disastrous blend of human-introduced avian diseases, too-rapid changes to their ecosystems, and hunting. Surely another factor was that their eggs were hard to ignore for anyone tempted by so much easily obtained protein in one place, as these were about 150 times more voluminous than those of chickens. Almost nothing is known about their nests, but they definitely were not located in trees, which left their eggs quite vulnerable to any creatures with opposable thumbs that were able to easily carry them away.

Until just recently—that is, in historical times—hefty birds also lived on New Caledonia, Malta, Cuba, the Hawaiian Islands, and an island of the Fiji archipelago. New Caledonia, a Pacific island with a main landmass that was connected to Australia and New Zealand during the Mesozoic, had both *Sylviornis neocaledoniae*—a galliform bird, related to chickens, turkeys, and other ground-dwelling fowl, but standing 1.5 m (5 ft) tall and weighing about 25 kg (55 lbs)—and *Megapodius molistructor*, a megapode ("large footed" bird) that weighed about 3 kg (6.6 lbs), three times larger than any living megapode. Both birds were likely mound nesters, and fortunately some trace evidence supported this idea. Archaeologists at first thought huge earthen mounds on New Caledonia—measuring as much as 40 m (130 ft) wide and 5 m (16 ft) tall—were human burial grounds, similar to those made by Native Americans. The problem with this hypothesis was that the mounds lacked an essential component of human burial grounds: bones and artifacts. This is when scientists instead thought "really big avian traces," linking these hillocks to *Sylviornis* and *Megapodius* as more likely tracemakers for these landscape-altering features.

Other sizeable birds on islands included: the giant swans of Malta (*Cygnus falconeri*); the moa-nalo, consisting of four species of ducks in the Hawaiian Islands; the Viti Levu giant pigeon (*Natunaornis gigoura*), which was on one of the Fiji islands; and the giant cursorial owl (*Ornimegalonyx* sp.) of Cuba. All of these birds were flightless, and as one might guess from the adjective "giant" applied to their common names, they were significantly larger than any of their living relatives; for instance, imagine a 1.2 m (4 ft) tall owl prowling the forests of Cuba. All of these birds had something else in common, which was their rapid extinction soon after humans came in contact with them, found them delicious, introduced egg predators, and changed their habitats. Fortunately, traces again have helped fill in a few details about the behaviors of these vanished birds. For instance, moa-nola diets are known through their coprolites, which showed these were important grazers in Hawaiian ecosystems that lacked mammals as herbivores.

Although all of these recent birds were impressively sized, none were potential predators of people. So perhaps the closest situation comparable to *Jurassic Park*, in which theropods could have preyed on people only half their size on a remote island, was during the Pleistocene Epoch and on the island of Flores, Indonesia. Bones from about 12,000 to 100,000 years ago there include remains of both the largest known stork (*Leptoptilos robustus*) and a diminutive species of hominin (human relative), *Homo floresiensis*, nicknamed the "hobbit." The size disparity between these two species was like that between *Utahraptor* and a modern *Homo sapiens*. The storks were just shy of 2 m (6.6 ft) tall, whereas adult *H. floresiensis* stood only a little more than 1 m (3.3 ft) off the ground, and stork tracks were probably twice as long as the hominin tracks.

Also, storks, despite being associated with heart-warming fables of delivering babies, are voracious carnivores. Hence, these big storks, rather than fulfilling fables, would have been more interested in consuming warm hearts. So far, though, no one has noted any trace fossil evidence of storks preying on or scavenging *H. floresiensis*, such as beak marks on bones or coprolites holding the remains of furry feet. More likely menu items for these storks would have been rodents of unusual size, such as the Flores giant rat (*Papagomys armandvillei*), which was more than double the length of the biggest urban rats.

However, Pleistocene trace fossils from South Africa do show evidence of a bird attacking a small hominin. These consist of holes in the skull of the "Taung child," which belonged to a juvenile *Australopithecus africanus*. At first interpreted as leopard toothmarks, these holes are now regarded as beak and talon marks from an eagle, neatly matching traces left on modern monkey bones by African crowned eagles (*Stephanoaetus coronatus*). Some paleoanthropologists even speculate that frequent eagle attacks would have selected for more cooperative behavior and larger body size in hominins, thus deterring these predators and contributing to greater stature.

As one might have figured out by now, islands do funny things with the evolution of birds, especially related to body size.

Paleontologists and biologists have noted two seemingly opposite trends: smaller animals that colonize islands tend to get larger over time, whereas larger animals get smaller. Both can be summarized as the "island rule," although this can be split into "island gigantism" and "island dwarfism," respectively. Given the right ecological conditions and enough time, natural selection on an island can drive an animal's lineage down either path. For instance, if what we would consider today as "normal"-sized swans, ducks, pigeons, or owls had flown to islands during the Pleistocene, were genetically isolated from others of their species over several hundred generations, had no natural predators, and plenty to eat, then their descendants might have become much larger. However, with great size came great responsibility: to gravity, that is. Part of this natural selection toward gigantism in birds would have reduced any energetically expensive traits, such as wings. Consequently, wings shrank as weight and height went up in these island birds, rendering most of them flightless.

Speaking of islands, if you've ever fantasized about ones where dinosaurs continued to evolve after the Cretaceous, then look no further than New Zealand. Back in the Cretaceous, plate-tectonic shifting caused the main landmasses of New Zealand to split from eastern Australia, taking its dinosaurs with it. However, very few mammals came along for the ride, so geographic isolation combined with the end-Cretaceous extinctions to favor the survival and evolution of birds over non-avian dinosaurs and mammals. As a result, New Zealand experienced a grand evolutionary experiment, answering the question: What would have happened to dinosaurs over the last 65 million years if there had been no pesky mammals competing with them? More important, as an ichnologist I might also ask what sorts of unique traces might have been made as a consequence of such evolutionary innovations?

The variety of birds and the ecological niches they filled in New Zealand were remarkable. For example, before humans showed up about eight hundred to a thousand years ago, Haast's eagles (*Harpagornis moorei*) were the top carnivores in New Zealand, predatory theropods that delivered death from above to other large theropods. Like some other birds, female eagles were larger than males, weighing about 12 to 15 kg (26–33 lbs) and with wingspans of 2.5 to 3 m (8–10 ft). Although the biggest eagles today have comparable wingspans, they are all less than half the weight of Haast's eagle. And just what other large theropods were these eagles hunting in New Zealand? Moas, which were among the most famous and diverse of recent avian dinosaurs that remind us of non-avian dinosaurs.

Fortuitously enough, paleontologist Sir Richard Owen—who coined the word "dinosaur"—first studied these fabulous flightless birds in the early 19th century after British explorers brought moa bones to the U.K. Since then, paleontologists figure there were about nine species, which ranged from 1 to 3 m (3.3–10 ft) tall. Moas were grazing and browsing herbivores, basically filling the ecological role of ornithomimids, ornithopods, small ceratopsians, and other mid-sized Cretaceous dinosaurs in New Zealand, yet were alive when Geoffrey Chaucer wrote *Canterbury Tales*. Considering their sizes, abundance, and wide geographic ranges, moas must have worn deep trails throughout the landscapes of New Zealand and left easily visible browse lines along forest borders. Thanks to their coprolites, we even know what plants they ate and in which ecosystems, described in a 2013 study done by Jamie Wood and other scientists. Like the elephant birds in Madagascar, though, they fell victim to the ways of humans, and were extinct by the end of the 15th century.

Luckily, New Zealand still holds some examples of unique avian legacies, including some of the strangest birds in the world and some unusual avian traces. Among these birds are the flightless kiwis, consisting of five species of *Apteryx*. These birds are nocturnal foragers, eating a wide variety of plant materials, invertebrates, and small vertebrates, which they locate with nostrils on the ends of their beaks; no other bird has such an unusual adaptation. Female kiwis also stand out from other avians by having two ovaries, sharing nesting burrows with males for as long as twenty years, and laying a single egg that can take up one-third of their body volume and one-fourth of their weight.

Two other New Zealand birds, the kea (*Nestor notabilis*) and the kakapo (*Strigops habroptilus*), are examples of parrots that evolved in unexpected ways. Although we normally think of parrots as subtropical–tropical birds flitting about in rainforests, keas live in alpine environments—hopping along easily on icy glaciers—and kakapos are flightless, nocturnal, and the largest of all parrots. Kakapos also have odd mating rituals that involve traces. Male kakapos dig a series of half-meter (20 in) wide bowl-like depressions that they use like megaphones to project their mating calls. They further link these depressions by making tens-of-meters-long trails between them—made by their compulsive need to clean out their "amplifiers," which prompts them to walk constantly between them.

The point of this all-too-short stroll through the evolutionary history of birds and a glimpse at some of their ichnology is to emphasize how birds, despite the extinction of their theropod relatives 65 million years ago, continued to be dinosaurs. Accordingly, they also made traces that overlapped in size with those of Mesozoic dinosaurs, such as the tracks and nests of theropods and ornithopods. Moreover, these bird traces reflect myriad behaviors that also may have been shared by their non-avian predecessors, giving us search images for trace fossils that might be used to interpret behaviors currently unknown in dinosaurs. For instance, imagine the incredible coolness of finding a Cretaceous trace fossil consisting of a series of depressions linked by a network of trails and realizing that it might be analogous to a kakapo wooing trace, perhaps providing a key to understanding a dinosaur's mating behavior.

Thus, knowing about this evolutionary history of avians helps us observers of present-day traces better appreciate that every bird track, nest, probe, splatter, or other mark reflects a rich evolutionary history linked to a dinosaurian past that is minimally 160 million years old. Yes, birds have certainly changed and diversified enormously since the Jurassic, as well as spread throughout the world, occupying land, sea, and air. Yet modern birds also still hold insights to their Mesozoic origins, and the behaviors and traces we observe today might help us to better understand dinosaurs of the Mesozoic.

Tracking Birds, on the Ground and in the Air

As mentioned before, tracking is a longtime practical activity connected to hunting, and as a science has great applications toward understanding animal presence and behavior. Despite all of this, tracking is normally applied to mammals, not birds. Just to put this in perspective, watch the difference in people's reactions if you say "I'm going to go track a deer!" versus "I'm going to go track a robin!" I predict the latter will result in arched eyebrows, double-takes, and quizzical laughter. After all, the conventional wisdom is that most birds are too small to track, they fly so often that they do not leave many tracks, and their tracks all look alike.

Wrong, wrong, and wrong. Granted, the tiniest of birds—such as the bee hummingbird (*Mellisuga helenae*) of Cuba, which weighs only about 2 grams (0.07 oz)—would be very difficult to track. But nearly all other flighted birds, from wrens to condors, come to earth and make abundant and easily identifiable tracks. Of course, some flightless birds—such as rheas, emus, ostriches, and cassowaries—make the largest amount of tracks, and the largest tracks period. But many flighted birds regularly come in contact with the ground as part of their everyday lives and are thus capable of leaving tracks, too.

Bird tracks can be divided into four main categories based on overall form, all of which are derived variations of theropod dinosaur feet: anisodactyl, palmate, totipalmate, and zygodactyl. Anisodactyl is the easiest of these to link to Mesozoic dinosaurs, as it consists of three forward-pointing toes and sometimes a more backwardly pointing one (digit I). In some birds, though, this toe also may be reduced in size or absent. All songbirds have anisodactyl feet, as do chickens, turkeys, vultures, and raptors, as well as most wading birds and shorebirds, such as herons and plovers respectively. Palmate is a condition in which a basic anisodactyl foot has webbing between its main three digits, like on the feet of ducks, geese, and gulls. Totipalmate feet take this form a little further with webbing between all four toes, making for impressively broad tracks, such as those left by pelicans. Zygodactyl feet have their digits arranged with two pairs of toes and each bunched together to form either a K or X pattern. These feet are typical of owls, roadrunners, and woodpeckers.

Each bird-foot form reflects adaptations selected over thousands of generations of birds, meaning that each track made by a bird also reflects its evolutionary history and adaptations. For example, the extremely long, widely spaced, and thin toes of some wading birds, such as moorhens, are well suited for walking on lily pads in freshwater ponds. The three forward-pointing and single backward-pointing toes of songbirds are excellent for grasping and perching on branches. The webbed feet of ducks, geese, and pelicans aid in their surface swimming and diving in either freshwater or marine environments. Woodpecker feet work quite nicely for moving rapidly up and down vertical tree trunks. Raptor and owl feet are very good at grabbing and pinning down struggling prey. As a result, many bird species can be identified from their tracks, which can be further used to interpret their probable lifestyles.

Figuring out which bird made which track, however, does require careful measurements of footprint lengths and widths, toe lengths and widths, angles between toes, as well as basic knowledge about where birds normally live and when certain birds might be in the neighborhood for a visit. For one, I will not be adding secretary bird (*Sagittarius serpentarius*) tracks to my list of possible trackmakers while doing field work in North America, as this bird is restricted to Africa. Yet migratory birds, such as sandhill cranes (*Grus canadensis*), might stop by for brief cameo appearances in places where they typically do not hang out. Hence, people who track birds also should be aware of which birds are migrating and when.

Bird trackway patterns fall into five behavioral groupings: diagonal walking (or running), hopping, skipping, standing, and flying. Of course, birds are not necessarily locked into making just one type of trackway while doing their business. Robins, for example, are great little runners when on the ground, zipping from one place to another while looking for earthworms or other invertebrate treats. However, they can also switch from running to skipping, in which both of their feet leave and hit the ground at nearly the same time, but with one foot slightly ahead of the other. (Hopping differs from skipping by being done with both feet together, side-by-side.) Also, just before changing from running to skipping, robins might stop with their feet next to one another. Moreover, robins are very good at taking off from a standing start; so paired tracks might be the last ones seen in a trackway. Conversely, paired tracks might be the first in a robin trackway, telling exactly where a robin landed. Many a time I have followed a bird trackway and along its length seen such shifts in its behavior, both understated and overt, recorded all throughout: a script narrating a scene that is normally much more detailed than if I had actually watched the bird make the tracks.

Diagonal walking is an easy pattern to understand because it is so similar to our own walking pattern: right, left, right, and so on, in which a diagonal line can be drawn from one track to the next. I have seen this trackway pattern made by birds as small as sanderlings (*Calidris alba*) to as large as cassowaries and emus, and by birds with anisodactyl, palmate, totipalmate, or zygodactyl feet. However, the biggest difference between bird and human diagonal-walking pattern is how bird trackways typically have very narrow straddles, as if they are walking on a tightrope. For example, if some pranksters decided to create fake moa tracks in New Zealand by wearing big three-toed "feet" and wanted to make the trackway look more convincing to experts, they would need to swing their hips, placing one foot almost directly in front of the other. On the other hand (or foot, rather), walking normally would cause a wider-straddle trackway, which would be a dead giveaway that someone wearing oversized three-toed shoes had fabricated them. (However, the fact that a few idealistic people would happily accept such tracks as evidence that 3-m-tall flightless birds are somehow winning a centuries-long game of "hide-and-seek" is another matter.)

The most important variation on the diagonal-walking trackway pattern is running, in which the footprints are farther apart from one another, reflecting increased stride length. With this increase in speed, trackway straddles become even tighter, and individual tracks may show signs of claws digging in deeper, and sand or mud having been pushed behind where the feet registered. In this respect, nearly every non-avian theropod and most ornithopod trackways bear similar basic patterns or features like those made by equivalently sized modern birds. Thus, despite having anatomies that differ from Mesozoic dinosaurs, it is no wonder that dinosaur ichnologists still turn to birds—especially large flightless ones—as their default models for how bipedal dinosaurs moved and made tracks.

Has a bipedal dinosaur trackway—whether from a theropod or ornithopod—ever shown hopping and skipping patterns like those made by some modern birds? Not yet, and no one is holding their breath in anticipation of finding these in the fossil record. Even so, non-avian feathered theropods might have been capable of short flights, such as the Late Jurassic *Anchiornis* or Early Cretaceous *Microraptor*, and thus could have made hopping or skipping tracks to get themselves aloft.

In a 2013 study, paleontologists using CT scans of non-avian and avian theropod skulls (including that of *Archaeopteryx*) suggested that some non-avian dinosaurs had brains well suited for the complexities of flight, with a few better than *Archaeopteryx*. Based on other anatomical traits, like long fingers with claws, other small feathered theropods, such as the Late Jurassic *Scansoriopteryx* (*Epidendrosaurus*) and *Epidexipteryx* of China, look like they were better adapted for climbing trees and hanging on branches than being on the ground. This means a better place to look for their trace fossils might be on petrified logs in same-aged strata, or at least sedimentary rocks that originally formed near forests. However, a few feathered theropods were far too big to have either flown or climbed trees, such as the Early Cretaceous *Yutyrannus huali*, which was close to 9 m (30 ft) long and weighed more than a ton. For such weighty theropods, there was no hopping, skipping, jumping, or tree climbing, unless they did these as small tykes.

Thus it might behoove paleontologists who are interested in learning more about the origins of bird flight to pay attention to flying tracks associated with modern birds. After all, a continuing controversy in dinosaur paleontology—a real one, not a fake one like the old "Was *Tyrannosaurus* a predator or scavenger?" argument—was how self-powered flight evolved in non-avian theropods. Granted, nobody denies that self-powered flight provided some great advantages for those dinosaurs. For one, it took them to far more places than running, swimming, or gliding, but while using less energy than those means of transportation. Moreover, those places may have offered more choices in food, mates, nest sites, and habitats for raising offspring. This ability especially came in handy for migrating, in which these dinosaurs could more easily switch locations with seasonal changes or severe alterations of local climates. These are big questions that might be helped by looking at little tracks.

Flying tracks are by far my favorite of all bird tracks to find, and many other people must share this feeling, as the more frequently forwarded photos I receive from ichnologically inclined fans are of these. One of these photos shows a snowy vista punctuated by repeating sets of four mouse tracks, which end abruptly as they coincide with the feathered outline of an owl. The tale is all there in the tracks: a small mouse galloping across the snow, knowing that it is risking its life by being out in the open; an owl spotting it from its roost, then taking off with a whispery flap; a glide down with talons extending and grasping the mouse just as it landed on the snow; a beat of the owl's wings against the snow surface to continue its forward momentum and become airborne again, but this time carrying a little treat. Sadly for ichnologists, though, such detailed and evocative traces are more likely to be made in snow, which, once warmed, has a rude habit of melting and thus erasing all of the evidence.

In my experience, the best places to look for flying tracks are in soft mud or sand along a seashore, lake margin, or river floodplain.

While examining these tracks, watch for the paired ones, and for gaps in their trackway patterns. Nothing quite says "flight" like a right–left pair of bird tracks with no other tracks in front of or behind them. Do you see no tracks behind them, followed by a normal trackway? These are landing tracks. Do you see a normal trackway that ends with two tracks, and nary a track after that? These are take-off tracks. Other details to note in such tracks are linked to whether a bird was trying to control its descent or begin its ascent. For instance, in landing tracks, birds with rearward-pointing toes (digit I) on each foot direct these forward, leaving long claw-marks while "putting on the brakes." As the other forward-pointing digits contact the ground, these will push against the mud or sand, forming mounds in front of those digits as the entire foot comes to a halt. For take-off tracks, these reveal whether a bird left the ground instantly—with a burst of wing-driven power, it is aloft—or needed a little more forward momentum, such as a running, skipping, or hopping start, aided by much flapping. In such trackways, distances between the sets of tracks become greater, reflecting how lift forces gradually took the bird farther upward into the wild blue (or mild gray) yonder.

The controversy about the evolution of flight in dinosaurs is not whether or not it happened, but how it happened. At first only two hypotheses were offered, with pithy summary titles: "ground up" and "trees down." The "ground up" hypothesis says that certain lineages of non-avian theropods at first were ground dwellers but later evolved flight through a combination of natural selection for longer arms, fast running, lighter body frames, and flight feathers. The "trees down" hypothesis states that those same originally ground-dwelling dinosaurs climbed trees and otherwise became more arboreal, and at first were gliders. But then natural selection favored those that could flap under their own power instead of just glide.

Fortunately for scientists who do not like dichotomous arguments, a third hypothesis was offered just recently, and it synthesized elements of the other two with the special bonus of sounding more technical. Called the "wing-assisted incline running" (WAIR) hypothesis, the researchers who developed this model—mostly Kenneth Dial and his colleagues—noticed how baby birds, with only half-formed wings, flap these while running up inclines, including vertically oriented tree trunks. Thus, flapping combined with running helps the birds to get up those surfaces much more easily than if they did not flap at all. Half-formed wings also come in handy when baby birds fall out of nests, as flapping slows them down enough that they are more likely to survive once they strike the ground.

This is where trace fossils could help to test and otherwise augment these hypotheses, especially in Middle through Late Jurassic rocks, which is when paleontologists suspect that birds began separating from dinosaurs, evolutionarily speaking. Track evidence supporting ground-up flight would show small theropod tracks with take-off patterns similar to those we see in modern birds, but probably more typical of those that need a little more of a running or hopping start before becoming airborne. Track evidence for trees-down or incline-assisted flight is a little more challenging from a trace fossil standpoint, as this would require paleontologists to recognize scratch marks on fossil tree trunks and branches, and to distinguish scratch marks made by climbing versus running while flapping. Although not completely impossible, such evidence would have a much lower preservation potential than those made on muddy or sandy surfaces. The best trace fossils related to this hypothesis, of course, would be those showing the landing marks of an awkward baby dino-bird—complete with feather impressions—that also show it walking away from its crash site, successfully surviving a fall from a nearby tree.

Have non-avian dinosaur tracks indicating flight, however brief, been recognized from the fossil record? Not yet. However, some Early Cretaceous (about 120-million-year-old) small-shorebird tracks from Korea, reported by Amanda Falk and others in 2009, bear some of the same traits I just described for landing and take-off tracks. In 2013, Pat Vickers-Rich, Tom Rich, and I interpreted two of three closely associated anisodactyl tracks from Early Cretaceous rocks of Victoria, Australia as bird tracks, with the third coming from a non-avian theropod. These turned out to be the oldest bird tracks known in Australia, dating from about 105 million years ago. One of the tracks, made by a bird's right foot, also had an elongated mark left by digit I, matching modern examples I've seen formed by herons in their landing tracks. Unfortunately, the slab of rock was broken along the front edge of this track, so we could not further test our provocative hypothesis by checking whether another track was paired with it, and bearing the same marks of flight. When we published our results, we encouraged other paleontologists to start thinking about evidence for flight whenever studying Cretaceous tracks, whether these were from birds or non-bird theropods.

One last point I would like to make about bird tracks, and a surprising one to many people, is how these traces can actually alter environments. For instance, if you ever see mudcracks in an area frequented by birds, take a closer look at the geometry of those cracks and you may see the familiar three- or four-toed patterns of bird tracks in them. I first noticed this phenomenon during a hot summer in 2004 while walking along the edge of a pond on San Salvador Island, Bahamas. Yellow-crowned night herons (*Nyctanassa violacea*) and black-necked stilts (*Himantopus mexicanus*), both long-legged birds with thin toes, had been hunting for crabs in the muddy areas around the pond, and both species made hundreds of tracks. Wherever these birds' feet had punctured the muddy surfaces, sunlight dried the mud exposed along the edges of the footprints, which started cracks that grew and radiated from these edges. The mudcracks grew enough that they sometimes joined, especially where the birds' tracks were closer together (a result of slowing down while stalking crabs). From then on, I've looked for avian-caused mudcracks elsewhere and was gratified to find them in Arctic environments. In the summer of 2007, while doing field work at a dinosaur dig site on the North Slope of Alaska, I noticed bird tracks—this time from seagulls, geese, swans, and plovers—in the muddy patches of a river floodplain. Sure enough, wherever the muddy surfaces had dried, mudcracks developed, and nearly all connected perfectly to the bird footprints. The extended sunlight of a polar summer—with the sun setting for only a few hours during each 24-hour cycle—accelerated the drying after birds had pierced the muddy surfaces with their feet.

Did non-avian theropods ever cause mudcracks from their tracks? Yes, indeed. Among the extensive collections of dinosaur tracks at the Amherst Museum of Natural History in Massachusetts are slabs of Early Jurassic sandstones with gorgeously defined natural casts of mudcracks. Some of these mudcracks have theropod tracks in the middle of them, with cracks joining theropod clawmarks. The same apparent causal relationship between theropod tracks and mudcracks shows up in Early Jurassic rocks of southwestern Utah, in the same strata holding thousands of dinosaur swim-tracks, and is one of the better surviving examples of such intersections between dinosaur trace fossils and their correspondingly altered landscapes.

Nests, Empty and Otherwise

Bird nests are extremely variable as traces, ranging from simple scrapes in the ground to some of the most elaborate structures made by any living vertebrates. Examples of scrape nests are those of penguins or some shorebirds, such as oystercatchers (species of *Haematopus*) or plovers (mostly species of *Pluvialis* and *Charadrius*), which use their feet to scratch out slight depressions into which they deposit a few eggs. Among the latter are the massive individual nests of bald eagles (*Haliaeetus leucocephalus*), as well as the communal nests of the sparrow-sized sociable weaver birds (*Philetairus socius*). The largest bald-eagle nest yet measured was 3 m (10 ft) wide, 6 m (20 ft) deep, and weighed more than two tons. These nests are usually constructed in trees sturdy enough to hold such nests or on the tops of rocky cliffs. Sociable weaver nests of South Africa are among the most spectacular of all avian-made traces. Unlike the individual typical cup- or saucer-shaped nests people may normally see in urban settings, these nests are made collectively, shared, and reused by hundreds of weaver birds. Placed in stout trees or on telephone poles, weaver nests are composed of multiple chambers and entrances, are as much as 3 to 7 m (10–23 ft) in outline, and are used for both nesting and shelter by non-nesting adults. They are so massive, other bird species nest in or on them.

Not to be outdone, the mallee fowl (*Leipoa ocellata*) of Australia constructs gigantic incubation mounds for its eggs. These mounds can be as much as 4 m (13 ft) tall and 10 m (33 ft) wide, representing the movement of more than 200 m^3 (7,000 ft^3) of soil. For incubation, the parents let anaerobic bacteria do all of the work for them, breaking down organic material in the mounds and generating enough heat to keep the eggs toasty. In short, birds can make modest nests, incredible nests, and everything in between.

Although bird nests are almost as diverse as birds themselves, they can be placed into eight basic kinds, based on where they are located or their overall shape: scrape, platform, crevice, cup (or saucer), spherical, pendant, cavity, burrow, and mound. Of these, probably the one most people see are cup nests in trees, in which various songbirds arranged sticks or other vegetative debris—perhaps cemented by mud, bird spit, or feces—into a semi-circular bowl that keeps eggs from falling out. Indeed, of the nest categories, most of these are in trees, although a few are on or in the ground, such as scrape, platform, burrow, and mound nests.

Based on what we know about the evolution of birds from dinosaurs, bird nests probably started on the ground, but with underground as another possibility. The earliest known dinosaur nests, made by the Early Jurassic sauropodomorph *Massospondylus* of South Africa, were definitely on ground surfaces. At the other end of the Mesozoic Era, Late Cretaceous *Maiasaura*, *Troodon*, and titanosaur nests were bowl-like depressions surrounded by raised rims. A burrow entombing the small Cretaceous ornithopod *Oryctodromeus* and two of its offspring may or may not have been a nest site, although it was likely used as a den for raising its young. Of these dinosaurs, *Troodon* was the most closely related to birds, with at least one species of feathered troodontid known. Moreover, based on dinosaur egg porosities, nearly all were likely partially buried, whether in sediment or vegetation. So for now, paleontologists assume that nearly all dinosaurs nested on or in the ground.

Yet at some point in the Mesozoic Era—probably in the Early Cretaceous—nests began going up and into trees. This behavioral and evolutionary innovation likely happened in many places with multiple lineages of theropods. However, the proliferation of nesting must have been hosted in forested ecosystems, which provided plenty of opportunities for small theropods to go up and into trees, using these as safe havens from predators looking for eggs, baby theropods, or theropod parents. Also, just to show the plasticity of such behavior, some modern birds that we may normally think of as ground nesters—such as wild turkeys (*Meleagris gallopavo*)—may use trees on an *ad hoc* basis. In one instance, egg predation from feral hogs on a Georgia barrier island was so pervasive that wild turkeys began nesting in the trees there. In other words, trying circumstances can sometimes prompt birds to resort to behaviors that might have been deeply buried in their evolutionary histories.

As one might imagine, nests consisting of collections of sticks, leaves, feathers, and mud—however artfully or systematically arranged—had poor fossilization potential. The possibility of winning the fossilization sweepstakes shrank even more if these nests were high up in trees, away from entombing sediments. Not surprisingly, then, stick nests attributable to non-avian or avian dinosaurs have not yet been interpreted from the Mesozoic Era. Much more likely to be preserved are massive nest mounds, like those made by modern mallee fowls or some of the extinct birds of New Caledonia.

Nevertheless, even though trace fossils do not yet tell us exactly when non-avian or avian theropods first started nesting in trees, we do know when they bored into tree trunks to make cavity nests, like those made today by woodpeckers. At least one vase-shaped structure described from a petrified log from central Europe, dating from the Miocene Epoch (about 23 to 5 *mya*), is remarkably similar to woodpecker nests. Because this was well after the dinosaurs were gone, and no older ones have been found yet, this fossil cavity nest implies that the exploitation of tree trunks for nesting might have been a totally Cenozoic innovation in birds. Yet all we need to start a good argument on that point is a trace fossil of a Late Cretaceous cavity nest.

Burrows: The Avian Underground

Some birds are impressive burrowers. In my experience, this statement surprises many people until you say the words "burrowing owls" (*Athene cunicularia*), which elicits big smiles and vigorous head nodding in appreciation of these adorable birds. Yet these owls are not the only birds that use burrowing as part of their normal lifestyles. In North America, belted kingfishers (*Megaceryle alcyon*), bank swallows (*Riparia riparia*), and rough-winged swallows (*Stelgidopteryx serripennis*) burrow into soft, sandy bluffs adjacent to rivers or other bodies of water for their nesting and raising of young. In Europe, Africa, and elsewhere in the world, all species of bee-eaters make bank burrows, sometimes by the thousands at nesting sites. The Atlantic puffin (*Fratercula arctica*) of North America digs lengthy burrows, some of which are 3 m (10 ft) long and 1 m (3.3 ft) below the ground surface, a large burrow for a small bird. Some species of penguins, such as the little penguin (*Eudyptula minor*) of Australia and New Zealand, as well as the Magellanic penguin (*Spheniscus magellanicus*) of southern South America, also excavate burrows.

In all instances, birds use these burrows for nesting. Nonetheless, burrows also have the great multi-purpose advantage of protecting chicks and adults from predators while providing a place that maintains a near-constant temperature year-round. The latter especially comes in handy in high-latitude environments, where trying to stay warm during wintertime without shelter could quickly sap energy reserves.

Because birds do not have shovels, nor arms well adapted for holding shovels, they must use a combination of their beaks and rear legs to excavate a burrow. Imagine carrying out a mouthful of sediment at a time or scratching with your feet to make a tunnel many times longer than your body: like *The Shawshank Redemption*, but without even the benefit of a little rock hammer. These burrows are not just tubular, either, but also expand at their ends to include a nesting chamber. These chambers must be large enough to accommodate eggs, chicks, and at least one adult to tend to the eggs and feed their offspring, which have the annoying habit of growing bigger with multiple feedings.

An important point to know about burrowing birds, however, is that they may reuse or steal burrows, then—like squatters taking over a house—remodel and otherwise modify them to suit their own needs. For example, burrowing owls may take over appropriately sized mammal or gopher-tortoise burrows. Rough-winged swallows might usurp kingfisher burrows, and vice-versa. Some birds, though, may treat their burrows more like vacation homes, coming back to them at certain times each year. A male–female pair of Atlantic puffins, which normally mate for life, might return to the same burrow they used the previous year for nesting, then do some home improvement to prepare the nursery for their next brood. If the old burrow collapsed or otherwise is in bad shape, they may dig a new one. Because puffins are quite sociable, their togetherness results in huge colonies; this means they dig many burrows and many generations of burrows, which often intersect with one another, making for massive composite traces below ground. Complicating these underground neighborhoods is another bird, the Manx shearwater (*Puffinus puffinus*), which also nests in burrows among all of the puffin burrows. This implies that what might be classified as a "bird-burrow complex" in the geologic record might actually be the result of multiple species living in close relation to each other.

As mentioned before with the example of *Oryctodromeus* and other burrows credited to small dinosaur burrows in the geologic record, burrows conferred many advantages favoring dinosaur survival, or at least gave them enough time to mate and breed. However, one big question remains about burrowing dinosaurs and birds: When did theropods start going underground? This inquiry might be answered by learning more about the forms of modern bird burrows, their environments, and possible ways these traces could be preserved in the fossil record. Given such tools, paleontologists are much more likely to find Mesozoic burrows that can then be linked to either birds or non-avian theropods.

Probing Questions: Beak Marks and Other Feeding Traces

Like all other animals, birds have to eat. But when compared to most animals, they also must eat a greater proportion of food compared to their body weights, making the idiomatic expression "eat like a bird" (applied to someone who doesn't eat very much) quite incorrect. This energy debt is owed because of generally high metabolisms, especially for birds that fly vigorously or over long distances. Extreme examples are hummingbirds, some of which can produce thousands of wing beats per minute, or Arctic terns (*Sterna paradisaea*), which cover more than 40,000 km (25,000 mi) in an annual migration. Fortunately for ichnologists, one of the consequences of birds needing to eat frequently is the multitude of traces this produces. Starting with their mouths and ending with their cloacas, bird-feeding traces can be found wherever birds live.

Cough pellets, gastroliths, and feces have already been discussed in loving detail, so this section will focus on a few of their other feeding traces. Among these are beak marks, and in case researchers need an excuse to go to the beach, the best places to see these traces are along sandy shorelines. In such environments, shorebirds from small to tall stick their beaks into the sand to probe for, pry out, and procure foodstuff. Much of this sustenance consists of small clams, snails, and crustaceans, which burrow into the sand in a vain attempt to avoid predation. (As far as these invertebrates are concerned, shorebirds might as well be tyrannosaurs.)

When used in combination with tracks, beak-probe patterns can even be attributed to specific birds, such as sandpipers, plovers, or sanderlings. For instance, sandpipers do double taps—bang-bang—which leave distinctive two-holed patterns. In contrast, plovers keep their beaks on a beach surface while running along at high speed, making a more continuous and connected series of holes.

With clams, oysters, and other bivalves, more subtle traces of bird predation might be on the shells themselves. Look for chipping along shell edges, revealing where shorebirds wedged in their beaks and then opened them, an action called gaping; this exhausts an assaulted clam, which eventually can't resist any longer and opens its valves, revealing its soft goodies inside. Further evidence of gaping might be apparent where clams are still in their burrows but all that is left are two empty valves. Sometimes these valves—still joined at their hinges—have been pulled out of their burrows and left lying on a beach surface like open books, perhaps with a valve holding a small pool of clam juice. The final piece of ichnological and forensic evidence linking a shorebird with these molluscan molestations are their tracks, which help to identify which species pried into these clams' personal lives, then ended them.

Do these methods seem a little cruel to us? Sure, but they work great for birds. In fact, here's something even more brutal: Some of them pick up their intended victim—either a thick-shelled clam or whelk—fly up to a height of 5 to 10 m (16–33 ft), and drop it onto concrete or hard-packed sand. Thick shell or not, few molluscans can withstand this aerial-assisted assault, and their shells will either break or at least the animals will be stunned enough by the impact that they open up. I have often seen traces of this predation along the Georgia coast, in which a seagull prowled a sandflat at low tide, extracted a clam or whelk from its burrow, held on to it with its beak, took off, and then let go. In many instances, the bird left tracks directly in front of the clam or whelk's burrow, followed by a take-off pattern. When I look a little farther away, a bounce mark and broken shell show exactly where it impacted, and trampling around the shell pieces tells of how the gull successfully dined on its gravitationally prepared meal.

Birds may also take advantage of seashores with good rock outcrops or paved beachside parking lots, using these for opening their naturally packaged seafood. I've watched American crows do this in the Florida Everglades, in which they drop freshwater snails onto paved roads to break them open. In urban environments of Japan, carrion crows (*Corvus corone*) learned a variation of this technique, used with walnuts, in which they place these on the crosswalks of busy streets for cars to run over and break. They then wait for pauses in traffic provided by stoplights, fly down to quickly eat their meals, then take off just before the light turns green.

Beak marks made by birds living away from shorelines, especially in forests or plains, are a little different, but mostly because these are preserved in different substrates. Insect-eating birds, such as grackles or starlings, systematically insert their beaks into soil or ground vegetation to find their food. Northern flickers in particular are both persistent and insidious in their ground probing. These birds find an ant nest, insert their beaks into the main burrow shaft, and open wide (gape). This causes enough of a disturbance to encourage ants to run directly toward its mouth, which is where its long tongue picks them off, rapid-fire; for dessert, it may then drill further down to get some delectable ant larvae. Either way, this probing and gaping results in a conical hole at the top of the ant nest, and its depth corresponds roughly with the length of the bird's bill. Seeing that ant eating has been proposed as a lifestyle for some theropods, such as the Late Cretaceous *Xixianykus zhangi* of China, trace fossils resembling those formed by gaping might also have been preserved in fossil ant nests.

Meanwhile, above the ground and in the trees, woodpeckers and related birds either drill into trees or peel off bark to expose bountiful supplies of wood-boring insects. Bark peeling from birds can be considerable; I have seen large dead pine trees ringed by thick piles of bark on the ground, the cumulative work of insect-eating birds and not just decay. Also, seeing that eggs are such rich sources of nutrition, beak marks also may be in eggshells of other birds. Yet predatory birds do not just limit themselves to eggs for breakfast, nor do they necessarily stick with "non-avian only" policies while hunting. Sure signs of bird-on-bird predation are neat piles of feathers below roost spots, which a raptor plucked while enjoying a once-chirping treat. These piles may also have the distinctive white fecal sprays of raptors. Of the preceding traces, though, only the drill holes in trees are likely to have fossilized, and thus far nothing like these has been interpreted from the geologic record.

Less subtle raptor feeding traces are decapitated bodies of birds, or their counterparts, severed heads. One of the most unnerving predation traces I have ever seen, belonging to a bald eagle (*Haliaeetus leucocephalus*), was along the shoreline of a Georgia barrier island. The trace consisted of the isolated head of a laughing gull (*Leucophaeus altricilla*) that was lying just above the high tide mark. All of its feathers were still there and its eyes were open, while its neck had been sheared cleanly from its torso, which was nowhere in sight. No tracks were around the head, either, which made me wonder if the eagle performed its "Red Queen" impression while airborne. Regardless, it was a vivid reminder of how dinosaurs are still killing one another, and sometimes in brutally efficient ways.

Dinosaurian Bachelor Pads and Sexy Time

In evolution, not all selection is natural: much of it is sexual. Every second of every day, mates are chosen or rejected on the basis of whether or not they possess attractive qualities to justify exchanging gametes. For males, ensuring that their sperm fertilize females' eggs sometimes means doing more than just looking good, but also requires more effort, such as courtship. In birds, courtship may involve impressive feats of singing, or flashy displays of feathers combined with dancing, or showing up rival males in a competition; in other words, not so different from humans. So as a male, anything that attracts and convinces a female that your genes are the best (or at least will do for now) provides an advantage over other males and will be selected. But could such efforts actually result in traces that are later recognizable as the efforts of desperate male birds attempting to entice female birds? Furthermore, could courtship traces be used as models for detecting analogous trace fossils, whether made by non-avian dinosaurs or birds?

Some tracks could qualify as traces that might be connected with pre-mating rituals. For example, nearly everyone knows about peacocks and their exhibitions of glorious tail feathers, but perhaps fewer think about what the peacocks are doing with their feet. Peacocks, in order to show off the full extent of their tail feathers, perform a slow pirouette, often defining a tight circle. This results in many overlapping tracks in the same small, semi-circular area; whether tracks overlap from right to left or left to right could be used to figure out whether the peacock turned clockwise or counterclockwise, respectively. Male plovers also make distinctive pre-mating trackways by high stepping (also called "marking time") and placing one foot directly in front of the other. The trackways from this behavior are downright weird when compared to their normal walking trackways, consisting of parallel sets of tracks in which each track is arranged end-to-end. If another male plover is in the area, a race might happen in which the two contenders run parallel to one another, back and forth, in front of a female. The tracks from this behavior are distinct from high stepping, as the tracks would be much farther apart from one another but still would consist of two parallel trackways.

Other than tracks, male birds of different species make many additional traces that tell us they were looking for love, perhaps in all the right places. We also already learned about male kakapos and their ground amplifiers, which they use as the avian equivalents of "Mr. Microphone" systems to advertise their availability. But how about birds that go the extra distance by making something special for that little ladybird in their life? Indeed, a few birds go the route of becoming artists and engineers, going for visual impact by constructing elaborate, shiny structures that impress females with their ingenuity.

Meet the bowerbirds of Australia and New Guinea. Sometimes nicknamed the "amorous architects," male bowerbirds build a variety of structures—bowers—that are intended to catch the eye of a prospective female. Bowers are placed into three basic categories: mats, maypoles, and avenues. Mats are like ground canvases adorned with meticulously placed leaves, stones, shells, or even beetle wings. Maypoles are vertical structures, composed largely of sticks arranged around a small tree, but can also resemble teepees. Avenues, which consist of parallel walls made of leaves, sticks, and flowers, leading to a nuptial meeting spot, seem like runways that lead to a heart-shaped bed at their ends. The intricacy of most bowers is astounding, and may include shiny human-made objects such as items made of glass, metal, or plastic. Some bowerbirds even use perspective in their handiwork, placing smaller pieces of glass or rocks at the front of a bower and larger bits toward the rear: the only examples of such artistic renderings known outside of humans. For further enticement, in an attempt to seal the deal, bowerbirds usually throw in a dance or two.

Sadly, like stick nests, these monuments to mating have extremely low preservation potential in the fossil record. Besides, an exquisitely arranged collection of leaves, flowers, and sticks will probably not retain its original integrity after weathering, erosion, and burial for it to be interpretable as a bower; however, it might excite paleobotanists. The bowers most likely to be recognized as trace fossils are those in which their makers used rock or shell collections. After all, a large grouping of perfectly identically sized and shaped stones (and ones that clearly are not gastroliths), along with snail shells and beetle wings, should catch the attention of a paleontologist encountering these in the geologic record, especially if associated with a dinosaur skeleton, or (even better) two of the same species.

Fortunately, a lack of evidence for Mesozoic bowers did not stop the imaginations of screenwriters and computer-graphics artists for the Discovery Channel, in the TV series *Dinosaur Revolution* (2011), from recreating a Cretaceous scene inspired by bowerbirds. This scene starred a computer-generated image of the large feathered theropod *Gigantoraptor*, an oviraptorosaur from the Late Cretaceous (70 *mya*) of Mongolia. In it, a male with gaudily colored feathers, as well as a bright blue-and-red wattle (reminiscent of a cassowary), had cleared out a shallow circular depression and lined it with sticks and stones. Once a more drably colored (yet comely) female *Gigantoraptor* approached with some interest in what he had to offer, he commenced showing off his wattle and feathers and began to dance within the bower-like structure. (Amusingly enough, the dance was scored with a flamenco-like piece of music, underlining its seductive qualities.) As an added ichnological bonus, a small-mammal burrow complex was under the bower, with its mammals increasingly disturbed by the commotion above.

Unfortunately for both the male *Gigantoraptor* and the mammals, the bower soon collapsed under the weight of the dancing *Gigantoraptor*; this was not a surprising development, as these dinosaurs weighed about 1.5 tons, or ten times as much as an adult cassowary. This accident brought his performance to an embarrassing end, and his love interest quickly retreated, presumably to find a mate with better bower-building abilities. So not only did he not get the girl theropod, but he also ended up being a home wrecker.

Given this basic knowledge about modern bird courtship traces and what to look for, a huge advantage of interpreting trace fossils of these is that they are tied to gender; few trace fossils, other than some insect and sea-turtle nests, allow for such specificity. Thus, I have a dream that some day kinky-minded paleontologists—which is to say, virtually all paleontologists—will recognize the fossilized remains of theropod or bird courtship tracks, amplifiers, bowers, or other such wooing traces.

Dust Baths, Sun Baths, and Other Traces of Dirty Birds

Feathers require maintenance for several reasons. For one, they provide nice hiding places for skin parasites or other microbiota that might like to hang out underneath them. For another, birds have glands that secrete oil onto their feathers, an especially important accessory for birds as a form of waterproofing. This means that birds must preen and otherwise groom themselves regularly by pulling out or repairing damaged feathers, or spreading oil more equally. However, sometimes grooming is not enough, so birds seek treatments to rid themselves of too much oil in some spots, or reduce the nasty effects of lice or other arthropods that like to latch on to them for free meals.

Two self-administered therapies used by birds to rid themselves of unwanted passengers—therapies that also leave traces—are dust baths and sun baths. Dust baths are done through the following: finding a place with plenty of exposed dry clay and silt; getting a pit started by scratching with their feet; hunkering down in this pit; and rapidly fluffing feathers while throwing wings out, up, and back to toss fine-grained sediment onto their backs. Sometimes they dip into and shake their heads in the dust, too, just to make sure all feathers are covered. These actions help to break down any oil building up on their feathers, as the dust adheres to the oil and makes it easier to shed. Moreover, fluffing uncovers skin under the feathers, which allows dust to reach those areas and choke out parasites. The trace that results from a dust bath is a shallow semicircular depression slightly wider than the wingspan of the bird; sparrows make ones that are only about 10 to 15 cm (4–6 in) across, whereas turkeys' dust baths are more than 50 cm (20 in) wide. These hollows may also hold feather impressions or tracks, but if the substrate is too dry and collapsed on itself, these details may not be preserved.

Dust bath traces are surprisingly common traces, even in urban environments. For instance, in Atlanta, Georgia, I have seen dozens of closely spaced dust baths next to bus stops. Wherever people have worn down the grass next to those stops, sparrows took advantage of these easily available dusty patches to clean themselves by getting dirty. On the opposite end of the size spectrum from sparrows, ostriches in Africa make dust baths by sitting down, scratching out a hollow with their feet, and using both their short wings and long legs to heap dirt onto their feathers. Just to make sure they don't miss a spot, they also roll their long necks onto the ground, leaving what must look like snake-like traces in front of their main body impressions.

Sunbathing accomplishes some of the same goals as dust baths but uses solar power. When sunbathing, some birds simply lie down on the ground and soak up rays, whereas others sit upright on branches or other perches. What they have in common is that the birds spread their wings out to their sides. Why? When I've seen black vultures (*Coragyps atratus*) and turkey vultures (*Cathartes aura*) adopt this posture on chilly winter mornings, I always assumed they did this to warm up. But it may not be just the vultures that are getting warmed up but their parasites, which start moving once they reach a certain temperature. This makes it easier, then, for a bird to find and evict them.

Ornithologists have noted avian sunbathing since the 1950s, but paleontologists have seemed less aware of how this behavior might connect to non-avian dinosaurs. However, paleontologist Darren Naish, in a 2013 blog post, made such a link by reviewing sunbathing in modern birds while also encouraging people to visualize and artistically depict sunbathing theropods. Naish's post was accompanied by photographs of various species of birds—including an owl—lying on the ground with wings spread out, basking in direct sunlight. Of these, the most striking photo from an ichnological standpoint was of a secretary bird (which, incidentally, are wicked predators) lying on its tummy and with its head up, wings out, and long tail feathers behind it. Upon seeing this photo, I promptly imagined the bird disappearing, leaving only impressions of its torso, wings, and tail feathers. I then wondered how such a resting trace might act as a model for finding similar traces made by predatory feathered theropods, whether in Jurassic or Cretaceous rocks. Trace fossils like these would not only be very nice finds indeed, but also potentially interpretable as something more than just "resting."

Of course, many small songbirds bathe in what we regard as a conventional way by using water. In such instances, the bodies of water used for bathing might vary from puddles to ponds to lakes to oceans. Anyone who owns a birdbath or has watched songbirds around puddles knows how they partially immerse themselves, and then shake their bodies to ensure that all of their feathers get wet. Although the body of water itself will not preserve traces of this activity, moist mud or sand on their banks and shallow bottoms might record a curious series of tracks, perhaps with splatter marks on emergent areas next to the water.

So now think of, say, a 1.5-ton *Gigantoraptor* from the Late Cretaceous, one perhaps rejected by a potential mate. In an attempt to better its health and appearance, it decides to take a dust, sun, or water bath. Now envisage the marks these respective behaviors might have produced. A *Gigantoraptor* dust bath would be a relatively shallow semicircular structure, but enormous compared to those of all modern birds, probably more than 5 m (16 ft) wide. Linking this wallow to a large theropod could be tough, but doable if it also rolled its neck and head on the ground outside of the main depression, giving more anatomical clues. In contrast, sunbathing would have left body, wing, and tail impressions, with some movement blurring their outlines but not as much as in a dust bath. A water bath would perhaps show a series of walking tracks going from less saturated to more saturated (gooey) sediments, a stopping pattern, then lots of tracks in a small area caused by shuffling of the feet; water droplets on the shoreline and on top of the tracks there would complete the picture. Hopefully this mental exercise serves as yet another example of how modern bird traces can act as predictors for what trace fossils might be out there waiting to be recognized, and the behaviors that might be divined from them.

Bird Intelligence and Tools as Traces

At one time, the phrase "bird brain" was an insult hurled at someone perceived to lack common sense. Nowadays, it should be taken as a compliment. Paralleling the extreme makeover for dinosaurs showing they were not slow reptilian-like dullards, research on bird behavior in the past thirty years or so has completely changed our view of birds as simplistic automatons obeying their genetic codes. Instead, we are increasingly seeing birds more as sophisticates with their own complicated individual and social lives, language, tools, and even culture, in which parent birds actually pass down information to their young.

Nevertheless, if you are skeptical about the term "bird intelligence" and need examples, here are a few. How about recognized gradations of bird language, in which birds can tell one another through alarm calls that a possible threat is coming from above (a hawk) or on the ground (a fox)? This was verified with chickens (*Gallus gallus*), which are often impugned as the dumbest of all birds and thus deserving of roasting pans. Not complex enough behavior for you? Okay, how about when birds inform one another that not only is a human approaching but a specific individual who harassed them several years before? Or that they then teach this to their children? This sort of learning, recall, and teaching ability has been documented in American crows. Other birds that can learn individual human faces include pigeons, magpies (*Pica pica*), and northern mockingbirds (*Mimus polyglottus*); the latter can learn to associate specific faces as threats within about thirty seconds. Still not impressed? How about superb fairy-wrens (*Malurus cyaneus*) giving their chicks a "password" (a single note) as a cue for feeding, but while the chicks are still snugly inside their eggs? Researchers who verified this found that the fairy-wrens had likely evolved this behavior as a defense against feeding the hatchlings of cuckoo birds.

Given these samples, it should not be such a stunning revelation that birds are also among the few animals that use and make tools. Nonetheless, when this was first documented starting in the 1960s, it was an eye-opener for behavioral biologists. After all, the conventional wisdom at that time was that only primates used or shaped implements from their environments to accomplish a task. Although bird tools are oftentimes quite understated as traces (how many people can tell whether a wren used a cactus spine to pry out an insect?), it is still good to know about these and add them to their bird-trace checklists.

Tool-using birds include, at a minimum, New Caledonian crows (*Corvus moneduloides*), woodpecker finches (*Cactospiza pallida*) of the Galapagos Islands, Egyptian vultures (*Neophron percnopterus*) of Africa, bristle-thighed curlews (*Numenius tahitiensis*) of a few Pacific islands, the palm cockatoo (*Probosciger aterrimus*) of Australia, brown-headed nuthatches (*Sitta pusilla*) and burrowing owls of North America, and various herons, egrets, and seagulls. Of these birds, New Caledonian crows are the most impressive of tool users and problem solvers. These crows carefully choose or shape sticks, leaves, or feathers into practical objects, which they use to acquire food. In laboratory settings, they have even used metal hooks provided by researchers, and sometimes bent these into usable shapes. Woodpecker finches latch on to cactus spines or twigs with their beaks and then manipulate these like fine surgical tools to extract insect larvae from tight spots in trees. Like New Caledonian crows, they also may modify these items—such as shortening them—to improve their utility. Egyptian vultures and bristle-thighed curlews share the practice of grabbing a rock with their beaks and throwing these at eggs to break them open. These vultures do this to ostrich eggs, whereas the curlews crack albatross eggs; intriguingly, the vultures seem to pick rocks that are egg-shaped. Male palm cockatoos grasp sticks in their beaks, which they then drum against trees to alert females that they are in the area, echoing the mating habits of beat poets who used bongos in 1950s coffeehouses. Brown-headed nuthatches acquire short pieces of bark or sticks, which they lever against the bark on tree trunks to expose insects. Perhaps my favorite tool-using bird, though, is the burrowing owl. These actually use mammal feces as a tool, by picking up pieces of bison or cattle dung, placing these in front of their burrows, and waiting for dung beetles to arrive, which they then happily devour. Lastly, a few species of herons, egrets, and seagulls go fishing by employing feathers, berries, and even bread as lures. They either dangle these items from their beaks above a water surface or drop them onto the water to attract fish, which they then nab.

So do birds have memes—culturally acquired behaviors that modify over time—as well as genes? Although acquired knowledge and tool use is certainly shared among contemporary peers and with offspring in some birds, behavioral biologists still have not verified that this knowledge is being passed down or changed over many generations. Nonetheless, our awareness of bird culture and tool use lends to some fun (and not so crazy) speculation about whether or not Mesozoic theropods used tools, had their own way of communicating life-saving information to one another, and whether any of these resulted in trace fossils. Of course, the likelihood of recognizing tools or related traces would be minuscule, unless paleontologists found them in the mouth or hands of a dinosaur, or they noticed an appropriately sized and shaped rock sticking out of the side of a dinosaur egg. Nevertheless, all of these insights on bird behavior make us look at dinosaurs with a little more imagination, allowing us to wonder how many of their behaviors might have been shared with birds' Mesozoic ancestors and contemporaries, and their traces.

Thus while keeping in mind bird traces as a bridge to understanding dinosaurs of the past, as well as a few examples of how birds as modern dinosaurs affect the world today, it is time to think on a grander scale with our ichnological perspectives. In the next chapter, we will explore how some dinosaur traces probably changed Mesozoic landscapes, as well as how bird traces are currently changing our landscapes. Indeed, in some instances these changes may have even affected the evolution of modern ecosystems, implying that dinosaurs left a lasting imprint that still surrounds us and likely will persist well into our future as a species.

CHAPTER 11

Dinosaurian Landscapes and Evolutionary Traces

The Living Trace Fossil

We all live in a dinosaur trace fossil. It's not a trace fossil in the conventional sense, like an ankylosaur trackway in Bolivia, a prosauropod nest in South Africa, toothmarks on a bone from Canada, or a sauropod coprolite in India. Nor is it a trace made by modern cryptic dinosaurs, which no doubt will be reported breathlessly by actors playing scientists on a made-for-TV "documentary." It's not something more human-made, either, such as the worldwide cultural trace of dinosaurs related to their enduring and multigenerational popularity. Instead, the trace fossil is infused in the totality of terrestrial environments, what we sense from those environments, and even some of the earth resources we use. Where we live and what we do today is all somehow related to the former existence of Mesozoic dinosaurs and the continued presence of Cenozoic dinosaurs, birds.

With regard to resources and a cultural presence of dinosaurs, in the 1960s I grew up hearing the statement "Oil is made from dinosaurs." Sinclair Oil Corporation encouraged this illusion by sponsoring dinosaur exhibits at the Chicago and New York World's Fairs in 1933–34 and 1964–65 respectively, in which they overtly connected "dinosaurs" and "oil" in the public mind. (As a legacy of the 1964–65 Fair, statues of *Tyrannosaurus* and *Apatosaurus* constructed for it can still be seen in Dinosaur Valley State Park, Texas, less than a kilometer from real Early Cretaceous theropod and sauropod tracks.) Sinclair even adopted a green *Brontosaurus* as a symbol of its company, using this logo on service station signs and in magazine ads, while also selling plastic dinosaurs at their service stations. The plastic, of course, was also partially made of petroleum, which in retrospect seemed as if Sinclair Oil was into recycling long before it was hip.

Naïvely, I accepted the adage "dinosaurs make up oil" as true until a few science classes in college—particularly those in geology—straightened me out. It turns out that nearly all petroleum is from algae, most of which were deposited and buried in marine environments; no dinosaurs contributed their bodies to the original organic matter, and they had no role in helping to bury it, let along mature the organic compounds sufficiently that these later became oil and gas deposits. Indeed, some of the most prolific petroleum reservoirs in the world are filled with oil that post-dates the end-Cretaceous extinction of dinosaurs. Given all of these revelations, I had learned a lesson in not blindly accepting popular assumptions no matter how much we want to believe them, and to beware of the power wielded by smart, pervasive advertising.

Yet it was not until I became a geologist, paleontologist, and ichnologist that my perspective started coming back to this childhood thought and I wondered how, in some small part, it could be justified as true. Sure, dinosaurs did not directly contribute their remains to petroleum reserves. My mind is not going to change on that point. Furthermore, some petroleum deposits definitely formed millions of years after the last of the non-avian dinosaurs had left their traces. But did dinosaurs somehow change environments globally so that algae—which did contribute their bodily remains to oil—became more prolific in the world's oceans during the Mesozoic Era? Did they alter their local environments so that rivers changed their courses, which affected the locations of river deltas where many oil reservoirs are located? Did dinosaurs affect the evolution of terrestrial ecosystems and their organic productivity so much that marine ecosystems were impacted by these landscapes, thus affecting what happened in ocean waters, shallow and deep?

Up until now, we've learned that dinosaur ichnology applies to dinosaur trace fossils like tracks, nests, burrows, gastroliths, toothmarks, and coprolites, ranging in scale from two-meter-wide sauropod tracks to microscopic scratch marks on dinosaur teeth. Yet dinosaur ichnology also could be expanded to a more global view. Going back to a basic definition—that a trace is any indirect evidence of behavior aside from body parts—this concept can be taken further. For instance, to use a well-documented phenomenon, global climate change today is largely a human-caused trace. Did dinosaurs affect the world in a similar (albeit non-industrial) way? Could it be that the burning of fossil fuels today is really a composite trace, one that would not be happening if it were not for dinosaurs changing the earth to one conducive for making those fuels?

Maybe not. But let's explore anyway. The worst that will happen is to learn something new, while also expanding our perspectives by considering how dinosaurs may have been the original "ecosystem engineers" of terrestrial environments, altering them in ways that never would have happened without them and their behaviors and resulting traces. We will also take a look at how these alterations constitute dinosaurian traces that still affect us in significant ways today, and how these traces will continue to influence our future.

That One's Going to Leave a Mark: Dinosaur Trails and Their Effects on Landscapes, Rivers, and Ecology

I'd seen plenty of large sauropod tracks in the western U.S. and parts of Europe, but never ones this big. I tried to informally measure a few of the larger ones by making a circle with my arms above them. But my hands were always wide apart, making only semi-circles. Had I been doing ballet, I would have failed to complete the first position *bras au repos,* meaning the tracks were well over a meter wide. Once recognized, they were easily visible along the seashore as shallow rounded or oblong pits in the reddish Cretaceous sandstone exposed there. Once my wife Ruth and I picked out a few as search images, hundreds revealed themselves, accentuated by indirect light as the sun began to set over the ocean. It was a dinosaur-trampled mess, and a glorious one.

Although the marine platform was heavily eroded, a few of the flat sandstone bedding surfaces were continuous enough for trackway patterns to emerge. With one, a sauropod had made a "narrow gauge" diagonal-walking trackway, and one where its rear feet stepped directly on top of its front footprints. In other places, though, tracks were paired and closely spaced, either offset or overlapping. These were front- and rear-foot impressions, with the offset ones reflecting an understep (slow walking) pace. We could even see some of the sauropod tracks in vertical sections of the coastal outcrops. The normally near-horizontal layering of the sandstones had been distorted and contorted, showing where massive dinosaur feet had deeply compressed soft sandy layers about 130 million years before we were there. Sprinkled between the sauropod-made pits on the marine platform were three-toed theropod tracks. These seemed minute in comparison to the sauropod footprints, but were still 30 to 40 cm (12–16 in) long, indicating theropods with hip heights of about 1.4 to 1.6 m (4.6–5.2 ft)—big enough to stare us in our faces had they come back to life just then.

It was May 2009, and Ruth and I were on vacation in Broome, Western Australia. We had just finished a week of field work in Victoria, and to celebrate we were fulfilling one of the items on our Australian checklist, which was to visit Broome. Although it's a long way from anywhere else, friends told us that it was a lovely place to visit, with a gorgeous beach, art galleries, cultural tours, and some quirky, unconventional touristy attractions such as an open-air theater and camel rides on the aforementioned beach. What about the dinosaur tracks? Well, okay, as a card-carrying ichno-nerd, I have to admit these factored into our decision, especially once I learned the dinosaur tracks at Broome were only a few kilometers outside town and publicly accessible at low tide.

I first heard about these tracks at a scientific meeting, the first International Palaeontological Congress, which was held in Sydney in 2002. At this meeting, Tony Thulborn—introduced previously as one of the original paleontologists to study the Lark Quarry tracksite—gave a talk simply titled "Giant Tracks in the Broome Sandstone (Lower Cretaceous) of Western Australia." The audience of 25 to 30 paleontologists attending his presentation was in for a treat. Along with some of the preliminary scientific findings—that the Lower Cretaceous Broome Formation held a huge number and variety of dinosaur tracks—Thulborn showed photographs of what were then known as the largest extant footprints made by any land animal in the history of the earth. Some of the sauropod tracks were nearly two meters across; I'd slept in beds smaller than these tracks. At the end of his presentation, he announced with rightful pride, "Mine's the biggest!" (Just for context, he was talking about the tracks.)

Additional photos shown by Thulborn effectively communicated another point he wanted to make, which was that the dinosaurs—which were mostly sauropods, but also included some large theropods—had literally impacted their environments. Through sheer quantity of footfalls, as well as those footfalls coming from massive animals, the sauropods—and to a lesser degree the theropods—had altered the surfaces of their landscape enough to change the topography of their local environments. In 2012, Thulborn elaborated on that idea in an article titled "Impact of Sauropod Dinosaurs on Lagoonal Substrates in the Broome Sandstone (Lower Cretaceous), Western Australia." In that paper, he provided evidence that the dinosaurs had stomped soft sediments along a lagoonal shoreline so much that they formed low-lying areas flanked by higher areas, like levees on either side of well-worn trails.

Furthermore, these trails may have been routes used habitually by the dinosaurs. Once established, they became paths of least resistance for moving about, as if they made their own highways. Photographs in Thulborn's article showed huge sauropod tracks in depressed areas, but no tracks on the elevated areas on either side. The sandstones also lack plant-root trace fossils or other evidence of fossil plants, so it either was an already clear area for the dinosaurs to saunter through there or they denuded it by stomping plants into submission, while also compacting the soils, which prevented further plant colonization.

So imagine an Early Cretaceous shore next to a lagoon, with these deeply impressed trails running parallel to the average high-tide mark along that shore. With no vegetation along the way, the sauropods would have had a clear view, just in case they needed advance warning of the big predatory theropods waiting for them out there. Occasionally one or several of these theropods came down to the shoreline too, hoping to pick off a straggling sauropod for a big score, but for the most part they stayed away; this was sauropod country, and the uneven ground made ambush hunting and quick pursuits problematic. If seen from a pterosaur's point of view, the coastal trails would have connected to more inland ones, criss-crossing the forested interiors and freshwater wetlands like a great spider web.

Such grand disturbances of pliable mud or sand, in which great numbers of overlapping footprints made by immense dinosaurs made trails or left churned messes in the geologic record, are sometimes called "dinoturbation." I personally dislike this term, because it literally means "terrible [or awe-inspiring] mixing." This handiwork, however, is not the exclusive domain of dinosaurs. After all, earthworms and ants also mix tons of sediment every day. This term also distracts from how a few modern vertebrates, such as elephants and hippopotamuses, are capable of doing their own awesome mixing of sediment, which we somehow manage to restrain ourselves from labeling "elephanturbation" or "hippoturbation." Semantics aside, huge-sized dinosaurs, which sometimes traveled together in herds like the proverbial ships passing in the night, would have left sedimentary wakes with their passage, massively disturbing and altering terrestrial and freshwater ecosystems wherever their feet landed.

As the largest living land animals, elephants are the first analogs ichnologists reach for when trying to estimate the potentially far-reaching ecological effects of dinosaur trails. Elephants consist of three species: the African bush elephant (*Loxodonta africana*), African savannah elephant (*L. cyclotis*), and Indian elephant (*Elephas maximus*). Of these, the African bush elephant (*L. africana*) is the largest, with males weighing more than 7 tons, but Indian elephant males can also reach 5 tons. Elephants of all three species travel extensively and migrate annually. They normally walk in groups led by an adult female (matriarch), although adult males will go off on their own to make their own tracks. Thanks to fossil trackways recently discovered in the United Arab Emirates which show a series of parallel and overlapping tracks (group behavior) crossed by one trackway (a lone male), we know elephants and their relatives have likely held these same behaviors minimally for the past seven million years.

These behaviors also imply that local vegetation is normally worn down and sediments compacted by groups of elephants, not individuals, and that once a path has been cleared, it will be used repeatedly, perhaps by generations of elephants. Elephants also need plenty of water, so they try to stay near rivers or ponds, which they often enter and exit to drink or bathe. These habits mean they wear down banks, form wide divots on those banks, and muck up water-body bottoms, especially if they start wallowing. Elephant trails can also form depressions deep enough for water to flow along them, creating canals that connect previously isolated rivers or ponds. Trails on riverbanks similarly allow easier passage for floodwaters to cut through levees and pour out onto floodplains, depositing sediment in what are called *crevasse splays*.

Modern hippopotamuses (hippos), despite being smaller than elephants—with adults weighing in at 2.5 to 4 tons—have an even larger impact on their aquatic environments, which is where they spend most of their time. In a study by geologist Daniel DeoCampo published in 2002, he documented how hippos in Tanzania made a 30 m (100 ft) wide and 2 m (6.6 ft) deep muddy wallow pond, which connected to 1 to 5 m (3.3–16 ft) wide trails that imparted radiating and branching patterns onto the surrounding landscape. Hippos made these trails by frequently moving into and out of the wallow pond to feed on nearby vegetation; their activities, combined with their bulks, compressed and otherwise altered sediments. Most important, hippo trails actually changed the direction for water flow in the area, in which channels followed the trails, a type of channel abandonment called an *avulsion*. This channelization via hippo traces, in which their trails eventually turn into new river channels, is also well documented in the Okavango Delta of Botswana.

So did dinosaurs affect the courses of rivers with their trails, carving out new routes for flowing water through avulsions, or connect previously isolated water bodies? Given the known effects of much smaller modern large animals on rivers, the probable effects of individual dinosaurs that weighed 10 to 20 tons or more, herd sizes of these dinosaurs, and their geological longevity, I would be extremely surprised if they did not. If so, these effects may be detectable by geologists by more closely examining Mesozoic river deposits that also contain plenty of dinosaur bones and trace fossils—such as those of the Late Jurassic Morrison Formation in the western U.S.—for sauropod-width incisions in ancient river levees. They might also reexamine crevasse splays to see whether these connect to such divots, and whether these contain sauropod or other dinosaur tracks. In this respect, in 2006, two geologists—Lawrence Jones and Edmund Gustason—did indeed propose that avulsion features in the Morrison Formation of east-central Utah were likely caused by sauropod trails that created "channels" for the flow of floodwaters. Indeed, some former river-channel sandstones in the Late Jurassic Morrison Formation have dinosaur tracks on their bottoms, which may have been made by theropods or sauropods crossing rivers.

Amazingly, geologists have also determined that some modern rivers have been flowing more or less in the same valleys since before and well into the Mesozoic Era. These rivers include the Nile in eastern Africa, the Amazon of South America, the Macleay and Murrumbidgee of Australia, and the Colorado River of the western U.S., among others. All of these rivers are in places where dinosaurs—including big sauropods—used to roam, meaning at some point in their geological histories dinosaurs and previous versions of these rivers intersected. Considering the substantial human populations that settled along these rivers (some for as long as 10,000 years or more) and have since depended on the rivers and their floodplains for their lives, it is humbling to think that their present-day locations may have been at least partially determined by dinosaurs. Also think of how the ecological communities of these rivers evolved in those river valleys, beholden to these indirect effects of dinosaurs.

Now this is where a paleo-curmudgeon might preemptively scold me (augmented with much finger wagging) by saying, "Correlation is not causation!" Yet it is also unrealistic to accept the notion that dinosaurs, through their habitual movements, had no effect whatsoever on these or any other rivers during their times. Consequently, a valid question to ask is not *whether* dinosaurs affected the course of rivers or other water bodies in the past, but rather *how much* did dinosaurs affect these and other rivers? Although perhaps unanswerable, this inquiry is worth reflecting upon.

So while pondering this concept of how dinosaurs changed the very landscapes on which they walked, keep in mind how some may have been obliged to tread where generations had walked before. In that respect, here are the final two sentences from Thulborn's 2012 paper about the sauropod tracks of Western Australia:

> *If sauropods were as wary as elephants in negotiating sloping terrain, they would naturally have tended to walk on the lower and safer ground—which, in practice, would be any area that was already trodden by earlier visitors. In doing so, they would automatically have followed, deepened, and widened the routes*

pioneered by their predecessors, thereby reshaping the topography of the landscape they inhabited.

In short, dinosaur trails influenced the behavior of dinosaurs, traces that affected their decisions in everyday life, perhaps extending back into the Jurassic but very likely by the Early Cretaceous. In turn, those dinosaur-made trails transformed the land and waterways for future generations, traces that extended well beyond the extinction of their species and descendants, affecting all life thereafter.

Dinosaur-Caused Avalanches

Did dinosaurs ever cause landslides? The short answer is "yes," although it should be followed up by another question: "What do you mean by 'landslide'?" In popular vernacular, this term refers to any movement of earth material, but most often conjures visions of a chaotic mass of rocks rapidly moving down a slope, perhaps toward a village, often preceded by a distant rumble and someone yelling dramatically, "Landslide!" or "Avalanche!" But geologists actually apply these two words to different mass movements of earth material: a slide moves along a defined plane, whereas an avalanche is a chaotic flow of rocks and air. Hence, the latter is the scientifically correct warning in this example, which is good to know as it prevents arguments with geologists who might object to your use of "landslide" as a descriptor while you are in a village being destroyed by an avalanche.

Anyway, back to dinosaurs and their effects on mass movements. In an article by geologist David Loope published in 2006, he showed how Early Jurassic dinosaurs caused sand-dune surfaces to collapse as they walked across them. These ancient wind-blown dunes are preserved in the Navajo Sandstone, an Early Jurassic (190–180 *mya*) formation that is well exposed in many national parks of Utah and other spots in the western U.S. This formation is best known for showing off its spectacular *cross bedding,* which is the internal structure of sand dunes formed by Jurassic winds, its beautifully curved, parallel, and intersecting lines only much later becoming inert subjects of clichéd landscape photography.

Fortunately for ichnologists, the Navajo Sandstone also holds tens of thousands of dinosaur tracks. In contrast, dinosaur bones are rare in the Navajo Sandstone, so these tracks show that dinosaurs were there, which ones were present, and what they were doing.

In Loope's study, he looked at an area on the Arizona–Utah border that was particularly rich in dinosaur tracks, most of which were from variously sized theropods but also included prosauropods. The tracks could be observed both on bedding planes and in vertical sections, which helped to define just how much the dinosaurs were disturbing the sand dunes as they walked on them. Perhaps the most surprising conclusion of his investigation was that many of the dinosaur tracks had been made on dry sand, which was always regarded as a poor medium for preserving tracks. Even better, because the footprints were made on dry sand, each spot where a dinosaur stepped became a potential avalanche, which multiplied with every subsequent step. Other geologists had reckoned that the climate for that area during the Early Jurassic was monsoonal, so Loope figured that the tracks were made and preserved during the dry season, which was likely in the winter.

For anyone who has walked on a sand dune, especially one with dry sand, what happens to the sand depends on where you are walking on the dune. For instance, if you confine yourself to the dune crests by just walking along their tops, this motion causes minimal movement of sand, and when you look back at your path you'll mostly see a softly defined track. But as soon as you start walking up or down the side of a dune, its sloped surface starts collapsing. As each foot pushes down into and against the dry sand and then withdraws, the sand grains become part of a miniature-scale avalanche, tumbling on top of one another as they move downslope. This sort of movement is called *grainflow,* in which sand grains flow as if they are in water, but in this instance are cushioned by air between the grains.

This phenomenon is best experienced while walking down a dune slope, in which one should note how each step causes sand to flow ahead of you, creating an apron of avalanched sand in front of where your next step will land. Once finished with a descent, take a look behind and you will see a wide, featureless area of disturbed sand, which probably will not show any clear outlines of your footprints. If you had not just done this personal experiment, you would have no idea if a person or other animal had caused the avalanches, or whether this change in the dune landscape was even caused by animals. Gravity happens, so dune slopes at high angles can also simply collapse under their own weight without the help of trackmakers.

So this is what the Navajo Sandstone dinosaurs did. As they walked down dry dune slopes, their feet initiated grainflows, which preceded each successive footfall. Step, grainflow, then step into a grainflow, which was repeated for however far the theropods or prosauropods walked down dune slopes. However, what the geologic record revealed, which was not readily apparent from contemporary examples of walking down dunes, is that outlines of the dinosaur feet were preserved below the surfaces of the original grainflows. In cross-sections of the sandstone, Loope found theropod and prosauropod tracks directly associated with grainflows, in which the dinosaurs caused downslope movements with each step, then stepped into the grainflow deposit they had just made. The dinosaurs had caused small avalanches, which then buried their tracks, leaving behind the most important clues needed to determine what caused these grainflows in a Jurassic dune field.

With this example in mind, we might wonder how else dinosaurs, through their walking on slopes, might have wrought other changes in their landscapes, such as triggering mass movements of materials in upland areas. For example, did dinosaurs that lived in mountainous regions ever start an avalanche by walking across a rocky slope? Did polar dinosaurs ever start snow avalanches? Both of such examples are extremely unlikely to have been preserved in the geologic record: for one, snow tends to melt, which was also true of Mesozoic snow. Nonetheless, sand dunes and other sedimentary deposits could reveal more secrets of how dinosaurs modified their environments, step by step.

Nesting Grounds: Turning Flat into Bumpy

As we learned from the Late Cretaceous titanosaur nests in Argentina, sauropod nesting grounds probably changed the appearances of their landscapes, especially because they kept coming back to the same area over generations. This meant that sauropod nests, which were probably made by scratch digging, were placed above the graves of former nests, eggs, and embryos. In Montana, the Late Cretaceous hadrosaur *Maiasaura* and theropod *Troodon* also seemed to nest in the same places over time, implying that these dinosaurs also may have habitually used nesting grounds. High concentrations of dinosaur eggs elsewhere in the world, such as in South Africa, Spain, and India, likewise point toward how dinosaur parents kept choosing the same places over generations for their nesting. All of this nesting over time meant many ground nests, some of which may have been reused; but once buried by a river flood, new nesting grounds would have been stacked on top of previous ones.

Although we do not have sauropods or ornithopods today that we can study for the effects of their nesting on local landscapes, we do have birds, or theropods, making vast nesting grounds. As mentioned in a previous chapter, many shorebirds make simple scrape nests, but they may also concentrate these nests in vast breeding colonies, congregating by the thousands along oceanic and lake shorelines or on small islands. In some instances, birds in breeding colonies, such as Australasian gannets (*Morus serrator*), make more prominent nests as raised platforms with a central bowl-like depression for holding eggs. They then place these nest structures close together—just slightly more than the body length of the gannet next door—which creates regularly bumpy and dimpled surfaces with hundreds or thousands of regularly spaced nests throughout their nesting grounds. While on a trip to New Zealand, I remember seeing and marveling at the gannet nest sites there, which were on a few isolated marine platforms just offshore. But while gazing at these busy colonies and their nesting traces, I also wondered whether Mesozoic theropods might have had similar structures, and over extensive areas.

Now, one would think researchers looking for examples of ground-nesting birds that modify their environments via nesting, and as analogs to Mesozoic dinosaurs, might be tempted to look first at ratites. But this would be a mistake, as ostriches, emus, cassowaries, and rheas do not form breeding colonies but instead act more like rugged individualists, albeit family-oriented ones. Rather, one of the best examples of breeding grounds in which nesting birds remade their environments, even affecting the locations of deltas and river channels, comes from those long-legged pink birds we revere as lawn-ornament idols, flamingos.

In an article published in 2012, Jenni Scott, Robin Renaut, and Bernhart Owen described the terrain-altering effects of flamingo nesting grounds along lakeshores in the Kenya Rift Valley of eastern Africa. Flamingos there consist of two species, the lesser flamingo (*Phoeniconaias minor*) and the greater flamingo (*Phoenicopterus roseus*). These birds gather and breed by the millions next to high-salinity lakes in this area. Feeding, flirting, and fornicating flamingos significantly trample the nearshore and shallow-water muds of these lakes, which these birds also use for bathing. They further contribute to this "flamingoturbation" by constructing thousands of nests out of gooey shoreline mud. Their nests look like miniature volcanoes, wider at their bases (30 cm, or 12 in) than at their tops (20 cm, or 8 in), and about 15 to 20 cm (6–8 in) tall, with a shallow bowl at the top to secure the egg clutch.

Flamingos mine the lakeshore mud, which they do by dragging it with their beaks or scooping up mud in their bills and spitting it onto the nest. These actions result in shallow rings or semi-circles ("moats") around each nest. Mixed in with the mud shaping the nests were feathers, vegetation, and (somewhat morbidly) the bones of dead flamingos. Insects may add their own traces to the nests, burrowing into and forming brooding cells in their sides, and plant roots sometimes invade abandoned nests. Based on their detailed descriptions of these modern nests, the researchers were confident that they could easily identify ancient examples. Not surprisingly, then, they promptly did this in the same study area, discovering flamingo-nest trace fossils that were more than 10,000 years old, and sometimes directly next to or beneath modern nests.

Probably the most important aspect of these nests, though, is how both extant and older nests affected lakeshore environments and river deltas emptying into the lakes. Nests were normally spaced 1 to 2 m (3.3–6.6 ft) apart, either forming lines or clustered. However, wherever nests were closer together, the "moats" between them joined, turning these low areas into small ponds or canals through which water could easily accumulate or flow, respectively. Both the trampling and nests also compacted lakeshore sediments so that these were less porous, acting more like cement. These "case-hardened" surfaces then resisted wave or stream erosion, which accordingly also influenced where water flowed in delta channels. Thus through a combination of activities, but especially their nesting, flamingo traces had significantly changed the environments around these lakes.

Were dinosaur ground nests ever so numerous and closely spaced that these formed nesting grounds and likewise changed their local environments? We don't know for sure yet. Jack Horner originally proposed that the Late Cretaceous hadrosaur *Maiasaura* of Montana may have had nesting colonies, but not enough of these were preserved to say whether or not nests affected nearby rivers, lakes, or other environments. Near the *Maiasaura* nests are *Troodon* nests, which are rimmed, bowl-like depressions that outwardly resemble flamingo nests. However, these nests are much lower and wider and not nearly as numerous.

Another intriguing idea about dinosaur nesting inspired by modern birds is that perhaps some made massive nest mounds for incubating their eggs. Remember the mallee fowls (*Leipoa ocellata*) of Australia and the 4-m-high hills they make with their nest mounds? How about the recently extinct mound-nesting birds of New Caledonia, *Sylviornis neocaledoniae* and *Megapodius molistructor*, which made mound nests so big that archaeologists at first mistook these for human burial mounds? Of these two birds, *Sylviornis* was the bigger one, weighing about 25 kg (55 lbs), but still much smaller than most Jurassic and Cretaceous theropods. In at least one recent documentary (*Dinosaur Revolution*, 2011), *Tyrannosaurus rex* was depicted as a mound nester with an accordingly huge mound for its eggs. Sadly, though, such nest mounds may have been made mostly of vegetation, not sediment, so these would have had a much lower chance of being preserved as trace fossils. Alternatively, if composed more of sediment, remnants of massive nest mounds like these might have been left behind, although if unaccompanied by dinosaur eggshells or embryos these might be tough to discern.

Still, there is much we need to learn about dinosaur nesting. For example, no one has found nests of stegosaurs, ankylosaurs, and many other dinosaurs, including those commonly preserved as adults, such as *Allosaurus* and *Triceratops*. Perhaps some of these dinosaurs—especially those for which we have evidence of gregariousness, like ceratopsians—had immense nesting grounds that played a major part in shaping their local environments. However, if such nesting grounds lacked eggshells and bones, geologists and paleontologists may not have recognized them for their true nature and merely regarded them as strangely bumpy ground of uncertain origin. So it is with some hope that an ichnological outlook and a little bit of luck may aid in the discovery of large-scale dinosaur nesting grounds, and the effects these grounds had on their local environments.

Dinosaur Flatulence and Global Warming

Farts do not fossilize. This is an extremely fortunate situation for geologists and paleontologists worldwide who otherwise would have to wear gas masks every time they cracked open a Mesozoic rock with their hammers. Let's face it, the Mesozoic was probably a very stinky time in earth history, given the considerable numbers and sizes of dinosaurs actively eating, belching, vomiting, pooping, peeing, and, yes, breaking wind in a way that likely generated its own measureable wind speeds.

Probably the only way for ichnologists to detect whether a specific dinosaur let one rip is to carefully examine the sediment surrounding the few theropod resting traces described from Early Jurassic rocks and see whether any unexplained structures might be in the former vicinity of the theropod trackmaker's cloaca. However, I am not hopeful about our finding such trace fossils, nor do I expect anyone to receive government-funded research grants to reproduce modern examples under laboratory conditions.

So although individual trace fossils of dinosaurs' gaseous emissions may not be recognizable, their collective effects likely were. Paleontologists, geologists, and *paleoclimatologists* (people who study ancient climates and how these changed through time) are considering the intriguing possibility that dinosaur farts may have contributed to global climate change.

The sudden release of gases from intestines, politely termed *flatulence*, can cause localized changes in atmospheric chemistry that may or may not be detectable through smell, but normally is preceded by sonic qualities. The mixture of gases, called a *flatus*, varies considerably within or between species of flatus emitters. For example, in humans, a flatus produced by swallowed air that made its way through the digestive tract approximates normal proportions of atmospheric elements and compounds, which is 78% nitrogen (N_2), 21% oxygen (O_2), and less than a percent of carbon dioxide (CO_2), methane (CH_4), and other gases. However, if originating as a by-product of digestion, the proportions may include as much as 30% carbon dioxide and/or 50% hydrogen and methane. Most of the unpleasant smells we associate with flatulence come from sulfurous gases, such as hydrogen sulfide (H_2S). In contrast, carbon dioxide is odorless, and methane, although detectable by smell, is not as offensive as its sulfurous companions. Most of these gases, methane included, are among the waste products of anaerobic bacteria and bacteria-like organisms (*archeans*) living in our guts.

Other than initiating temporary social discomfort, adverse olfactory sensations, embarrassment, or lowbrow humor, every flatus by all humans or other animals contributes gases to the global atmospheric budget. Although each emission is small, the joint effect of these gases from animal- and non-animal sources is what changes the earth's atmosphere and ultimately affects climate. As everyone now knows (with the exception of a few billionaires), the overproduction of certain gases—such as carbon dioxide and methane—has insulating qualities in our atmosphere, trapping infrared radiation (heat) and thus contributing to warmer global temperatures. This is not just a matter of quantity, but also the type of gas. Although carbon dioxide is often vilified for its role in climate change, methane in smaller concentrations has a greater impact, as it insulates much better than carbon dioxide—like the difference between using nylon or wool for a blanket.

In a 2012 paper, three scientists—David Wilkinson, Euan Nisbet, and Graeme Ruxton—asked themselves this: Did dinosaur-produced methane contribute to global warming? They then tried to answer it by applying a few biological principles, ecological concepts, atmospheric chemistry, and a whole bunch of math. In their model, they concluded that, yes indeed, sauropods alone could have produced almost as much methane in a year as is made by all sources (human and non-human) today: about 570 million tons versus 600 million tons, respectively. Among the factors they considered in their model were:

- Sauropod metabolism, such as whether they were endothermic or ectothermic;
- Mode of digestion, in which they assumed sauropods used gut bacteria or archeans to break down low-quality plant matter;
- An average sauropod size of 20 tons, which was typical for the Late Jurassic sauropod *Apatosaurus;*
- Estimated sauropod herd size;
- Land area needed to ecologically support this dinosaur biomass;
- High ecological productivity of land ecosystems during the Mesozoic, which was partly a function of how average global temperatures were much warmer than today.

So did Mesozoic global warming begin before dinosaurs added their gaseous contributions to the atmosphere, or afterwards? This is actually a tough question to answer, because unlike other times in earth's history, the Mesozoic lacked long periods of global cooling. Instead, it was warm throughout almost its entire 185 million years, from the start of the Triassic Period to the end of the Cretaceous Period. Dinosaurs were a part of this equation for the last 165 million of those years, but the grandest of herbivorous gas-bag dinosaurs—sauropods—did not show up until about the middle of the Jurassic, which is when they were joined by stegosaurs, ankylosaurs, and nodosaurs. Later, in the Cretaceous, sauropods were accompanied by great numbers of big ornithopods and ceratopsians. This means that a warm climate was started before the Late Triassic—which was when the dinosaurs arrived on the scene—but dinosaurs may have done their part to keep the home fires burning once they evolved, spread, and farted throughout the Jurassic and Cretaceous.

Nevertheless, as intriguing as this model of dinosaur-caused warming might sound, it's based on many assumptions and with lots of leeway granted for some of the factors. For instance, the jury is still out on whether or not some herbivorous dinosaurs—such as sauropods, ornithopods, stegosaurs, ankylosaurs, and ceratopsians—used fermentation in their hindguts to digest their food. As mentioned previously, gut-produced methane comes from anaerobic bacteria and archaeans, which together are called *methanogens*. Many modern animals, from termites to cattle, have this symbiotic relationship in which these methanogens live in their intestines and through fermentation break down otherwise low-quality plants (think "woody"). Although paleontologists know from coprolites that some dinosaurs had gut bacteria, the evidence is otherwise scanty for methanogens having shared time with dinosaur innards.

As a result, paleontologists sometimes have to point to an absence of trace fossils in order to say hindgut fermentation (and hence methane production) was common in large herbivorous dinosaurs. In this instance, the rarity of gastroliths in most herbivorous dinosaurs is now regarded as a good reason to think that sauropods and other hefty herbivores used hindgut fermentation. Gastroliths were always considered as one way for herbivorous dinosaurs with pencil- or leaf-like teeth to grind up their food, thus easing their digestion. But with an awareness that gastroliths are not as common in sauropods as once supposed—or used more as mixing stones, rather than grinding stones—and that they are almost unknown in stegosaurs, ankylosaurs, ceratopsians, and large ornithopods, other digestive aids have to be considered. Granted, some of these dinosaurs, such as ceratopsians and hadrosaurs, had terrific teeth for shearing and grinding plants. But no amount of chewing can break down cellulose, which requires biochemical intervention. So without gastroliths, these dinosaurs were likely getting help from their little anaerobic friends, which in turn released great volumes of methane.

Unfortunately, given this revelation about the probable role of dinosaurs in methane production and their contribution to Mesozoic atmospheres, a few climate-change deniers have latched on to it and made pithy statements such as, "Dinosaurs caused global warming, too. So what's the big deal now?" The big deal is in the numbers provided by Wilkinson and his colleagues: in a contest for methane production in which dinosaurs take on modern humans, humans would still win. This means that humans are unwittingly conducting an experiment to find out whether we can reproduce Mesozoic temperatures without dinosaurs. If we succeed by confirming the hypothesis—elevated levels of methane in our atmosphere cause significantly greater global temperatures—the consequences are dire for many species, including ours. Hence, this possible dinosaur trace, one that may have maintained globally high temperatures and therefore high sea levels—thus affecting ocean ecosystems—is a lesson in what trials of earth history do not need repeating during our time here.

Look at a Flower, See a Dinosaur

As hinted at in previous chapters, paleobotanists are often the Rodney Dangerfields of paleontology, garnering only slightly more respect than ichnologists. Fortunately, their relevance among vertebrate paleontologists—and especially dinosaur paleontologists—has been bolstered in the past thirty years or so through their pursuing the answer to a provocative question: Did herbivorous dinosaurs have something to do with the evolution of flowering plants? In short, are modern flowering plants an evolutionary trace of Mesozoic dinosaur behavior?

What is brilliant about this question is that, unlike many rhetorical inquiries with "yes" or "no" answers that are uttered with confident finality, the answers are probably "Yes, in part" or "No, not exactly." These uncertain responses appropriately lead to broader discussions of whether herbivorous dinosaurs were the primary movers and shakers in the evolution of land plants during the Mesozoic, or whether plants were in the driver's seat, or whether a combination of the two took place. In other words, more good science came out of following and testing this possible dinosaur–flower connection.

Robert Bakker was the first dinosaur paleontologist to suggest that dinosaur feeding might have "invented" flowering plants (*angiosperms*) from non-flowering seed plants (*gymnosperms*), such as cycads, ginkgoes, and pines. His reasoning was that during the Early Cretaceous (about 135 *mya*), grazing and low-level browsing ornithopods mowed down gymnosperms so much that angiosperms, with their low-lying flowers and fruits, were favored over these plants. Consequently, flowering plants proliferated, diversified, and took over terrestrial ecosystems, which they have ruled ever since. This was an imaginative idea, but Bakker was not a paleobotanist, so his supposed connection between herbivorous dinosaurs and flowering plants seemed tenuous to most other paleontologists.

Among the first paleobotanists to start examining whether flowering plant and land-dwelling vertebrate evolution went hand in hand (or rather, leaf in mouth) were Scott Wing and Bruce Tiffney. In several articles published in the late 1980s, they proposed that herbivorous dinosaurs and the earliest flowering plants had what they termed a "reciprocal relationship," which resulted in their coevolution. This relationship meant dinosaurs that ate these plants helped with dispersing their seeds (more on that topic later). Moreover, as angiosperm seeds were placed in new ecosystems, this affected their natural selection and ultimately their evolution throughout the rest of the Cretaceous Period. In turn, as new species of angiosperms arose, some produced more specialized flowers and fruits, or defenses against herbivory (such as toxins and spines), which contributed to the natural selection of herbivorous dinosaurs. In other words, the plants were partially responsible for driving the evolution of new species of herbivorous dinosaurs that were more specialized in their feeding. In contrast, the first herbivorous dinosaurs to eat angiosperms may have been more generalists, eating every item from their Early Cretaceous all-you-can-eat salad bars.

However, other paleontologists, including some paleobotanists, have critically examined and questioned a direct cause-and-effect of herbivorous dinosaurs on flowering plants. In a 2001 study done by dinosaur paleontologist Paul Barrett and paleoecologist Katherine Willis, wherein they examined how herbivorous dinosaurs and fossil angiosperms overlapped in time and space during the Cretaceous, they could not conclusively say "Dinosaurs invented flowers." Instead, they concluded that other animal–plant interactions—such as between insects (especially bees and wasps), mammals, and plants—probably had a greater impact in angiosperm evolution, as well as extended periods of global warming. Yet they also admitted that dinosaurs and flowering plants likely affected each group's evolutionary histories, even though the evidence supporting this was still fairly meager.

Similarly, another review done in 2008 by Graeme Lloyd and eight other paleontologists examined how what they called the "Cretaceous Terrestrial Revolution" implicated many groups of animals—not just dinosaurs—in the evolution of angiosperms. They also found that the supposed "boom" in the biodiversity of dinosaurs toward the end of the Cretaceous was likely a consequence of paleontologists looking more closely for dinosaur fossils in Late Cretaceous rocks; after all, the more you look, the more you find. (Granted, the Late Jurassic Morrison Formation has been studied intensively too, but its dinosaurs are still not as biodiverse as those from the Late Cretaceous.) They also pointed out that direct evidence of dinosaurs eating fruits or other parts of flowering plants is actually quite rare, and only represented by a few trace fossils such as dental microwear in hadrosaurs, an ankylosaur cololite, and ornithopod and sauropod coprolites.

So what evidence would be needed for those who find themselves romantically attached to the idea that dinosaurs had somehow contributed to Valentine's Day festivities and were behind the evolution of roses, tulips, petunias, azaleas, and other gorgeous flowers we see, smell, and appreciate today? As just mentioned, some Cretaceous dinosaur trace fossils gave paleontologists specific examples of how these dinosaurs interacted with plants in their ecosystems. Also, in places where herbivorous-dinosaur tracks are common but their bones are rare or absent, these trace fossils help paleontologists to at least say whether or not these dinosaurs were in the same places as flowering plants, which they could have eaten. But these few trace fossils are like single frames taken from a 135-million-year-long movie, not showing the fuller connections between herbivorous dinosaurs and flowering plants throughout the Cretaceous Period. As a result, more comprehensive trace fossil studies are needed, and perhaps from Cretaceous rocks in one area, better enabling paleobotanists, vertebrate paleontologists, and ichnologists to assess what happened over time.

Still, flowering plants as dinosaur fodder are only one factor to consider when thinking about their evolution. Sure, if angiosperms grew and reproduced quicker than gymnosperms after being chomped by sauropods, hadrosaurs, ankylosaurs, and ceratopsians, that was an important trait in their favor. Yet two other dinosaur behaviors must have affected plants and are related to dinosaur traces: stomping and pooping.

For the former, whatever low-lying plants dinosaurs were not eating, they were stepping on. Hence, flowering plants may have had traits that allowed or promoted their survival after dinosaur feet had compressed them. Some of these traits might include different root systems, which allowed for more clonal growth into new shoots post-trampling, or increased selection for *vegetative growth*, in which pieces of a flowering plant take root when dropped somewhere other than their original home. Dinosaur feet also might have carried flower pollen or seeds to other places, helping plants as either sex surrogates (pollination) or scattering their babies (seed dispersal), something today's animals still do *en masse* for flowering plants. Dinosaurs trampling around freshwater environments also would have disturbed those sediments sufficiently that some gymnosperms were excluded, but made conditions perfect for angiosperms to take hold as "weedy" species that thrived in mixed-up soils.

Lastly, with regard to dung and its effect on flowering plants, also think "fertilizers," and on an immense scale. As a hint of how dinosaur feces might have affected soils and plants during the Mesozoic, paleoecologists, in a 2013 study done on extinctions of large herbivorous mammals in the Amazon River Basin at the end of the Pleistocene Epoch (about 12,000 years ago), found that after these mammals died out, the soils there never quite recovered. The paleoecologists concluded that the magic ingredient missing from these soils was large-mammal excrement. The big herbivores—which included giant ground sloths, glyptodonts (armadillo-like animals the size of a compact car), and elephant relatives—had acted as agricultural agents, spreading the wealth (so to speak) and donating nutrients to soils wherever their droppings dropped. Also, because they were big animals, they traveled greater distances, meaning their dung was distributed far and wide. However, once this copious supply of enriched organic matter was gone, soils suffered, which accordingly meant plant communities grew less exuberantly and became less diverse.

The main implication of this research was that the presence of large defecating herbivores was very important for maintaining plant communities during the Pleistocene in that part of South America. Furthermore, their long-time presence probably affected the evolution of angiosperms there, and few places in the world are more famous for their floral diversity than the Amazon basin. Likewise, whenever great poopers die out, one should also mourn for dung beetles, which die with them; this meant that insect biodiversity also declined with the demise of the megafauna.

So now take this concept much further back into the geologic past, such as during the Cretaceous, and ponder the effects of herbivorous dinosaur feces on enriching nutrients in soils and plant growth. Although not all dinosaur feces survived to become coprolites, these traces surely contributed to the fruition of flowering plants, and thus the evolutionary legacy of flowering plants embodies these fecal traces.

Scared Green? The Possible Effects of Predatory Theropods on Vegetation

What about large predatory theropods, those poor neglected dinosaurs that almost no one seems to care about, nor remember? How did these big carnivores relate to the evolution of land plants, including flowering ones? Putting oneself in their places, thick vegetation either would have served as great cover for ambush predation or gotten in their way when chasing down their prey. But that's thinking too small. When viewed from an ecological perspective, one should rather imagine how abundant, large, healthy theropods probably maintained river–floodplain plant communities and facilitated the growth of low-lying vegetation and forests wherever they lived. In other words, large theropods, such as *Allosaurus*, *Acrocanthosaurus*, *Gigantosaurus*, and *Tyrannosaurus*, might have been the original "green" dinosaurs, saving plants wherever they stalked, and hence helping the evolution of their ecological communities.

This seemingly incongruous leap of logic is loosely based on recent research into the effects of apex predators on riverbanks (called *riparian zones*) and forest ecosystems, exemplified by wolves in the vicinity of Yellowstone National Park (Wyoming). Since the 1990s in Yellowstone, wildlife biologists and ecologists have examined the ecological effects of wolf reintroductions, in which wolf packs were put back in places where they had been locally extinct for a while. One of the most surprising results of these reintroductions was how riparian ecosystems in Yellowstone improved, with greater and more vigorous plant growth that approached their recent, pre-colonial state, when wolves naturally inhabited this area. Rapid stream erosion and flooding also lessened, both direct results of more plants growing along stream banks.

What did this have to do with wolves, or even carnivorous dinosaurs? First, with regard to wolves, their favorite item on the menu—one that they will pick nearly every time if given a choice—is elk (*Cervus canadensis*). With no major threat from predators in Yellowstone ecosystems, elk ran wild (more so), overpopulating and eating much of the vegetation, including young tender plants along riparian zones. Reduced numbers and heights of plants along streams meant fewer plant roots holding down the soils, which led to accelerated erosion and flooding around Yellowstone streams, making it tougher for new plant growth to take hold. But once wolves were back in the neighborhood, salads that were once taken leisurely plummeted, riparian plant communities bounced back, and streams no longer lost so much sediment or flooded with quite so much ferocity.

Part of this situation was because wolves killed and ate some of these gluttonous plant munchers, but they also managed to exert a sort of mind control over their prey. For instance, once wolf packs had hunted elk over several generations in this area, these herbivores restricted themselves to eating vegetation only in certain places, and skittishly, with the threat of death as a big motivator for not hanging out in any one place and browsing too long. This fear factor even caused elk to have smaller families, as the added stress of possible predation triggered hormones that decreased female-elk fertility. With fewer elk, and elk eating less, plant communities expanded and became more contiguous. This effect even helped wolves' super-friends, grizzly bears, which then had lots more berries to eat during lean times. In short, wolves helped to change the ecosystems around and in Yellowstone National Park for the better, and as a result riparian plants there breathed a sigh of oxygen-laden relief.

Now imagine this situation with theropods as the predators—whether as pack hunters, or carrying the biomass of a dozen wolf packs in a single body—and herds of big herbivores as prey. Transfer these same concepts to their Mesozoic ecosystems, in which certain predators kept certain herbivores in check, preventing them from staying in any one place and eating too many plants. Consider all of the healed bite marks, other toothmarks, gut contents, coprolites, and other trace fossils that tell us about predator–prey relations between dinosaurs at different times during the Mesozoic. Then multiply these trace fossils by millions to recreate what must have happened over more than 150 million years, and envisage the aggregate effects of predators on herbivores in their plant-filled ecosystems.

Also contemplate how Mesozoic streams may have changed their dynamics—flow patterns, erosion rates, and flooding—according to a presence or absence of predatory theropods. Lastly, visualize how riparian plants, flowering or otherwise, then thrived and were more able to pass on their genes to future generations. If any or all of these scenarios happened, then these are additional subtle but large-scale dinosaur traces, ecological echoes of the interplay between dinosaurs seeking their respective sustenance.

Worldly Traces: Birds, Pollination, Seed Dispersal, and Hitchhiking Animals

As modern dinosaurs, birds have surely changed the world in small ways through their extremely varied behaviors and their resultant traces. Bird tracks, nests, burrows, beak probes, drillholes, cough pellets, gastroliths, feces, and tools certainly constitute bird calling cards, letting you know that individual birds have visited almost everywhere you look. Yet bird behaviors and their vestiges have also changed the face of terrestrial environments, resulting in the ultimate trace of their reign as Cenozoic dinosaurs.

Glance at nearly any landscape, and then look more carefully for a flowering plant in it, which you will likely find without much searching. (Hint: all grasses are flowering plants.) Chances are good that a bird was somehow involved in the evolutionary heritage of that plant for at least the past 65 million years. Now think about how flowering plants range in habitat from seashores to mountains, from deserts to freshwater ponds, and from Arctic tundra to tropical rainforests. Also consider how flowering plants often dominate those ecosystems through sheer numbers, or play integral roles as keystone species: remove certain flowering plants, and some ecosystems become ghosts of their former selves.

How did birds influence this situation, helping flowering plants to live almost everywhere on the land? Much of this world-altering activity came about through the special relationship between birds and these plants. Coincidentally (or not), early birds and flowering plants expanded and diversified at about the same time, which was in the middle of the Cretaceous Period (about 100–125 *mya*). Although paleontologists cannot say for sure that birds helped with the spread and evolution of flowering plants during the last half of the Cretaceous, or that flowering plants aided bird evolution, or that a combination of the two happened, the ecologically tight relationship we see today between these two suggests that they did indeed co-evolve.

To be sure, insects—especially pollinators, like bees, wasps, beetles, and others—were a big part of this picture, too. But once theropods, both non-avian and avian, began climbing trees and using powered flight, they must have also sought food resources in those trees, which surely included fruit. In some of the earliest studies done of this phenomenon, ecologists estimated that about half to 90% of all fruited trees of modern forests are adapted for birds and mammals to eat them. Similarly, dinosaurs, including birds, must have been powerful change agents, spreading flowering plants to places they never could have reached through other means. These actions even brought about changes that later benefited the evolution of tree-dwelling mammals, including those in our own lineage. Look at a friend or relative, then yourself, and thank a dinosaur for helping to shape the ecosystems that aided your shared ecological and evolutionary legacy.

The way Mesozoic non-avian dinosaurs and birds accomplished this momentous task, which was carried on as an evolutionary tradition by birds throughout the Cenozoic Era, was through their poop. Very simply, flowering plants produce seeds covered by delicious and nutritious fruit. Birds are among the animals that eat these fruits, seeds and all, which they carry in their bellies. Birds then later deposit the seeds somewhere else, while helpfully covering them with a nice mix of nitrogen- and phosphorus-rich fertilizer.

For an extreme example, recall the previously mentioned cassowaries of Australia and New Guinea, big flightless birds that eat the fruits of more than a hundred species of flowering plants and later dump the seeds of these plants, which are ensconced in voluminous piles of feces. Now apply this same thought to small, flight-worthy birds that ingest seeds in berries or other fruits, then fly away from those plants to drop seeds tens of kilometers away. No big deal, you might think: gravity would have done the same thing, through fruit just falling off plants, rolling a little bit downhill, or perhaps was aided by wind or water, which, after tens of millions of years, would have very gradually extended the geographic range of those plants. The same would have happened with islands, in which ocean currents or storms would have given these seeds a one-way ticket to a new home. Who needs birds?

Flowering plants do. The huge difference between "pre-birds" and "post-birds" for flowering plants was in rates of long-distance dispersal, which with the assistance of birds became hundreds of times faster and much more regular than relying on mere chance. Multiply these faster rates by the cumulative effects of generations of birds and angiosperms, and then add the effects of migrations to this equation. For instance, if Cretaceous birds started to move great distances annually, including over mountains and seaways that previously were barriers to plants and their seeds, the geographic spread of angiosperms would have accelerated dramatically. Birds aiding the long-distance dispersal of flowering-plant seeds—through eating fruit, carrying seeds, and defecating—must have transformed landscapes in an astonishing way over the last half of the Cretaceous Period. This Mesozoic crap was evolutionary gold.

How do ingested seeds resist digestion? Most have a hard coat that enables them to pass through the harshest of acidic digestive systems unscathed. This even happens in the guts of alligators and crocodiles, some of which eat a surprising amount of fruit. The bonus for these seeds is that they temporarily stay in a warm, moist place—namely, an animal's digestive tract—exit that place with a good amount of high-quality plant food on top, and grow up in a place different from where their parents lived. Birds make this happen more than any other animals. Granted, mammals do their part in playing the role of Johnny Appleseed, too, as a huge number of mammals are fruit eaters and very good at taking seeds to new places. For instance, fruit bats, justifying their common names, eat fruits, fly to other places, and defecate seeds covered with wondrous bat guano.

However, the big difference between birds and most mammals (including bats) is in their evolutionary histories. Throughout much of the Mesozoic Era, mammals were subordinate to dinosaurs in their shared ecosystems, and as far as we know no mammals evolved powered flight until nearly 15 million years after the end of the Cretaceous. On the other hand, the earliest birds that evolved from non-avian dinosaurs likely witnessed the first flowers. Although flying insects were also around then, they played more of a role in eating other plant parts and pollinating; in most instances, they would have been too small to carry seeds elsewhere. In other words, plants and birds have had a much longer time to get to know one another in an evolutionary sense, and this co-dependency is so deeply rooted that mammals have only added to the apple cart, not yet upsetting it.

Like all co-dependencies, though, a dark side emerges when one asks: What if I am a bird who does not carry out a plant's wishes? For plants, revenge is a dish best served fruity, as some flowering plants discourage seed eating by poisoning animals that dare to digest their seeds. For instance, apples and cherries are perfectly fine foods for humans and many other animals. But do not chew and swallow, say, a cup of apple seeds: every seed contains a small amount of cyanide, a very nasty toxin that interferes with oxygen absorption. Likewise, cashew nuts, which are the seeds inside the fruits of cashew trees (*Anacardium occidentale*), have poisonous shells, so every nut must be extracted from its shell before enjoying them. Hence, flowering plants used a "carrot and stick" approach in their co-evolution with birds. First, reward animals that eat your fruit, carry your children to a new, far-away land, and "plant" it with droppings there. Second, punish animals that try to take their hunger one step further by eating your children.

The role of birds in specially delivering plant seeds to novel places is now well documented, particularly for islands. Charles Darwin even thought of this, as he wondered how the isolated Galapagos Islands off the coast of South America had managed to acquire such thriving plant communities. Ocean currents and winds—which can transport seeds long distances—certainly played a role. But this assumes too much: after all, not every seed floats, nor do all seeds survive being immersed in salty water for long journeys, nor does each seed stay aloft once airborne. They needed help, and birds stepped in to do this, probably starting in the Cretaceous.

Yet birds did not just ferry about the potential offspring of plants to islands and other places: they also carried animals. Given this thought, it might be tempting to conjure an image of an avian airlift in which squadrons of cranes and storks on a continental landmass each grasp a small mammal or lizard, and one by one fly them to the nearest island and drop them off. They would then repeat this over generations, but taking these vertebrates to different islands, and farther afield. Assuming that each payload included both males and females of those animals, the islands eventually all got colonized. End of story.

Well, not quite. If birds want to go long distances without exhausting themselves, they are much better at handling very small passengers, ones they do not even notice as stowaways. These hitchhikers also can ride in larger numbers, which improves their chances of reproductive success when they arrive at their new destinations. Animal immigrants could include members of these birds' microbiomes, such as lice and other skin parasites, but other accidental tourists—such as snails, larval insects, or larval crustaceans—also can attach to bird feet. The best feet for these animals to latch onto are webbed ones, which have the most surface area; moreover, webbed feet tend to step into environments with lots of aquatic larvae. This means that ducks, geese, seagulls, pelicans, and other birds with webbed feet (palmate and totipalmate) are among the best candidates for taking off with the highest number of inadvertent travelers. The likelihood of this scenario is further improved if those bird feet step into mud, which acts as a temporary glue for sticking seeds and tiny invertebrates onto their pilots.

Based on fossil tracks from Korea, we know that palmate bird feet had evolved by about 120 *mya* (Early Cretaceous). Moreover, webbed bird tracks became more common and bigger throughout the rest of the Cretaceous. This meant that more birds evolved to shoreline habitats, and hence were more capable of picking up little invertebrates on their feet, taking them to ecosystems and places their ancestors had never before experienced. All of this implies that perhaps the largest living traces of dinosaurs, and ones we still live with every day, are these bird-assisted patterns of biogeography.

Darwin, Hitchcock, and the Dinosaur-Bird-Trace Connection

The idea of modern avian dinosaurs and their predecessors dispersing both plants and animals seems so brilliantly modern, one might wonder how scientists figured this out. Perhaps they tagged birds with GPS chips and then tracked their movement in real time with satellites, mapping and otherwise analyzing their routes with computers. Even better, the invertebrates were probably identified from afar by scanning the birds with lasers, and the scans were then converted to 3-D images, enlarged, and reproduced on a 3-D printer. Or maybe the researchers used other high-tech tools, all of which made science reporters giddily compare these to something they once saw on *Star Trek*, regardless of whether it was the original series, *The Next Generation*, *Deep Space Nine*, or *Voyager*. (We will not speak of *Enterprise*.)

But if you ever play a word-association game and the words "brilliantly modern" are referred to a concept associated with evolution and island biogeography, it is best to just answer "Darwin." Yes, that's right, Charles Darwin thought of flighted birds taking both seeds and small invertebrates to new homes, and repeating these actions over many generations. He even set up a few experiments to test this idea, none of which involved using GPS-enabled devices, satellites, computers, lasers, 3-D printers, or emergency medical holograms. For one experiment, he simply dipped duck feet into a pool of water with aquatic snails, watched them crawl onto the feet, then took the duck feet out of the water to see how long the snails stayed attached. In short, he just observed, questioned, tested, observed more, and then wrote a carefully worded conclusion based on the preceding. From such methods, he not only devised some of the most important tenets of modern evolutionary theory, but also explained how flowering plants, freshwater invertebrates, and marine invertebrates were able to travel to faraway places through the power of birds.

What Darwin did not know, though, was that through his study of birds and their long-distance movement of seeds and animals, he was also studying the ichnology of dinosaurs. In the late 19th century, Darwin was aware of the few Jurassic and Cretaceous non-avian dinosaurs that had been discovered in the United Kingdom and elsewhere, as well as the Late Jurassic *Archaeopteryx* from Germany. But he did not know, nor could he have suspected at the time, that birds are dinosaurs.

Interestingly, fossil theropod and ornithopod tracks were already known during Darwin's lifetime, but their makers had been misidentified. For instance, in 1845 he corresponded with ichnologist Edward Hitchcock of Amherst College in Massachusetts, who surmised that Late Triassic and Early Jurassic dinosaur tracks from the Connecticut River Valley were oversized bird tracks. In that letter to Hitchcock, in which Darwin refers to these dinosaur tracks as "footsteps," he mused about the meaning of these tracks with relation to birds:

> *In my opinion these footsteps (with which subject your name is certain to go down to long future posterity) make one of the most curious discoveries of the present century & highly important in its several bearings. How sincerely I wish that you may live to discover some of the bones belonging to these gigantic birds: how eminently interesting it would be (to) know whether their structure branches off towards the Amphibia, as I am led to imagine that you have sometimes suspected.*

Had both Hitchcock and Darwin lived to today, they would have been disappointed to learn that this wistful hope of finding bones of those "gigantic birds" (dinosaurs, as it were) would have been largely unfulfilled. As of this writing, despite continued discoveries of theropod and ornithopod tracks, almost no dinosaur bones are known from the Connecticut River Valley. Furthermore, body fossils of birds are unknown from rocks older than the Late Jurassic anywhere in the world.

Nonetheless, even though Darwin and Hitchcock missed the connection between dinosaur tracks, the evolution of birds, and biogeography, they each in their own way linked these seemingly disparate subjects. What Hitchcock got right, but without knowing it, was how the close resemblance of theropod tracks to those of modern birds did indeed reflect their evolutionary relatedness. What Darwin also got right was how birds changed the face of our modern ecosystems. Darwin had also very likely seen tracks of rheas (*Rhea pennata and R. pennata*) during his travels in South America, which closely resembled the fossil tracks Hitchcock described. Most important, though, both of their initial studies cleared intellectual trails, on which subsequent generations of biologists and paleontologists, likewise making careful observations of tracks and other traces, led to broader and more eclectic views of how the earth's surface has been and will continue to be shaped by the traces of dinosaur behavior.

Regardless of how quickly our technological tools might serve us in such endeavors, the preceding methods and concepts can be done and understood by using the greatest scientific tools we already possess—our senses and cognition.

Dinosaurs Without Bones: Trace Fossils of the Future

So now that we know that we are living with dinosaur traces, both ancient and modern, we can also better appreciate the role of individual dinosaur trace fossils in our understanding of these iconic animals. But what then to do with this ichnologically bestowed enlightenment, a newly realized super-power that allows for recognizing how dinosaur trace fossils superbly augment and oftentimes surpass the paleontological worth of dinosaur bones in interpreting dinosaur behavior?

Obviously, we must apply what we've learned about dinosaur trace fossils of our paleontological past to those of the future. With that goal in mind, here are words of advice and predictions, given with full awareness that my own knowledge of dinosaur trace fossils, much like a landscape occupied by dozens of small burrowing ornithopods, is full of holes.

We'll find more dinosaur trace fossils, and they'll be different. One of the most excitedly received dinosaur books in recent years was *All Yesterdays* (2012), coauthored by paleontologists Darren Naish and C.M. Kosemen and co-illustrated by paleoartists John Conway and Scott Harman. Although only 100 pages long, and with its illustrations taking up more space than its words, its inventive artistic renderings and descriptions of unexpected behaviors in dinosaurs—such as tree-climbing ceratopsians and mud-wallowing sauropods—wowed many of its fans. (The book's subtitle very easily could have been *Dinosaurs Gone Weird.*) That book's popularity then led to a sequel published in late 2013, titled *All Your Yesterdays,* bearing contributions by a variety of artists in which they portrayed yet more unconventional dinosaur behaviors. This sort of cooperation between paleontologists and visual artists—including those who create computer-generated imagery—points toward a potential fountain of creativity that can expand our perception of dinosaurs as real, living animals.

The main point of these books was to promote thinking a little differently about dinosaur behaviors. In that spirit, I will now take it a step further in this book and ask us to think differently about trace fossils that could be made by different behaviors and different dinosaurs. Furthermore, because dinosaur trace fossils are so much more common than their bones in most places, finding odd trace fossils should be easier than finding bones of odd dinosaurs. Sure, the usual theropod, ornithopod, and sauropod tracks will continue to be discovered nearly every day. Yet I have a more ambitious wish list of trace fossils, some of which I sincerely hope will be uncovered by ichnologically adept paleontologists in upcoming years. These are a few of my favorite things:

- Ceratopsian and pachycephalosaur trackways that confirm these big-headed dinosaurs really did smash into one another with their bony accouterments. Or, more interesting, show that they did not, and instead got along famously and that those big heads and accompanying horns and shields were there for some other purpose.
- More examples of dinosaur urolites. After all, they all had to go some time.
- Clear, definitive dinosaur regurgitalites, ideally with partially digested food directly associated with a probable regurgitator nearby.
- Long, continuous trackways made by really big theropods. I'm talking about trackways made by 7–9 ton Cretaceous theropods like *Spinosaurus* or *Carcharodontosaurus* of northern Africa, *Gigantosaurus* of South America, or *Tyrannosaurus* of North America. Such trackways would give us much more information on how these dinosaurs really moved, while also reeking of awesomeness.
- Sauropod trackways that demonstrate whether or not these animals traveled in family structures like modern elephants, with mixtures of young and old.

- More detailed studies of microwear on ornithopod teeth, illuminating how these dinosaurs ate, and in some instances what they ate.
- Dinosaur toothmarks preserved in unusual substrates, like molluscan shells or eggshells, and not just bones.
- Closer looks at broken and healed lower limb bones in theropods to see if these were the marks of cranky stegosaurs and angry ankylosaurs who objected to unwanted attention from those theropods. Likewise, more healed tail spikes and clubs of these dinosaurs, further showing how they disciplined their oppressors.
- Dig marks in fossil termite or ant nests that match the dimensions of therizinosaur hands or the limbs of other unusual dinosaurs.
- More dinosaur ground nests and nesting grounds. So few of these are known as trace fossils, yet more must be out there, waiting to be recognized.
- Burrows made by small theropods or other dinosaurs. Why restrict all of the underground fun to small ornithopods?
- More gastroliths, whether in or out of their dinosaur hosts: who really had them, who really lacked them?
- Enterolites or coprolites that show clear evidence of insectivory or omnivory in dinosaurs. They weren't just all "meat eaters" or "plant eaters" (how boring).
- Dinosaur sex traces. (Need I say more?)

Although paleontologists may find none of these, or only stumble upon a few of them, if any are found we should all celebrate their success with alacrity. What is also gratifying to know is that some of the people who find these may not even be professional paleontologists, but people who happened to be looking in the right place at the right time and with the right search image in mind. Paleontology, archaeology, and astronomy are among the few sciences in which amateurs regularly make important contributions through their numerous eyes on the ground (or on the stars). Let the games begin.

Nevertheless, we should describe and interpret these trace fossils with care. One of the best facets of the fossil record is that it gets better every day, and trace fossils are a big part of this daily improvement. Yet of the dinosaur trace fossils found, many either remain undescribed or (far worse) have been under-described, without enough meaningful attention to detail to be faithful records of dinosaur behavior. If dinosaur trace fossils are treated like curios that are simply named "track," "toothmark," or "coprolite," for instance, and linked to a possible dinosaur, they will become as dead as their makers. Technology will certainly play a part in better describing trace fossils. For example, 3-D images and 3-D printers of dinosaur tracks have already allowed researchers on other sides of the globe to work cooperatively, or to add to one another's observations.

Remember that dinosaurs are still making traces. With 10,000 species of extant birds, all of which are making traces, we have no shortage of modern analogs for dinosaur behaviors. Granted, these dinosaurs are limited to avian theropods, but their tracks, nests, burrows, gastroliths, feces, and other traces supply many opportunities to think about how dinosaurs both avian and non-avian might have behaved, and whether they made similar traces. Furthermore, although crocodilians are only distant relatives of dinosaurs, these big archosaurs are wonderful examples of big scaly four-legged toothy tracemakers that fit as tracemaking surrogates for a few dinosaurs and their Mesozoic relatives.

Become ichnologically imaginative. As one might have intimated from the wish list given above, I am encouraging everyone interested in dinosaurs to speculate about behaviors dinosaurs may have done and the traces that resulted from these behaviors. In the spirit of leading by example, I've done a few flights of reasonable fancy in this respect throughout the book, with the hope that more people will do the same. As a result, I fully expect dinosaur enthusiasts, armed with a heightened awareness of trace fossils and their importance in our understanding of dinosaurs, to make and share a surfeit of their own creative traces, both whimsical and serious.

So as we all make our own traces in our everyday lives, also imagine which ones will outlast us all. Will any of our traces also survive longer than the tracks, nests, burrows, and toothmarks of dinosaurs? Or will dinosaur trace fossils somehow surpass our own meager ichnological record? They already have had a 230-million-year head start, and with thousands of living descendants sharing the earth now, the odds are stacked against us. But no matter. This renewed appreciation for dinosaurs is now permanent, and one that recognizes another dimension of them beyond their bones. Place your hand on a dinosaur track, and you connect with the breathing essence of its maker, leaving your own fingerprints on it: life traces intersecting through time.

Notes

CHAPTER 1: SLEUTHING DINOSAURS

p. 10 "It turns out a good number of *Triceratops* head shields, which are composed of paired parietal and squamosal bones, bear deformities in the squamosals." Farke, A.A., Wolff, E.D.S., and Tanke, D.H. 2009. Evidence of combat in *Triceratops*. *PLoS One*, 4(1): e4252. doi:10.1371/journal.pone.0004252.

p. 10 "Ceratopsians, a group of dinosaurs that includes *Triceratops* and related horned dinosaurs, also made tracks, which are preserved in Cretaceous rocks from about 70 million years ago. . . ." Ceratopsian tracks are still quite rare, but a few are known from Colorado, Utah, and Alaska: (1) Lockley, M.G., and Hunt, A.P. 1995. Ceratopsid tracks and associated ichnofauna from the Laramie Formation (Upper Cretaceous: Maastrichtian) of Colorado. *Journal of Vertebrate Paleontology*, 15: 592-614; (2) Milner, A.C., Vice, B.S., Harris, J.D., and Lockley, M.G. 2006. Dinosaur tracks from the Upper Cretaceous Iron Springs Formation, Iron County, Utah. *In* Lucas, S.G. and Sullivan, R.M. (editors). 2006 Late Cretaceous vertebrates from the Western Interior. *New Mexico Museum of Natural History and Science Bulletin*, 35: 105-113.

p. 10 "In the Cretaceous Period, about 95 million years ago and on a lakeshore in what is now Queensland, Australia, nearly a hundred small, two-legged dinosaurs ran in the same direction and at high speed." Although this interpretation has been disputed in the past few years (see Chapter 3), the original study of these tracks is: Thulborn, R.A., and Wade, M., 1979. Dinosaur stampede in the Cretaceous of Queensland. *Lethaia* 12: 275-279.

p. 11 "One grouping of swim tracks, made by seven separate theropods, is preserved in Early Cretaceous rocks (110 million years ago) of Spain. Many more swim tracks are in Early Jurassic rocks from about 190 million years ago in southwestern Utah." Two articles on dinosaur swim tracks came out in quick succession in 2006 and 2007: (1) Ezquerra, R., Doublet, S., Costeur, L., Galton, P.M., and Pérez-Lorente, F., 2007. Were non-avian theropod dinosaurs able to swim? Supportive evidence from an Early Cretaceous trackway, Cameros Basin (La Rioja, Spain). *Geology*, 35, 507-510. (2) Milner, A.R.C., Lockley, M.R., and Kirkland, J.I. 2006. A large collection of well-preserved theropod dinosaur swim tracks from the Moenave Formation, St. George, Utah. *In* Harris, J.D., *et al.* (editors), *The Triassic-Jurassic Terrestrial Transition. New Mexico Museum of Natural History and Science Bulletin* 37: 315-328.

p. 11 "In 2007, I helped two other paleontologists document the first known burrowing dinosaur (*Oryctodromeus cubicularis*, a small ornithopod) from Cretaceous rocks (95 million years old) in Montana." Varricchio, D.J., Martin, A.J., and Katsura, Y. 2007. First trace and body fossil evidence of a burrowing, denning dinosaur. *Proceedings of the Royal Society of London, B*, 274: 1361-1368.

p. 12 "Two years later, I interpreted similar burrows in older Cretaceous rocks (105 million years old) of Victoria, Australia." Martin, A.J., 2009. Dinosaur burrows in the Otway Group (Albian) of Victoria, Australia, and their relation to Cretaceous polar environments. *Cretaceous Research*, 30: 1223-1237.

p. 12 "A different type of digging by other dinosaurs also has been inspired by unusual trace fossils found in Late Cretaceous rocks (about 75 million years ago) of Utah in 2010." Simpson, E.L., Hilbert-Wolf, H.L., Wizevich, M.C., Tindall, S.E., Fasinski, B.R., Storm, L.P., and Needle, M.D. 2010. Predatory digging behavior by dinosaurs. *Geology*, 38: 699-702.

p. 12 "These nests also contained clutches of paired eggs, which were arranged vertically in the nests by one or both of the parents after egg-laying." (1) Varricchio, D.J., Jackson, F., Borkowski, J., and Horner, J.R. 1997. Nest and egg clutches of the dinosaur *Troodon formosus* and the evolution of avian reproductive traits. *Nature*, 385: 247-250. (2) Varricchio, D.J., Jackson, F., and Trueman, C.N. 1999. A nesting trace with eggs for the Cretaceous theropod dinosaur *Troodon formosus*. *Journal of Vertebrate Paleontology*, 19: 91-100.

p. 12 "Similarly, a spectacular find of Late Cretaceous nests in Argentina from 70 to 80 million years ago and attributed to gigantic sauropods called titanosaurs shows that dinosaurs other than *Troodon* made ring-like enclosures for their eggs." (1) Chiappe, L.M., Coria, R.A., Dingus, L., Jackson, F., Chinsamy, A., and Fox, M. 1998. Sauropod dinosaur embryos from the Late Cretaceous of Patagonia. *Nature*, 396: 258-261. (2) Chiappe, L.M., Schmitt, J.G., Jackson, F., Dingus, L., and Grellet-Tinner, G. 2004. Nest structure for sauropods: sedimentary criteria for recognition of dinosaur nesting traces. *Palaios*, 19: 89-95.

p. 13 "What has surprised paleontologists in recent years, though, is the realization that a few theropods, a group of dinosaurs once assumed to have been exclusively carnivorous, also have these 'stomach stones.'" Wings, O. 2007. A review of gastrolith function with implications for fossil vertebrates and a revised classification. *Acta Palaeontologica Polonica*, 52: 1-16.

p. 13 "Although no one has yet found gastroliths directly associated with *Struthiomimus*, some of its relatives, collectively called ornithomimids ('ostrich mimics'), do have them." Kobayashi, Y., Lu, J.-C., Dong, Z.-M., Barsbold, R., Azuma, Y., and Tomida, Y. 1999. Herbivorous diet in an ornithomimid dinosaur. *Nature*, 402: 480.

p. 13 "This is backed by healed toothmarks caused by a large predatory theropod preserved in a few bones of *Edmontosaurus*, including at least one with a smoking gun (or tooth, as it were) linking it to *Tyrannosaurus* or its close relatives." Carpenter, K. 2000. Evidence of predatory behavior by carnivorous dinosaurs. *Gaia*, 15: 135-144.

p. 14 "*Triceratops* bones also bear toothmarks that could only have been made by tyrannosaurs, including those that mark the front of the face. . . ." At the time of this writing, this research had only been reported through an abstract, so hopefully a paper is out now: Fowler, D., Scannella, J., Goodwin, M., and Horner, J. 2012. How to eat a *Triceratops*: large sample of toothmarks provides new insight into the feeding behavior of *Tyrannosaurus*. Journal of Vertebrate Paleontology [Supplement to 3], Society of Vertebrate Program and Abstracts, October 2012, p. 60.

p. 14 "Amazingly, not one but two colossal coprolites attributed to tyrannosaurs have been documented, each with finely ground bone and one containing fossilized muscle tissue." (1) Chin, K., Tokaryk, T.T., Erickson, G.M., and Calk, L.C. 1998. A king-sized theropod coprolite. *Nature*, 393: 680-682. (2) Chin, K., Eberth, D.A., Schweitzer, M.H., Rando, T.A., Sloboda, W.J., and Horner, J.R. 2003. Remarkable preservation of undigested muscle tissue within a Late Cretaceous Tyrannosaurus coprolite from Alberta, Canada. *Palaios*, 18: 286-294.

p. 14 "From coprolites, we also suspect that at least some Late Cretaceous hadrosaurs ate rotten wood." Chin, K. 2007. The paleobiological implications of herbivorous dinosaur coprolites from the Upper Cretaceous Two Medicine Formation of Montana: why eat wood? *Palaios*, 22: 554-566.

p. 14 "We even figured out from dinosaur coprolites that at least a few animals—namely, dung beetles—depended on dinosaur feces as 'manna from heaven' to ensure their survival." Chin, K., and Gill, B.D. 1996. Dinosaurs, dung beetles, and conifers: participants in a Cretaceous food web. *Palaios*, 11: 280-285.

p. 14 "Trace fossils, such as dinosaur tracks and burrows in sedimentary rocks from formerly polar environments, tell us that they likely stayed put during the winters." (1) Martin, A.J., 2009. Dinosaur burrows in the Otway Group (Albian) of Victoria, Australia, and their relation to Cretaceous polar environments. *Cretaceous Research*, 30: 1223-1237. (2) Martin, A.J., Rich, T.H., Hall, M., Vickers-Rich, P., and Vasquez-Prokopec, G. 2012. A polar dinosaur-track assemblage from the Eumeralla Formation (Albian), Victoria, Australia. *Alcheringa: An Australasian Journal of Palaeontology*, 36, 171-188.

p. 14 "In one recent study, the earliest ancestors of dinosaurs were proposed on the basis of not-quite-dinosaur tracks in 245-million-year-old rocks in Poland from the earliest part of the Triassic Period." Brussatte, S.L., Niedźwiedzki, G., and Butler, R.J. 2010. Footprints pull origin and diversification of dinosaur stem lineage deep into the Early Triassic. *Proceedings of the Royal Society of London, B*, 278: 1107-1113.

p. 15 "This relatedness has been certified through many lines of evidence, including fossilized feathers directly associated with the skeletons of more than thirty species of theropods." Godefriot, P., Cau, A., Dong-Yu, H., Escuillié, F., Wenhao, W., and Dyke, G. 2013. A Jurassic avialan dinosaur from China resolves the early phylogenetic history of birds. *Nature*, doi:10.1038/nature12168.

p. 15 "A similar behavior can be inferred from tracks in Early Cretaceous rocks of east Texas, in which the footprints of a large theropod paralleled and then crossed those of a sauropod, apparently shadowing it." Farlow, J.O., Chapman, R.E., Breithaupt, B., and Matthews, N. 2012. The scientific study of dinosaur footprints. *In* Brett-Surman, M.K., Holtz, T.R., Jr., and Farlow, J.O. (editors), *The Complete Dinosaur* (2nd Edition). Indiana University Press, Bloomington, Indiana: 713-759.

p. 15 "Other compelling theropod trackways include some from the Cretaceous of China that tell of six theropods, equally spaced and all moving in the same direction, which very much looks like evidence of pack hunting." Li, R., Lockley, M.G., Makovicky, P.J., Matsukawa, M., Norell, M.A., Harris, J.D., and Liu, M. 2008. Behavioral and faunal implications of Early Cretaceous deinonychosaur trackways from China. *Naturwissenschaften*, 95: 185-191.

Chapter 2: These Feet Were Made for Walking, Running, Sitting, Swimming, Herding, and Hunting

p. 17 "Thus far, dinosaur tracks have been found in eighteen states of the

U.S. and on every continent except for Antarctica, with thousands of newly discovered ones each year." Although a little dated, the following three books give a geographic sense for where one might find dinosaur tracks: (1) Lockley, M.G. 1991. *Tracking Dinosaurs: A New Look at an Ancient World.* Cambridge University Press, Cambridge, U.K.: 267 p. (2) Lockley, M.G., and Hunt, A.P. 1995. *Dinosaur Tracks and Other Fossil Footprints of the Western U.S.* Columbia University Press, New York: 338 p. (3) Lockley, M.G., and Meyer, C. 2000. *Dinosaur Tracks and Other Fossil Footprints of Europe.* Columbia University Press, New York: 360 p.

p. 18 "Ideally, then, each recognizable dinosaur bone can be correlated with about six broad groups of dinosaurs. . . ." The latest two publications explaining dinosaur classifications in detail are *The Complete Dinosaur* and *The Dinosauria*, both in their second editions: (1) Brett-Surman, M.K., Holtz, T.R., Jr., and Farlow, J.O. (editors), *The Complete Dinosaur* (2nd Edition), Indiana University Press, Bloomington, Indiana: 1112 p. (2) Weishampel, D.B., Dodson, P., and Osmólska, H. (editors). 2004. *The Dinosauria* (2nd Edition). University of California Press, Berkeley, California: 861 p.

p. 19 "For example, the hypothetical 'first dinosaur,' which would have evolved about 235 *mya* and was the common ancestor to both saurischians and ornithischians. . . ." Figuring out what animal constituted the "first dinosaur" is likely impossible, but two of the earliest known dinosaurs are *Eodromeus* and *Eoraptor*, discussed here: Martinez, R.N., Sereno, P.C., Alcober, O.A., Colombi, C.E., Renne, P.R., Montañez, I.P., and Currie, B.S. 2011. A basal dinosaur from the dawn of the dinosaur era in southwestern Pangaea. *Science*, 331: 206-210.

p. 19 "For the major evolutionary groups of dinosaurs, the following modes of movement and digit numbers, with only a few exceptions, can be applied to help with identifying their tracks." Farlow, J.O., Chapman, R.E., Breithaupt, B., and Matthews, N. 2012. The scientific study of dinosaur footprints. *In* Brett-Surman, M.K., Holtz, T.R., Jr., and Farlow, J.O. (editors), *The Complete Dinosaur* (2nd Edition). Indiana University Press, Bloomington, Indiana: 713-759.

p. 20 "These were unusual theropods, so unusual that no self-respecting dinosaur paleontologist discussing them can complete a sentence with these as the subject without also saying 'strange'. . . ." Although slightly out of date now, a good overview of therizinosaurs is here: Clark, J.M., Maryanska, T., and Barsbold, R. 2004. Therizinosauroidea. *In* Weishampel, D.B., Dodson, P., and Osmólska, H. (editors), *The Dinosauria* (2nd Edition). University of California Press, Berkeley, California: 151-164.

p. 20 "As of this writing, a few therizinosaur tracks are known, with the most astonishing recently found in Cretaceous rocks of Alaska." Fiorillo, A.R., and Adams, T.L. 2012. A therizinosaur track from the Lower Cantwell Formation (Upper Cretaceous) of Denali National Park, Alaska. *Palaios*, 27: 395-400.

p. 21 "*Dromaeosaurids* . . . were bipedal theropods with three toes retained on their rear feet, but only two of those digits contacted the ground." One of the best summaries of theropod traits, including those of dromaeosaurids, is: Holtz, T.R., Jr. 2012. Theropods. *In* Brett-Surman, M.K., Holtz, T.R., Jr., and Farlow, J.O. (editors), *The Complete Dinosaur* (2nd Edition), Indiana University Press, Bloomington, Indiana: 347-378. Once you've read that, you'll be better prepared to focus on the traits of dromaeosaurids: Norell, M.A., and Mackovicky, P.J. 2004. Dromaeosauridae. *In* Weishampel, D.B., Dodson, P., and Osmólska, H. (editors), *The Dinosauria* (2nd Edition). University of California Press, Berkeley, California: 196-209.

p. 21 "*Pachycephalosaurs* . . . often nicknamed 'bone-headed dinosaurs' because of their thick, bony skulls, share a common ancestor with those other big-headed dinosaurs, ceratopsians." For seeing how big-headed ceratopsians and pachycephalosaurs were related through their anatomy: Mackovicky, P.J. 2012. Marginocephalia. *In* Brett-Surman, M.K., Holtz, T.R., Jr., and Farlow, J.O. (editors), *The Complete Dinosaur* (2nd Edition). Indiana University Press, Bloomington, Indiana: 524-549.

p. 22 "But the easiest way to tell the difference between a theropod track and an ornithopod track is to apply three criteria. . . ." (1) Lockley, M.G. 2009. New perspectives on morphological variation in tridactyl footprints: clues to widespread convergence in developmental dynamics. *Geological Quarterly* 53: 415-432. (2) Also see Farlow *et al.* (2012).

p. 22 "However, many prosauropod tracks also show them walking on their rear feet only." Rainforth, E.C. 2003. Revision and re-evaluation of the Early Jurassic dinosaurian ichnogenus Otozoum. *Palaeontology*, 46: 803-838.

p. 22 "So where we originally had none, we now frolic in the land of plenty, as sauropod tracks have been found on all continents except for Antarctica, and in rocks ranging from the Late Triassic (230 *mya*) through the Late Cretaceous periods (65 *mya*)." (1) Lockley, M.G., Wright, J.L., Hunt, A.P., and Lucas, S.G. 2001. The Late Triassic sauropod track record comes into focus. New Mexico Geological Society Guidebook, 52nd Field Conference: 181-190. (2) Carrano, M.T., and Wilson, J.A. 2001. Taxon distributions and the tetrapod track record. *Paleobiology*, 27: 564-582.

p. 23 "For example, the first undoubted stegosaur tracks were not found until 1994, in Middle Jurassic (about 170 *mya*) rocks of England." Whyte, M.A., and Romano, M. 1994. Probable sauropod footprints from the Middle Jurassic of Yorkshire, England. *Gaia Revista de Geociencas,* 10: 15-26.

p. 23 "Now stegosaur tracks are becoming more readily recognized, also having been found in Spain, Portugal, Morocco, and the exotic far-off land of Utah." (1) Pascual, C., Canudo, J.I., Hernández, N., Barco, J.L., and Castanera, D. 2012. First record of stegosaur dinosaur tracks in the Lower Cretaceous (Berriasian) of Europe (Oncala group, Soria, Spain). *Geodiversitas,* 34: 297-312. (2) Mateus, O., Milàn, J., Romano, M., and Whyte, M.A. 2011. New finds of stegosaur tracks from the Upper Jurassic Lourinhã Formation, Portugal. *Acta Palaeontologica Polonica,* 56: 651-658. (3) Belvedere, M., and Mietto, P. 2010. First evidence of stegosaurian *Deltapodus* footprints in north Africa (Iouaridène Formation, Upper Jurassic, Morocco). *Palaeontology,* 53: 233-240. (4) Milàn, J., and Chiappe, L.M. 2009. First American record of the Jurassic ichnospecies *Deltapodus brodricki* and a review of the fossil record of stegosaurian footprints. *Journal of Geology,* 117: 343-348.

p. 23 "Some of these tracks even include skin impressions, the first known glimpse at the scaly feet of stegosaurs." Mateus, O., Milàn, J., Romano, M., and Whyte, M.A. 2011. New finds of stegosaur tracks from the Upper Jurassic Lourinhã Formation, Portugal. *Acta Palaeontologica Polonica,* 56: 651-658.

p. 23 "Now their tracks are documented from places as widespread as Bolivia, British Columbia (Canada), Colorado (USA), and elsewhere. . . ." McCrea, R.T., Lockley, M.G., and Meyer, C.A. 2001. Global distribution of purported ankylosaur track occurrences. *In* Carpenter, K. (editor), *The Armored Dinosaurs.* Indiana University Press, Bloomington, Indiana: 413-454.

p. 24 "The current claim for 'oldest dinosaur from the fossil record' . . . lies with *Eodromeus* ('dawn runner')." Martinez *et al.* (2011).

p. 25 "Other dinosaur fossils from rocks of nearly the same age in Argentina includes one other theropod, *Herrerasaurus,* a basal sauropodomorph, *Eoraptor,* and a primitive prosauropod, *Panphagia.*" Martinez, R.N., and Alcober, O.A. 2010. A basal sauropodomorph (Dinosauria: Saurischia) from the Ischigualasto Formation (Triassic, Carnian) and the early evolution of Sauropodomorpha. *PLoS One,* 4: e4397. doi:10.1371/journal.pone.0004397. A new article by some of the same authors also concluded that *Eoraptor,* long considered as a primitive theropod, was actually a sauropodomorph: Sereno, P.C., Martínez, R.N., and Alcober, O.A. 2013. Osteology of *Eoraptor lunensis* (Dinosauria, Sauropodomorpha). Basal sauropodomorphs and the vertebrate fossil record of the Ischigualasto Formation (Late Triassic: Carnian-Norian) of Argentina. *Journal of Vertebrate Paleontology Memoir,* 12: 83-179 DOI: 10.1080/02724634.2013.820113

p. 25 "Sure enough, three- and four-toed tracks similar to those predicted for primitive dinosaurs are fairly common in some Middle Triassic rocks." (1) Demathieu, G.R. 1989. Appearance of the first dinosaur tracks in the French Middle Triassic and their probable significance. *In* Gillette, D.D., and Lockley, M.G. (editors), *Dinosaur Tracks and Traces.* Cambridge University Press, Cambridge, U.K.: 201-207. (2) A good, critical review of Early and Middle Triassic dinosaur-like tracks in the U.K. was provided by: King, M.J., and Benton, M.J. 1996. Dinosaurs in the Early and Mid Triassic? The footprint evidence from Britain. *Palaeogeography, Palaeoclimatology, Palaeoecology,* 122: 213-225.

p. 26 "Some of this disrespect for all things ichnological was allayed in 2010 when a team of paleontologists, led by Stephen Brussatte. . . ." Brussatte, S.L., Niedźwiedzki, G., and Butler, R.J. 2010. Footprints pull origin and diversification of dinosaur stem lineage deep into the Early Triassic. *Proceedings of the Royal Society of London, B,* 278: 1107-1113. An update on their research into dinosauromorph tracks in the Triassic rocks of Poland is here: Niedźwiedzki, G., Brusatte, S.L., and Butler, R.J. 2013. *Prorotodactylus* and *Rotodactylus* tracks: an ichnological record of dinosauromorphs from the Early-Middle Triassic of Poland. *Lyell Collection, Special Publications of the Geological Society of London,* 379, published online April 23, 2013: doi: 10.1144/SP379.12.

p. 26 ". . . footprints are often ignored or largely dismissed by workers focusing on body fossils, and are rarely marshaled as evidence in macroevolutionary studies of the dinosaur radiation." Brussatte *et al.* (2010) discuss this oversight of ichnological evidence, which is also addressed in Carrano and Wilson's (2001) paper.

p. 26 "A *ghost lineage* is one for which we have evidence that ancestral members of a clade and their descendants lived at a certain time in the geologic past. . . ." Cavin, L., and Forey, P.L. 2007. Using ghost lineages to identify diversification events in the fossil record. *Biology Letters,* 3: 201-204.

p. 27 "However, a few dinosaurs mixed it up, switching from bipedal to quadrupedal and back again. . . ." Wilson, J.A., Mariscano, C.A., and Smith, R.M.H. 2009. Dynamic locomotor capabilities revealed by early dinosaur trackmakers from southern Africa. *PLoS One,* 4(10): e7331. doi:10.1371/journal.pone.0007331

p. 27 "In a dinosaurian sense, though, a change from a four-legged to a two-legged gait meant that a dinosaur was *facultatively bipedal* (became bipedal when it wanted) and a normally two-legged dinosaur going on all fours was—you guessed it—*facultatively quadrupedal.*" See Farlow *et al.* (2012).

p. 28 "We also can get a better understanding of gaits by measuring distances between alternating feet (*pace*), between the same foot (*stride*), and the width of the trackway (*straddle*)." Basic tracking terms can be very confusing when applied by different trackers, but are approaching some standardization. Two scientifically well-done books on mammal tracking that clearly define these and other terms are: (1) Halfpenny, J., and Biesiot, E. 1986. *A Field Guide to Mammal Tracking in North America*. Johnson Publishing, Boulder, Colorado: 161 p. (2) Elbroch, M. 2003. *Mammal Tracks and Sign of North America*. Stackpole Books, Mechanicsburg, Pennsylvania: 778 p.

p. 29 "Speaking of sauropod trackways, their patterns can be further placed into two categories based on their widths: *narrow gauge* and *wide gauge*." (1) Farlow, J.O. 1992. Sauropod tracks and trackmakers: integrating the ichnological and skeletal records. *Zubia*, 10: 89-138. (2) Wilson, J.A., and Carrano, M.T. 1999. Titanosaurs and the origin of "wide-gauge" trackways: a biomechanical and systematic perspective on sauropod locomotion. *Paleobiology*, 25: 252-267.

p. 29 "Take the length of a theropod or ornithopod track and then multiply it by four. The resulting number gives the approximate hip height of the dinosaur. . . ." Henderson, D.M. 2003. Footprints, trackways, and hip heights of bipedal dinosaurs: testing hip height predictions with computer models. *Ichnos*, 10: 99-114.

p. 30 "Originally devised in 1976 by a paleontologically enthused physicist, R. M. Alexander. . . ." Alexander, R.M. 1976. Estimates of the speed of dinosaurs. *Nature*, 261: 129-130.

p. 31 "Olympic racewalkers regularly exceed 15 kph, which they can keep up for 20 km (12.4 mi)." U.S. and world records for racewalking are at the following USA Track and Field (USATF) site: http://www.usatf.org/Sports/Race-Walking/Records.aspx

p. 31 "Not surprisingly, then, other paleontologists have come up with their own formulas for estimating dinosaur speeds." The most-often cited alternative to Alexander's formula is here: Thulborn, R.A. 1982. Speeds and gaits of dinosaurs. *Palaeogeography, Palaeoclimatology, Palaeoecology*, 38: 227-256.

p. 31 "A few have even tried to say that some dinosaurs were not capable of running at all. . . ." Mallison, H. 2011. Fast-moving dinosaurs: why our basic tenet is wrong. *Journal of Vertebrate Paleontology* [Supplement], Program and Abstracts for Society of Vertebrate Paleontology meeting, Las Vegas, Nevada: 150.

p. 32 "In other words, tracks and these structures caused by the applied and released pressure—which some trackers call *pressure-release structures*, *pressure releases*, or *indirect features*. . . ." (1) Brown, T., Jr. 1999. *The Science and Art of Tracking*. Berkley Books, New York: 240 p. (2) Gatesy,

S.M. 2003. Direct and indirect track features: what sediment did a dinosaur touch? *Ichnos*, 10: 91-98. (3)

p. 33 "All of these factors culminate in what paleontologists consider as 'track tectonics'. . . ." Graversen, O., Milàn, J., and Loope, D.B. 2007. Dinosaur tectonics: a structural analysis of theropod undertracks with a reconstruction of theropod walking dynamics. *Journal of Geology*, 115: 641-654.

p. 33 "Consequently, a common way for dinosaur tracks to have made it into the fossil record was as *undertracks*." Milàn, J., and Bromley, R.G. 2006. True tracks, undertracks and eroded tracks, experimental work with tetrapod tracks in laboratory and field. *Palaeogeography, Palaeoclimatology, Palaeoecology*, 231: 253-264.

p. 35 "One of these, preserved in Late Cretaceous rocks of Bolivia, was made by an ankylosaur—a big, armored dinosaur—which must have looked like a living tank as it ambled along." Meyer, C.A., Hippler, D., and Lockley, M.G. 2001. The Late Cretaceous vertebrate ichnofacies of Bolivia: facts and implications. *Asociación Palaeontológica Argentina, Publicación Especial* 7: 133-138.

p. 35 "The first discovered running-dinosaur trackways were from a site in Texas, where at least three theropods moved at high speed." Farlow, J.O. 1981. Estimates of dinosaur speeds from a new trackway site in Texas. *Nature*, 294: 747-748.

p. 35 "To put it into a bipedal-human perspective, the top speed recorded by Usain Bolt over 200 m (656 ft) during the 2012 Olympics was also 27 mph. . . ." Hernández-Gómez, J.J., Marquina, V., and Gómez, R.W. 2013. On the performance of Usain Bolt in the 100 m sprint. *European Journal of Physics*, 34: 1227.

p. 35 "Theropods and humans alike, though, would be humbled by the top speed recorded by a cheetah (*Acinonyx jubatus*). . . ." Sharp, N.C.C. 1997. Timed running speed of a cheetah (*Acinonyx jubatus*). *Journal of Zoology*, 241: 493-494.

p. 35 "Nonetheless, a dinosaur tracksite in Queensland, Australia outdoes the Texas tracksite for sheer numbers of running dinosaurs." This tracksite was first interpreted as a "dinosaur racetrack" by: Thulborn, R.A., and Wade, M. 1979. Dinosaur stampede in the Cretaceous of Queensland. *Lethaia*, 12: 275-279. However, it was also recently reinterpreted as a "dinosaur natatorium" (swimsite) by: Romilio, A., Tucker, R., and Salisbury, S.W. 2013. Reevaluation of the Lake Quarry dinosaur tracksite (late Albian-Cenomanian Winton Formation, central-western Queensland, Australia): no longer a stampede? *Journal of Vertebrate Paleontology*, 33: 102-120.

p. 36 "For example, the fastest bipedal land animals today are ostriches (*Struthio camelus*), which have maximum speeds of 45 mph (72 kph)."

Alexander, R.M., Maloiy, G.M.O., Njau, R., and Jayes, A.S. 1979. Mechanics of running of the ostrich (*Struthio camelus*). *Journal of Zoology*, 187: 169-178.

p. 36 "One paleontologist, Jim Farlow, and two other colleagues figured out that given the average mass of an adult *T. rex*. . . ." Farlow, J.O., Smith, M.B., and Robinson, J.M. 1995. Body mass, bone "strength indication," and cursorial potential of *Tyrannosaurus rex*. *Journal of Vertebrate Paleontology*, 15: 713-725.

p. 37 "What they found was that a 45-mph-running *T. rex* would have required about 85% of its entire body mass concentrated in its legs. . . ." Hutchinson, J.R., and Garcia, M. 2002. *Tyrannosaurus* was not a fast runner. *Nature*, 415: 1018-1021.

p. 37 "Both of these skeletons belong to the same species of dinosaur, *Mei long* ('soundly sleeping dragon')." (1) Xu, X., and Norell, M.A. 2004. A new troodontid dinosaur from China with avianlike sleeping posture. *Nature*, 431: 838-841. (2) Gao, C., Morschhauser, E.M., Varricchio, D.J., Liu, J., and Zhao, B. 2012. A second soundly sleeping dragon: new anatomical details of the Chinese troodontid *Mei long* with implications for phylogeny and taphonomy. *PLoS One*, 7: e45203, doi:10.1371/journal.pone.0045203.

p. 38 "A few other dinosaur skeletons, such as those of the theropod *Citipati*, have been found preserved in sitting positions. . . ." (1) Norell, M.A., Clark, J.M., Chiappe, L.M., and Dashzeveg, D. 1995. A nesting dinosaur. *Nature*, 378: 774-776. (2) Varricchio *et al.* (1997).

p. 39 "Incidentally, dinosaur tail impressions are quite rare, with fewer than forty reported from the entire geologic record. . . ." Kim, J.Y., and Lockley, M.G. 2013. Review of dinosaur tail traces. *Ichnos*, 20: 129-141.

p. 39 "Reported in 2009 in southwestern Utah, this Early Jurassic trace fossil not only shows where a theropod approached a sitting spot and sat down. . . ." Milner, A.R.C., Harris, J.D., Lockley, M.G., Kirkland, J.I., and Matthews, N.A. 2009. Bird-like anatomy, posture, and behavior revealed by an Early Jurassic theropod dinosaur resting trace. *PLoS One*, 4: e4591. doi:4510.1371/journal.pone.0004591.

p. 40 "We were also testing an audacious claim that some of the wrinkle marks near the edge of the leg impressions were actually from feathers." Martin, A.J., and Rainforth, E.M. 2004. A theropod resting trace that is also a locomotion trace: case study of Hitchcock's specimen AC 1/7. *Geological Society of America Abstracts with Programs 36*(2): 96.

p. 41 "In short, this specimen records a full sequence of movement by the theropod and how it altered the ground beneath it. . . ." Martin and Rainforth (2004).

p. 42 "Only later did paleontologists realize this 'webbing' was actually a result of skin drying around its bones after the dinosaur had died." Manning, P. 2008. *Grave Secrets of Dinosaurs: Soft Tissues and Hard Science.* National Geographic Society, Washington, D.C.: 316 p.

p. 42 "A major flaw in this seemingly marvelous adaptation was that the hollow tube in the center of the crest, once studied in more detail later. . . ." Weishampel, D.B. 1981. The nasal cavity of lambeosaurine hadrosaurids (Reptilia:Ornithischia): comparative anatomy and homologies. *Journal of Paleontology,* 55: 1046-1057.

p. 42 "Once he investigated, he . . . made an astonishing discovery: the first known sauropod dinosaur tracks from the geologic record." Bird, R.T. 1985. *Bones for Barnum Brown: Adventures of a Dinosaur Hunter.* Texas Christian University Press, Ft. Worth, Texas: 225 p.

p. 43 "Later, a closer look at these tracks showed that the missing tracks in the sequence of steps could be attributed to differences in track preservation." Lockley, M.G., and Rice, A. 1990. Did Brontosaurus ever swim out to sea? *Ichnos,* 1: 81-90.

p. 43 "These tracks are also in rocks from near the start of sauropods in the fossil record (Late Triassic) to their very end (Late Cretaceous)." Lockley *et al.* (2001).

p. 43 "Yet Indian elephants (*Elephas maximus*) can swim as far as 25 miles (40 km), a feat far better than most humans are capable of." Johnson, D.E. 1980. Problems in the land vertebrate zoogeography of certain islands and the swimming power of elephants. *Journal of Biogeography,* 7: 383-398. However, on September 2, 2013, Diana Nyad successfully completed swimming 178 km (110 mi) from Cuba to Florida, a distance that would be the envy of all water-crossing elephants.

p. 44 "In fact, elephant swimming abilities show one of the probable ways mammoths dispersed to islands during the Pleistocene Epoch, where some isolated populations lasted until only about 4,000 years ago." (1) Vartanyan, S.L., Garutt, V.E., and Sher, A.V. 1993. Holocene dwarf mammoths from Wrangel Island in the Siberian Arctic. *Nature,* 362: 337-340. (2) Johnson (1980).

p. 44 "First, as early as 1980, a paleontologist interpreted swim tracks from Early Jurassic rocks of Connecticut as made by theropods. . . ." Coombs, W.P., Jr. 1980. Swimming ability of carnivorous dinosaurs. *Science,* 207: 1198-1200.

p. 44 ". . . in 2001, paleontologists working in separate studies and places (Wyoming and the U.K.) interpreted Middle Jurassic tracks as possible dinosaur swim tracks." (1) Kvale, E.P., Johnson, G.D., Mickelson, D.L., Keller, K., Furer, L.C., and Archer, A.W. 2001. Middle Jurassic (Bajocian and Bathonian) dinosaur megatracksites, Bighorn Basin, Wyoming, U.S.A. *Palaios,* 16: 233-254. (2) Whyte, M.A., and Romano,

M. 2001. A dinosaur ichnocoenoses from the Middle Jurassic of Yorkshire, UK. *Ichnos*, 8: 223-234.

p. 44 "Soon after that (2006), hundreds of much better examples were discovered and documented by Andrew Milner in Early Jurassic rocks of southwestern Utah. . . ." Milner, A.R.C., Lockley, M.R., and Kirkland, J.I. 2006. A large collection of well-preserved theropod dinosaur swim tracks from the Moenave Formation, St. George, Utah. *In* Harris, J.D., *et al.* (editors), *The Triassic-Jurassic Terrestrial Transition. New Mexico Museum of Natural History and Science Bulletin*, 37: 315-328.

p. 44 "The next year (2007), dinosaur swim tracks were again interpreted from long linear marks on an expansive surface of Early Cretaceous rock in Spain." Ezquerra, R., Doublet, S., Costeur, L., Galton, P.M., and Pérez-Lorente, F. 2007. Were non-avian theropod dinosaurs able to swim? supportive evidence from an Early Cretaceous trackway, Cameros Basin (La Rioja, Spain). *Geology*, 35: 507-510.

p. 44 "In 2013, yet more dinosaur swim tracks were reported from another Early Cretaceous site in Queensland, Australia." Romilio *et al.* (2013), and the controversy over this reinterpretation of what was regarded as a "dinosaur stampede" site is taken up in Chapter 3. Also in 2013, yet another example of dinosaur swim tracks was interpreted from China: Xing, L.D., Lockley, M.G., Zhang, J.O., Milner, A.R.C., Klein, H., Li, D.Q., Persons, W.S., and Ebi, J.F. 2013. A new Early Cretaceous dinosaur track assemblage and the first definite non-avian theropod swim trackway from China. *Chinese Science Bulletin*, 58: 2370-2378.

p. 45 "Not surprisingly, recreational purposes have never been suggested for swimming dinosaurs, but who knows whether an occasional dip might have also relieved any dinosaurs suffering from skin parasites or a hot day in the Mesozoic." While hiking by a lake in the Piedmont National Wildlife Refuge in mid-Georgia on a summer day, I was shocked to see a deer swimming vigorously around the lake edge. At some point it came up on the bank, and when I trained my binoculars on it, I could see its body was covered with hundreds of ticks. So it must have been desperate to either get rid of them or to at least feel better. I felt terrible for the deer, but also made a mental note that this was yet another reason why animals that don't normally swim would get into the water.

p. 46 "This site has trackways of more than twenty sauropods walking in the same direction and apparently made at about the same time." Castanera, D., Barco, J., Díaz-Martínez, I., Gascón, J., Pérez-Lorente, F., and Canudo, J. 2011. New evidence of a herd of titanosauriform sauropods from the Lower Berriasian of the Iberian Range (Spain). *Palaeogeography, Palaeoclimatology, Palaeoecology*, 310: 227-237.

p. 46 "A Late Jurassic sauropod tracksite near La Junta, Colorado, also shows the tracks of five sauropods moving in the same direction, spaced at regular intervals. . . ." (1) Lockley, M.G., Houck, K.J., and Price, N.K. 1986. North America's largest dinosaur trackway site: implications for Morrison Formation paleoecology. *Geological Society of America Bulletin*, 97: 1163-1176. (2) Lockley (1991).

p. 47 "The tracksite in the U.K., preserved in Middle Jurassic (about 165 *mya*) rocks, was likely made by dozens of sauropods moving together. . . ." Day, J.J., Norman, D.B., Gale, A.S., Upchurch, P., Powell, H.P. 2004. A Middle Jurassic dinosaur trackway site from Oxfordshire, UK. *Palaeontology*, 47: 319-348.

p. 47 "These include several sites from the Cretaceous of Korea, one of which has tracks of about twenty large ornithopods heading in the same direction, and another from the Cretaceous of Canada. . . ." (1) Lockley, M.G., Houck, K., Yang, S.-Y., Matsukawa, M., and Lim, S.-K. 2006. Dinosaur-dominated footprint assemblages from the Cretaceous Jindong Formation, Hallyo Haesang National Park area, Goseong County, South Korea: evidence and implications. *Cretaceous Research*, 27: 70-101. (2) Currie, P.J. 1983. Hadrosaur trackways from the Lower Cretaceous of Canada. *Acta Paleontologica Polonica*, 28: 62-74.

p. 47 "A few ankylosaur tracksites have parallel trackways, suggesting that at least two ankylosaurs were traveling together at the same time." McCrea *et al.* (2001).

p. 47 "Ceratopsian tracks are also uncommon enough to withhold judgment on that aspect of their lives too, although rocks bearing hundreds of bones of the same ceratopsian species tell us these dinosaurs were likely group-oriented also." (1) Wood, J.M., Thomas, R.G., and Visser, J. 1988. Fluvial processes and vertebrate taphonomy: the upper Cretaceous Judith River Formation, south-central Dinosaur Provincial Park, Alberta, Canada. *Palaeogeography, Palaeoclimatology, Palaeoecology*, 66: 127-143. (2) Qi, Z., Barrett, P.M., and Eberth, D.A. 2007. Social behaviour and mass mortality in the basal ceratopsian dinosaur *Psittacosaurus* (Early Cretaceous, People's Republic of China). *Palaeontology*, 50: 1023-1029.

p. 47 "At one site in Middle Jurassic (about 165 *mya*) rocks of Zimbabwe, trackways of at least five large theropods . . . were traveling together." Lingham-Solier, T., Broderick, T., and Ahmed, A.A.K. 2003. Closely associated theropod tracks from the Jurassic of Zimbabwe. *Naturwissenschaften*, 90: 572-576.

p. 47 "An Early Cretaceous (about 125 *mya*) site in China also shows six theropod trackways, equally spaced and pointing in the same direction. . . ." Li, R., Lockley, M.G., Makovicky, P.J., Matsukawa, M., Norell, M.A., Harris, J.D., and Liu, M. 2008. Behavioral and faunal implications of Early Cretaceous deinonychosaur trackways from China. *Naturwissenschaften*, 95: 185-191.

p. 48 "This trackway, discovered by paleontologist Roland Bird in 1938, was in a limestone bed cropping out in the Paluxy River." Bird, R.T. 1985. *Bones for Barnum Brown: Adventures of a Dinosaur Hunter.* Texas Christian University Press, Ft. Worth, Texas: 225 p.

p. 49 "So now the more reasonable explanation is that, yes, the theropod might have been stalking the sauropod but did not jump onto it there." Farlow, J.O., O'Brien, M.O., Kuban, G.J., Datillo, B.F., Bates, K.T., Falkingham, P.L., Piñuela, L., Rose, A., Freels, A., Kumagi, C., Libben, C., Smith, J., and Whitcraft, J. 2012. Dinosaur tracksites of the Paluxy River Valley (Glen Rose Formation, Lower Cretaceous), Dinosaur Valley State Park, Somervell County, Texas. *Actas de V Jornadas Internacionales sobre Paleontología de Dinosaurios y su Entorno,* Salas de los Infantes, Burgos: 41-69.

p. 49 "Glen Kuban, a paleontologist who has studied and mapped the Paluxy River dinosaur tracks for more than twenty years. . . ." Kuban, G. 1989. Elongate dinosaur tracks. *In* Gillette, D.D., and Lockley, M.G. (editors), *Dinosaur Tracks and Traces.* Cambridge University Press, Cambridge, U.K.: 57-72.

p. 50 "Researchers who worked on this site had also found sauropod tracks, showing they also lived in the area, but found ominous toothmarks on the bones along with shed teeth of juvenile and adult *Allosaurus.*" Jennings, D.S., and Hasiotis, S.T. 2006. Taphonomic analysis of a dinosaur feeding site using geographic information systems (GIS), Morrison Formation, southern Bighorn Basin, Wyoming, USA. *Palaios,* 21: 480-492.

p. 51 "In a 2009 journal article with the beguiling title of 'Dinosaur Death Pits from the Jurassic of China'. . . ." Eberth, D.A., Xing, X., and Clark, J.M. 2009. Dinosaur death pits from the Jurassic of China. *Palaios,* 25: 112-125.

p. 52 "Sure enough, a few fossils, such as snails and clams, found in dinosaur tracks were crushed underfoot. . . ." In the Morrison Formation of Colorado—at the same tracksite with herding sauropods—Lockley *et al.* (1986) found crushed clams in sauropod footprints and interpreted this as the sauropods having "killed" them. But I've also stepped on and crushed many clams that were dead long before I got there. Hence I'm a little skeptical that sauropods were responsible for their deaths: this could have been post-mortem smashing. Similarly, I've seen snail shells in Late Jurassic sauropod tracks in Switzerland, and again can't say for sure whether they were dead or alive when these dinosaurs put their feet down on them.

p. 52 "One theropod trackway in the Late Jurassic in Utah leaves no doubt that its maker had a tough time walking. . . ." Lockley, M.G., Hunt, A.P., Moratalla, J.J., and Matsukawa, M. 1994. Limping dinosaurs? trackway evidence for abnormal gaits. *Ichnos,* 3: 193-202.

p. 53 "One such trackway is from the Early Jurassic of Massachusetts, in which three tracks in sequence—right, left, right—have a perfectly fine three-toed theropod track on the left, but a two-toed one on the right, missing its innermost digit." Lockley (1991).

p. 54 "It also left lots of other evidence for its impact, not least of which is a huge crater of the right size and age in the Gulf of Mexico next to the Yucatan peninsula of Mexico." The original article that started the "meteorite killed the dinosaurs" hypothesis was: Alvarez, L.W., Alvarez, W., Asaro, F., and Michel, H.V. 1980. Extraterrestrial cause for the Cretaceous-Tertiary extinction. *Science*, 208: 1095-1108. (2) A good book written for a lay audience that summarizes the scientific story of the hypothesis, and written by one of the discoverers of the evidence for the meteorite impact is: Alvarez, W. 2008. *T. rex and the Crater of Doom*. Princeton University Press, Princeton, New Jersey: 216 p.

p. 55 "One such site in Utah has hundreds of deeply impressed tracks preserved in strata at multiple levels, one of which is only a few meters below the boundary." Dilfey, R.L., and Ekdale, A.A. 2002. Footprints of Utah's last dinosaurs: track beds in the Upper Cretaceous (Maastrichtian) North Horn Formation of the Wasatch Plateau, central Utah. *Palaios*, 17: 327-346.

p. 55 "A single—but huge—theropod track from New Mexico, attributed to *Tyrannosaurus rex*, also came from a layer just below the boundary." Lockley, M.G., and Hunt, A.P. 1994. A track of the giant theropod dinosaur *Tyrannosaurus* from close to the Cretaceous/Tertiary Boundary, northern New Mexico. *Ichnos*, 3: 213-218.

p. 55 "In Spain, abundant dinosaur tracks, some ascribed to hadrosaurs and sauropods, are preserved only a few meters below strata that preserved fossils of Paleogene fish and mammals." Riera, V., Oms, O., Gaete, R., and Galobart, A. 2009. The end-Cretaceous dinosaur succession in Europe: the Tremp Basin record (Spain). *Palaeogeography, Palaeoclimatology, Palaeoecology*, 283: 160-171.

p. 55 "At one site, more than forty horizons contained hadrosaur and sauropod tracks below the boundary, and paleontologists estimated that some tracks were made only 300,000 years before the end of the Cretaceous." Villa, B., Oms, O., Fondevilla, V., Gaete, R., Galobart, A., and Riera, V. 2013. The latest succession of dinosaur tracksites in Europe: hadrosaur ichnology, track production and paleoenvironments. *PLoS One*, 8: e72579. doi:10.1371/journal.pone.0072579

Chapter 3: The Mystery of Lark Quarry

p. 61 "As is still typical for fossil finds in many parts of the world, it was spotted by a sharp-eyed amateur, cattle station manager Glen Seymour, who lived and worked in the area." The best historical account

I've read about the discovery and uncovering of the Lark Quarry tracksite was in the annual magazine of the Australian Age of Dinosaurs (Winton, Australia): Meiklejohn, D., and Elliott, J. 2004. The ghosts of Lark Quarry. *Australian Age of Dinosaurs*, Issue 2: 18-31.

p. 61 "Otherwise, nothing much in a scientific sort of way happened until 1971, when Mr. McKenzie and Mr. Knowles took paleontologists to the site. . . ." Although this part of the story is again related in Meiklejohn's (2004) *Australian Age of Dinosaurs* article, the part where Pat Vickers-Rich and Tom Rich accompanied the others to the site was told to me personally by none other than Pat Vickers-Rich and Tom Rich.

p. 63 "Based on Thulborn and Wade's analysis of the site, the area was a lakeshore that had been submerged regularly by a nearby stream emptying into it." (1) Thulborn, and Wade. (1979).

p. 64 *"Persuasive circumstantial evidence leads us to conclude that they represent a stampede—that is, a wild, unreasoning and panic-stricken rush to escape the threat of danger. What could have caused such presumed panic?"* Thulborn and Wade (1979).

p. 65 "The second, published in 1984, was a much longer and more detailed report modestly titled 'Dinosaur Trackways in the Winton Formation (Mid-Cretaceous) of Queensland.'" Thulborn, R.A., and Wade, M. 1984. Dinosaur trackways in the Winton Formation (mid-Cretaceous) of Queensland. *Memoirs Queensland Museum*, 21: 413-517.

p. 67 "The fresh hypothesis states that the 'dinosaur stampede' was not triggered by the arrival of a predator, and no stalking of other dinosaurs by a voracious predator happened either." Romilio, A., and Salisbury, S.W. 2011. A reassessment of large theropod dinosaur tracks from the mid-Cretaceous (late Albian–Cenomanian) Winton Formation of Lark Quarry, central-western Queensland, Australia: a case for mistaken identity. *Cretaceous Research*, 32: 135-142.

p. 69 "After all, hypotheses are only accepted conditionally, and then are subject to further testing so we can find out whether or not they still hold up to scrutiny." Cleland, C.E. 2001. Historical science, experimental science, and the scientific method. *Geology*, 29: 987-990.

p. 70 "Why identifying the maker of these big tracks is so difficult is mostly attributable to the tracks only having three toes, a trait also known as *tridactyl*." Farlow *et al.* (2012).

p. 70 "On average, theropod tracks are longer than they are wide, whereas ornithopod tracks are wider than they are long." (1) Lockley (2009); (2) Farlow *et al.* (2012).

p. 71 "In an article published in 1988, they used a sample of 66 Early Cretaceous tridactyl dinosaur tracks from Spain, all of which had been identified confidently as either ornithopod or theropod tracks on the basis of their qualities." Moratalla, J.J., Sanz, S.L., and Jimenez, S. 1988. Multivariate analysis on Lower Cretaceous dinosaur footprints: discrimination between ornithopods and theropods. *Geobios*, 21: 395-408.

p. 73 "Skeletal remains of this ornithopod were discovered in 1963 near the small town of Muttaburra in central Queensland, northeast of Lark Quarry. . . ." Bartholomai, A., and Molnar, R.E. 1981. *Muttaburrasaurus*: a new Iguanodontid (Ornithischia:Ornithopoda) dinosaur from the Lower Cretaceous of Queensland. *Memoirs of the Queensland Museum*, 20: 319-349.

p. 73 "In contrast, bones of only one large predatory theropod . . . *Australovenator wintonensis*, lovingly nicknamed 'Banjo'. . . ." Hocknull, S.A., White, M.A., Tischler, T. R., Cook, A.G., Calleja, N.D., Sloan, T., and Elliott, D. A. 2009. New mid-Cretaceous (Latest Albian) dinosaurs from Winton, Queensland, Australia. *PLoS One*, 4: e6190.

p. 74 "Interestingly, objections to this identification, voiced formally by Romilio and Salisbury, were not new, as a few dinosaur paleontologists questioned it soon after Thulborn and Wade's second article came out in 1984." (1) Paul, G. 1988. *Predatory Dinosaurs of the World*. Simon and Shuster, New York: 464 p. (2) Lockley and Hunt (1994).

p. 74 "I say the last facetiously, but a small troupe of creative folks actually did write and perform a short musical number retelling the original story of the dinosaur stampede. . . ." The performance was recorded, placed on YouTube on September 29, 2012, and is titled *Dinosaur Stampede MASA2012*: http://www.youtube.com/watch?v=midGaNNHhbM

p. 77 "In this manuscript, which was posted online in the same journal (*Cretaceous Research*) that published the re-study. . . ." The article was initially available online November 22, 2011, then withdrawn sometime before May 2012. Apparently, the author was never notified by the journal of the withdrawal, and learned about its disappearance from others. The link to the former article and the "Withdrawn" notice are here: http://www.sciencedirect.com/science/article/pii/S0195667111001844

p. 78 "Hence, it did what very few academic journals do, which was pull the article offline." As part of an online scientific community, I recall the surprise and discussion that took place among paleontologists after this happened, with much speculation about why it was justified. Some of this is summarized by Brian Switek, along with his coverage of the new interpretation of the "stampede" tracksite as a "swimsite," here: http://phenomena.nationalgeographic.com/2013/01/11/swim-tracks-undermine-dinosaur-stampede/

p. 79 "Thulborn submitted a revised version of his article to the Australian paleontology journal *Alcheringa* . . . published in early 2013."

Thulborn, R.A. (2013): Lark Quarry revisited: a critique of methods used to identify a large dinosaurian track-maker in the Winton Formation (Albian–Cenomanian), western Queensland, Australia. *Alcheringa: An Australasian Journal of Palaeontology*, doi:10.1080/03115518.2013.748482

p. 79 "Then, also early in 2013, the aquamusical version was unveiled." Romilio, A., Tucker, R., and Salisbury, S.W. 2013. Reevaluation of the Lake Quarry dinosaur tracksite (late Albian-Cenomanian Winton Formation, central-western Queensland, Australia): no longer a stampede? *Journal of Vertebrate Paleontology*, 33: 102-120.

p. 80 "Just how did I get trapped in the building housing these fabled and controversial dinosaur tracks?" Much of this story I told before in an invited article for *Australian Age of Dinosaurs* magazine (2011 issue), written as a humorous end piece to the magazine (called the "Tail Bone") and titled *Sleeping with the Dinosaur Tracks*.

CHAPTER 4: DINOSAUR NESTS AND BRINGING UP BABIES

p. 86 "How did male dinosaurs get past the thick tails of some females to reach their goals? More intriguing, how did male dinosaurs get past the spiky or clubbed tails of some females?" Brian Switek, in his book *My Beloved Brontosaurus* (2013, Farrar, Straus, and Giroux, 256 p.), explores the intriguing topic of dinosaur sex with great humor and deeply penetrating prose in a chapter titled "The Big Bang Theory." In this chapter, he discusses the pros and cons of dinosaur positions, so with those in mind, it is even more fun to imagine the traces these would have made.

p. 87 "This discovery was of a small ceratopsian dinosaur, *Protoceratops*, on top of a dromaeosaur, *Velociraptor*, with both locked in an embrace but a pointedly deadly one." Although this deservedly famous pair of fossils has been much discussed since its discovery in 1971, the first peer-reviewed publication reporting it was: Barsbold, R. 1974. Saurornithoididae, a new family of theropod dinosaurs from Central Asia and North America. *Paleontologica Polonica*, 30: 5-22.

p. 87 ". . . but more emblematic of Tennyson's 'nature, red in tooth and claw.'" This famous phrase, with its allusions to Darwinian struggles (albeit ten years before publication of Darwin's *On the Origin of Species*) comes from section (Canto) 56 of Alfred, Lord Tennyson's epic poem, *In Memoriam A.H.H.*, finished in 1849. This poem is explored in depth by a Stephen Jay Gould essay, "The Tooth and Claw Centennial," in: Gould, S.J. 1995. *Dinosaur in a Haystack*. Harmony Books, New York: 63-75.

p. 87 "Paleontologists who studied these fossils agree that both dinosaurs must have been buried instantaneously. . . ." So much has been written about these dinosaurs and how they were buried, I won't even try to pick any one. But probably the best general explanation for how collapsing dunes (and wet sand) could have been responsible for the exquisite preservation of the Late Cretaceous Mongolian dinosaurs is: Dingus, L., and Loope, D. 2000. Death in the dunes. *Natural History*, 109: 50-57.

p. 90 "Nonetheless, eggs and eggshells are actually body fossils." Martin, A.J. 2006. *Introduction to the Study of Dinosaurs* (2nd Edition). Wiley-Blackwell, Oxford, U.K.: 560 p.

p. 91 "Vertebrates that lay eggs today, such as reptiles, birds, and monotremes (egg-laying mammals, such as the platypus and echidnas), are all classified as *amniotes*." Benton, M.J. 2005. *Vertebrate Palaeontology* (3rd Edition). Wiley-Blackwell, Oxford, U.K.: 455 p.

p. 91 "We currently have no reasonable evidence that dinosaurs gave birth to live young, but this trait (*viviparity*) showed up in ichthyosaurs. . . ." Many articles have been published on individual ichthyosaurs and their giving live birth. But for an exploration of how it evolved in Mesozoic marine reptiles, as well as an overview of the evidence for biological development in fossils, read: Sánchez, M.R. 2012. *Embryos in Deep Time: The Rock Record of Biological Development*. University of California Press, Berkeley, California: 265 p.

p. 91 "Although that strategy has worked very well for sea turtles for the past 100 million years. . . ." A fossil sea-turtle nest—from the Cretaceous Period, no less—was just recently described in detail, with its traits compared to those of modern sea turtles: Bishop, G.A., Pirkle, F.L., Meyer, B.K., and Pirkle, W.A. 2011. The foundation for sea turtle geoarchaeology and zooarchaeology: morphology of recent and ancient sea turtle nests, St. Catherines Island, Georgia, and Cretaceous Fox Hills Sandstone, Elbert County, Colorado. *In* Bishop, G.A, Rollin, H.B., and Thomas, D.H. (editors), *Geoarchaeology of St. Catherines Island, Georgia*. Anthropological Papers of the American Museum of Natural History, No. 94: 247-269.

p. 91 "Viviparity also occurs in some modern species of lizards and snakes, including sea snakes. . . ." Andrews, R.M., and Mathies, T. 2000. Natural history of reptilian development: constraints on the evolution of viviparity. *BioScience*, 50: 227-238.

p. 92 "The oldest known dinosaur eggs come from Early Jurassic rocks. . . ." Reisz, R.R., Evans, D.C., Roberts, E.M., Dieter-Sues, H., and Yates, A.M. 2012. Oldest known dinosaurian nesting site and reproductive biology of the Early Jurassic sauropodomorph *Massospondylus*. *Proceedings of the National Academy of Science*, 109: 2428-2433.

p. 92 "These categories, based on shell microstructure, arrangements of the pores, and overall forms, were given linguistically daunting names such as Spheroolithidae, Ovaloolithidae. . . ." Zelenitsky, D.K., Horner, J.R., and Therrien, F. 2012. Dinosaur eggs. *In* Brett-Surman, M.K., Holtz, T.R., Jr., and Farlow, J.O. (editors), *The Complete Dinosaur* (2nd Edition). Indiana University Press, Bloomington, Indiana: 613-620.

p. 93 "Paleontologists who studied *Troodon* egg clutches were surprised to notice that not only were eggs paired but also aligned vertically." Varricchio *et al.* (1997).

p. 93 "Thus far, we only know of one instance in which an egg assemblage contained embryonic bones of more than one species of dinosaur. . . ." This find was first described in 1994, then updated in 2009. (1) Norell, M.A., Clark, J.M., Dashzeveg, D., Barsbold, R., Chiappe, L.M., Davidson, A.R., McKenna, M.C., Perle, A., and Novacek, M.J. 1994. A theropod dinosaur embryo and the affinities of the Flaming Cliffs dinosaur eggs. *Science*, 266: 779-782. (2) Bever, G.S., and Norell, M.A. 2009. The perinate skull of *Byronosaurus* (Troodontidae) with observations on the cranial ontogeny of paravian theropods. *American Museum Novitates*, 3657: 51 p.

p. 94 "Sea turtles, for example, have a temporary extension of their beaks when born, which is applied like a can opener from the inside of the egg to open it." Ruckdeschel, C., and Shoop, C.R. 2006. *Sea Turtles of the Atlantic and Gulf Coasts of the United States*. University of Georgia Press, Athens, Georgia: 152 p.

p. 94 "Similar parts have been described from Late Cretaceous sauropod embryos in Argentina. . . ." Garcia, R.A. 2007. An "egg-tooth"-like structure in titanosaurian sauropod embryos. *Journal of Vertebrate Paleontology*, 27: 247-252.

p. 94 "Have hatchling trace fossils been identified in dinosaur eggs?" Mueller-Töwe, I. J., Sander, P.M., Schüller, H., and Thies, D. 2002. Hatching and infilling of dinosaur eggs as revealed by computed tomography. *Palaeontographica Abteilung A*, 267: 119-168.

p. 95 "Yet these trace fossils have not been discovered yet either, despite many decades of our demonizing mammals for eating too many dinosaur eggs. . . ." No one is quite sure when the "Mammals ate dinosaur eggs, so dinosaurs went extinct" idea originated, but it is a persistent myth. And myth it is, because dinosaurs and mammals lived together in the Mesozoic for about 150 million years, yet we still have no evidence—trace fossil or otherwise—for mammals eating dinosaur eggs, let alone eating enough of them to hasten dinosaur extinction.

p. 96 "Unidentified therizinosaurs from the Late Cretaceous of Mongolia, which may have even formed nesting colonies." This fantastic discovery of a probable therizinosaur nesting colony was announced in a talk at the Society of Vertebrate Paleontology meeting in October 2013: Kobayashi, Y., Lee, Y.-N., Barsbold, R., Zelenitsky, D., and Tanaka, K. 2013. First record of a dinosaur nesting colony from Mongolia reveals nesting behavior of therizinosauroids. *Journal of Vertebrate Paleontology*, Program and Abstracts, 2013: 155.

p. 96 "The ceratopsians *Psittacosaurus* from the Early Cretaceous in China, and *Protoceratops* from the Late Cretaceous in Mongolia. . . ." (1) Meng, Q., Liu, J., Varricchio, D.J., Huang, T., and Gao, C. 2004. Parental care in an ornithischian dinosaur. *Nature*, 431: 145-146. (2) Fastovsky, D., Weishampel, D., Watabe, M., Barsbold, R., Tsogtbaatar, K., and Narmandakh, P. 2011. A nest of *Protoceratops andrewsi* (Dinosauria, Ornithischia). *Journal of Paleontology*, 85 (6): 1035-1041.

p. 96 ". . . a nearly complete specimen of *Citipati osmolskae*—missing only its head—was found in a sitting position above a clutch of long oval eggs." (1) Norell, M.A., Clark, J.M., Chiappe, L.M., and Dashzeveg, D. 1995. A nesting dinosaur. *Nature*, 378: 774-776. (2) Clark, J.M., Norell, M.A., and Chiappe, L.M. 1999. An oviraptorid skeleton from the Late Cretaceous of Ukhaa Tolgod, Mongolia, preserved in an avian-like brooding position over an oviraptorid nest. *American Museum Novitates*, 3265: 1-36.

p. 97 "Indeed, the discovery of a dinosaur egg—containing embryonic bones, no less—in Late Cretaceous shallow-marine deposits in Alabama. . . ." Lamb, J.P., Jr. 2001. Dinosaur egg with embryo from the Cretaceous (Campanian) Mooreville Chalk Formation, Alabama. *Journal of Vertebrate Paleontology*, 21 (Supplement to 3): 70A.

p. 97 "If talking just about reptile nests, these can be summarized into two broad categories: ground nests and hole nests, with ground nests made on the ground surface and hole nests below the surface." Martin (2006).

p. 98 "Bird nests range from simple scrapes in the ground, to better defined excavations. . . ." Elbroch, M., and Marks, E. 2001. *Bird Tracks and Sign of North America*. Stackpole Books, Mechanicsburg, Pennsylvania: 456 p.

p. 98 "For instance, one good candidate for underground nesting would have been the small Cretaceous ornithopod *Oryctodromeus cubicularis*. . . ." Varricchio *et al.* (2007).

p. 98 "For arboreal nests, a few small feathered tree-climbing, gliding, and flying non-avian dinosaurs are known from Early Cretaceous rocks of China." (1) Czerkas, S.A., and Yuan, C. 2002. An arboreal maniraptoran from northeast China. *In* Czerkas, S.J. (editor), *Feathered Dinosaurs and the Origin of Flight*. The Dinosaur Museum Journal 1. The Dinosaur Museum, Blanding, Utah: 63-95. (2) Zhang, F., Zhou, Z., Xu, X., Wang, X., and Sullivan, C. 2008. A bizarre Jurassic maniraptoran from China with elongate ribbon-like feathers. *Nature*, 455: 1105-1108.

p. 99 "Fortunately, paleontologists who have interpreted the few indisputable dinosaur nests in the geologic record made a nice little checklist for the rest of us to follow. . . ." Chiappe *et al.* (2004).

p. 100 "At the time of this discovery, more than a hundred years had elapsed between paleontologists first linking fossil eggs to dinosaurs, which was in 1869." Matheron, M. 1869. Notice sur les Reptiles fossils des depots fluvio-lacustres Crétacés du basin a lignote de Fuveau. *Memoires l'Academie Impériale des Sciences, Belles-Lettres et Arts de Marseille*: 345-379.

p. 100 ". . . John ('Jack') Horner and his friend Robert ('Bob') Makela . . . found depressions filled with eggs and partly grown juveniles of the large ornithopod dinosaur *Maiasaura peeblesorum*." (1) Horner, J.R., and Makela, R. 1979. Nest of juveniles provides evidence of family structure among dinosaurs. *Nature*, 282: 296-298. (2) Horner, J.R. 1982. Evidence of colonial nesting and 'site fidelity' among ornithischian dinosaurs. *Nature*, 297: 675-676. (3) Horner, J.R. 1984. The nesting behavior of dinosaurs. *Scientific American*, 250: 130-137.

p. 101 "Recognized by David ('Dave') Varricchio, who was a Ph.D. student of Horner's in the early 1990s, these structures were the first dinosaur nests defined solely as trace fossils." (1) Varricchio *et al.* (1997, 1999).

p. 102 "Eggs in the clutch were oriented almost vertically, narrow ends down but also pointing toward the center." Varricchio *et al.* (1997).

p. 104 "The width of the nest, rim included, was about half the total body length of an adult *Troodon*. . . ." Varricchio *et al.* (1999).

p. 104 "Just to compare, though, crocodilians and sea turtles normally take just a few hours to dig a hole nest." (1) Kusklan, J.A., and Mazzotti, F.J. 1989. Population biology of the American crocodile. *Journal of Herpetology*, 23: 7-21. (2) Ruckdeschel and Shoop (2006).

p. 105 "Some of these mounds, such as nests of the Australian brush turkey (*Alectura lathami*), can be 2 m (6.6 ft) tall and 20 m (67 ft) wide." (1) Jones, D.N. 1988. Construction and maintenance of the incubation mounds of the Australian brush-turkey *Alectura lathami*. *Emu*, 88: 210-218. (2) Seymour, R.S., and Bradford, D.F. 1992. Temperature regulation in the incubation mounds of the Australian brush-turkey. *The Condor*, 94: 134-150.

p. 105 "These anatomical traits of *Troodon* mothers are interpreted on the basis of how eggs are paired." Varricchio *et al.* (1997, 1999).

p. 106 "Even better evidence of brooding behavior is the adult itself. . . ." Varricchio *et al.* (1997, 1999).

p. 106 "Further study showed that this *Troodon*, as well as *Citipati* and *Oviraptor*, were probably male, providing an important clue related to how dinosaur parents took care of an egg clutch." (1) Varricchio,

D.J., Moore, J.R., Erickson, G.M., Norell, M.A., Jackson, F.D., and Borkowski, J.J. 2008. Avian parental care had dinosaur origin. *Science,* 322: 1826-1828. (2) Schweitzer, M.H., Wittmeyer, J.L., and Horner, J.R. 2005. Gender-specific reproductive tissue in ratites and *Tyrannosaurus rex. Science,* 308: 1456-1460.

p. 106 "Not coincidentally, this same sort of paternal protection is a modern behavior likewise seen in nesting emus, ostriches, and rheas." Varricchio, D.J. 2011. A distinct dinosaur life history? *Historical Biology,* 23: 91-107.

p. 107 "Much study on the sedimentary rocks there, including their geochemistry, revealed that this part of Montana was probably warm and semi-arid 75 to 80 million years ago." Rogers, R.R. 1990. Taphonomy of three dinosaur bone beds in the Upper Cretaceous Two Medicine Formation of northwestern Montana: evidence for drought-related mortality. *Palaios,* 5: 394-413.

p. 107 "Forests were farther away from the nesting grounds, as was an active volcano that occasionally erupted and helped to preserve both the trees of these forests and dinosaur bones." Roberts, E.M., and Hendrix, M.S. 2000. Taphonomy of a petrified forest in the Two Medicine Formation (Campanian), northwest Montana: implications for palinspastic restoration of the Boulder Batholith and Elkhorn Mountains volcanics. *Palaios,* 15: 476-482.

p. 108 "This is a behavioral trait shared with nesting sea turtles, crocodilians, and birds, in which mothers come back repeatedly to the same nesting site. . . ." (1) Ruckdeschel and Shoop (2006). (2) Thorbjarnarson, J.B., and Hernández, G. 1993. Reproductive ecology of the Orinoco crocodile (*Crocodylus intermedius*) in Venezuela. I. Nesting ecology and egg and clutch relationships. *Journal of Herpetology,* 27: 363-370. (3) Hepp, G.R., and Kennamer, R.A. 1992. Characteristics and consequences of nest-site fidelity in wood ducks. *The Auk,* 109: 812-818.

p. 108 ". . . I later did a study with Varricchio on fossil insect nests near the *Troodon* nesting sites." Martin, A.J., and Varricchio, D.J. 2011. Paleoecological utility of insect trace fossils in dinosaur nesting sites of the Two Medicine Formation (Campanian), Choteau, Montana. *Historical Biology,* 23: 15-25.

p. 109 "Horner originally conjectured that these cocoons, which were also near the *Maiasaura* nest sites, were those of carrion beetles, which he imagined fed on hatched eggs, or dead eggs and hatchlings." Horner, J.R., and Gorman, J. 1988. *Digging Dinosaurs: The Search that Unraveled the Mystery of Baby Dinosaurs.* Workman Publishing Company, New York: 210 p.

p. 109 "Thus it was not surprising that one of the same people who studied the *Troodon* nests and eggs with Varricchio in Montana, Frankie Jackson, later noticed and defined similar structures. . . ." Chiappe *et al.* (2004).

p. 110 "Instead, some of these nests had about the same width and depth of the *Troodon* nests. . . ." (1) Varricchio *et al.* (1999). (2) Chiappe *et al.* (2004).

p. 110 ". . . the titanosaur nests were longer than they were wide, giving them oblong or kidney-bean profiles." Chiappe *et al.* (2004).

p. 111 "Complicating matters further, the expansion and contraction of soils around the eggs fractured and moved them. . . ." Jackson, F.D., Schmitt, J.G., and Oser, S.E. 2013. Influence of vertisol development on sauropod egg taphonomy and distribution at the Auca Mahuevo locality, Patagonia, Argentina. *Palaeogeography, Palaeoclimatology, Palaeoecology*, 386: 300-307.

p. 111 "Even more odd is how the ungual on the first digit of the rear foot of a sauropod . . . is significantly larger than all of the others." Fowler, D.W., and Hall, L.E. 2011. Scratch-digging sauropods revisited. *Historical Biology*, 23: 27-40.

p. 112 "This same skewed orientation shows up in well-preserved sauropod tracks. . . ." Fowler and Hall (2011).

p. 112 "Interpretations of these pedal oddities, which paleontologists had noted since the 19th century, have been quite varied and include. . . ." Fowler and Hall (2011).

p. 112 "This was an intriguing clue, because some tortoises, such as the gopher tortoise of the southeastern U.S. (*Gopherus polyphemus*), are magnificent burrowers. . . ." Martin, A.J. 2013. *Life Traces of the Georgia Coast*. Indiana University Press, Bloomington, Indiana: 692 p.

p. 112 "To get a better idea of how sauropods might have dug out their nests, Fowler and Hall read many accounts of tortoises digging nests and watched videos of tortoises making nests. . . ." Fowler and Lee (2011).

p. 113 "For example, paleontologists who studied the Argentine sauropod nests do not think these eggs were actively buried by the sauropods." Chiappe *et al.* (2004).

p. 113 "Still, other sauropod eggshells have the right kind of pore structure for having been buried and incubated under a layer of sediment, just like in tortoises." Chiappe *et al.* (2004).

p. 114 ". . . even though one sauropod urination trace fossil has been interpreted from Late Jurassic rocks in Colorado, explained in a later chapter." Sorry, you're going to have to hold your curiosity for now, but I promise to relieve you with that reference soon.

p. 114 "Bernat Vila, Frankie Jackson, and others looked at nine stratigraphic horizons with fossil eggs, which they classified as megaloolithid, an egg type affiliated with sauropods." Vila, B., Jackson, F.D., Fortuny, J., Sellés, A.G., and Galobart, Á. 2010. 3-D modelling of Megaloolithid clutches: insights about nest construction and dinosaur behaviour. *PLoS One*, 5: e10362. doi:10.1371/journal.pone.0010362

p. 115 "All the same, Fowler and Hall thought about this problem too, and conjectured that large unguals in the front feet of sauropods may have been used for getting a grip. . . ." Fowler and Hall (2011).

p. 115 "For instance, from a full standing position, a mother titanosaur's cloaca might have been as high as 4 to 5 m (13–16 ft) above the ground." These heights can be imagined easily by looking at a mounted skeleton of a titanosaur, in which the upright postures correspond with leg lengths, which in turn are 80–90% of cloacal height. Look just below the base of the tail, and that is where the cloaca would have been located.

p. 116 "Alternatively, these mothers may have had an anatomical attribute, such as an extendable tube that projected from the cloaca and thus brought the eggs closer to the ground without the sauropod mother having to squat." Fowler and Lee (2011).

p. 117 "We don't know an answer for sure yet, although some sauropod tracksites show a wide range of sizes in their tracks—from very small to extra large—suggesting that young sauropods and parents may have been in the same place and time." (1) Bird (1985). (2) Lockley (1991).

p. 118 "One of the most impressive dinosaur nest sites with other trace fossils to address these questions is in Early Jurassic rocks of South Africa from about 190 *mya*." Reisz *et al.* (2012).

p. 118 "*Massospondylus* was not a massive dinosaur, probably weighing about 130 to 150 kg (285–330 lbs). . . ." Seebacher, F. 2001. A new method to calculate allometric length-mass relationships of dinosaurs. *Journal of Vertebrate Paleontology*, 21: 51-60.

p. 118 "Paleontologists had known about *Massospondylus* embryos and eggs since at least the 1970s. . . ." Reisz, R.R., Evans, D.C., Sues, H.-D., and Scott, D. 2010. Embryonic skeletal anatomy of the sauropodomorph dinosaur *Massospondylus* from the Lower Jurassic of South Africa. *Journal of Vertebrate Paleontology*, 30: 1653-1664.

p. 118 ". . . summarized in a report published by Robert Reisz and four other paleontologists in 2011, were spectacular." Reisz *et al.* (2012).

p. 119 "The first embryos and eggs were discovered at this site in 1976 in loose rocks on the ground. . . ." Reisz, R.R., Scott, D., Sues, H.-D., Evans, D.C., and Raath, M.A. 2005. Embryos of an Early Jurassic prosauropod dinosaur and their evolutionary significance. *Science*, 309: 761-764.

p. 119 "The nests were on four horizons and close to one another, which the paleontologists interpreted as good reasons to infer site fidelity and nesting colonies." Reisz *et al.* (2012).

p. 120 "These four-toed tracks, preserved as natural casts on blocks of rock that fell from the roadcut, just happened to match the front and rear feet of a sauropodomorph like *Massospondylus.*" (1) Farlow *et al.* (2012). (2) Reisz *et al.* (2012).

CHAPTER 5: DINOSAURS DOWN UNDERGROUND

p. 123 "Alligators are represented by two species that live in widely separated places. . . ." (1) Ouchley, K. 2013. *American Alligator: Ancient Predator in the Modern World.* University of Florida Press, Gainesville, Florida: 160 p. (2) Thorbjarnarson, J., and Wang, X.M. 2010. *The Chinese Alligator: Ecology, Behavior, Conservation, and Culture.* Johns Hopkins University Press, Baltimore, Maryland: 288 p.

p. 124 "Both are burrowers, especially the Chinese alligators. . . ." Thorbjarnarson and Wang (2010).

p. 124 "In their evolutionary history, Chinese and American alligators shared a common ancestor not too far back in the geologic past. . . ." Brochu, C.A. 2003. Phylogenetic approaches to crocodilian history. *Annual Review of Earth and Planetary Sciences*, 31: 357-397.

p. 125 *"Furthermore, no convincing evidence has revealed that dinosaurs lived underground because no dinosaur has ever been found in a burrow, nor have any burrows been attributed to them."* Martin (2006). (P.S.: What an idiot.)

p. 128 "A few other dinosaurs had anatomical attributes that could have helped them to dig, such as the small Late Cretaceous theropod *Mononykus* of Mongolia. . . ." Senter, P. 2005. Function in the stunted forelimbs of *Mononykus olecranus* (Theropoda), a dinosaurian anteater. *Paleobiology*, 31: 373-381.

p. 128 "Moreover, one specimen of the Early Cretaceous ceratopsian, *Psittacosaurus*, was found with 34 juveniles. . . ." Meng *et al.* (2004).

p. 128 "More recently, in 2010, trace fossils attributed to predatory dromaeosaurs—such as *Deinonychus*—were interpreted as dig marks. . . ." Simpson *et al.* (2010).

p. 128 ". . . mammals that live part or most of their lives underground will result in hundreds of species . . . from Oldfield mice (*Peromyscus polionotus*) to grizzly bears (*Ursus arctos*)." Elbroch (2003).

p. 129 "For a few animals, such as lungfish or some toads and turtles, these burrows serve as aestivation or hibernation chambers. . . ." Hembree, D.I. 2010. Aestivation in the fossil record: evidence from ichnology. *Progress in Molecular and Subcellular Biology*, 49: 245-262.

p. 130 "Anyway, Bakker speculated that *Othnielia* and *Drinker* . . . must have been living in burrows of their own making." Bakker, R.T. 1996. The real Jurassic Park: dinosaurs and habitats at Como Bluff, Wyoming. *The Continental Jurassic: Museum of Northern Arizona Bulletin*, 60: 35-49. David Varricchio also saw Bakker present this idea in a talk and he was influenced by it (Varricchio, personal communication with me, November 2013).

p. 130 "As discussed earlier, most scientists now accept that a large meteorite hit the earth about 65 *mya*, landing in the area of what is now the Yucatan Peninsula." Alvarez (2008).

p. 131 "So large animals likely died the soonest after the impact, and yes, I am talking about dinosaurs. . . ." McKinney, M.L. 1997. Extinction vulnerability and selectivity: combining ecological and paleontological views. *Annual Review of Ecology and Systematics*, 28: 495-516.

p. 132 "Lastly, burrowing is a behavior present in every major group of vertebrates that made it past the mass extinction. . . ." Robertson, D.S., McKenna, M.C., Toon, O.B., Hope, S., and Lillegraven, J.A. 2003. Survival in the first hours of the Cenozoic. *GSA Bulletin*, 116: 760-768.

p. 133 "This region of Montana had outcrops of the Blackleaf Formation, which was composed of mudstones and sandstones. . . ." Dyman, T.S., and Nichols, D.J. 1988. Stratigraphy of mid-Cretaceous Blackleaf and lower part of the Frontier Formations in parts of Beaverhead and Madison Counties, Montana. *USGS Bulletin*, 1773: 1-27.

p. 134 ". . . expressed succinctly in a quote by famed science fiction writer Isaac Asimov: 'The most exciting phrase to hear in science'. . . ." I have to admit that, however much I love this statement attributed to Asimov, and as often as you will find it connected to him in Internet searches, I've had a tough time finding exactly when and where he originally said it. So for now, I'll enjoy its sentiment regardless of whether he said it or not.

p. 138 "For example, modern gopher tortoise burrows play host to 200 to 300 species of other animals. . . ." Kinlaw, A., and Grasmueck, M. 2012. Evidence for and geomorphologic consequences of a reptilian ecosystem engineer: the burrowing cascade initiated by the gopher tortoise. *Geomorphology*, 157-158: 108-121.

p. 138 "As a Russian saying goes, 'Trust, but verify. . . .'" Доверяй, но проверяй, or transliterated as "Doveryai, no proveryai," which also rhymes.

p. 140 "The study of modern traces—also known as *neoichnology*—had been kind to us, with the gopher tortoise burrows of Georgia supplying a sensible explanation. . . ." Martin (2013).

p. 142 "The latter synced beautifully with previous work done by Dave's advisor, Jack Horner, who had proposed that *Maiasaura* raised its young in their nests. . . ." (1) Horner and Makela (1979). (2) Horner (1982).

p. 142 "First, one of the bones integral to its shoulder—the scapulocoracoid—had attachment sites for large muscles." Varricchio *et al.* (2007).

p. 142 "Second, the bone on the front of its snout—the premaxilla—was fused, an unusual feature in a dinosaur." Varrichio *et al.* (2007).

p. 142 "Third, when compared to its closest relatives (other small ornithopods), its hip had an extra vertebra." Norman, D.B. 2004. Basal Iguanodontia. *In* Weishampel, D.B., Dodson, P., and Osmólska, H. (editors), *The Dinosauria* (2nd Edition). University of California Press, Berkeley, California: 393-412.

p. 143 "In a study published the same year the dinosaur was discovered (2005), a zoologist compared the cross-sectional areas of burrows made by a wide variety of modern burrowing animals. . . ." White, C.R. 2005. The allometry of burrow geometry. *Journal of Zoology, London*, 265: 395-403.

p. 144 "Granted, the burrow was a close fit, but this situation also applies to many modern burrowing and denning mammals. . . ." Varricchio *et al.* (2007).

p. 144 "Gopher tortoise burrows display exactly the same sort of strategy, in which these at first run straight down. . . ." Martin (2013).

p. 147 "For example, in a paper published in late 2006, the author David Loope proposed small dinosaurs as possible burrowmakers. . . ." Loope, D.B. 2006. Burrows dug by large vertebrates into rain-moistened, Middle Jurassic dune sand. *Journal of Geology*, 114: 752-763.

p. 150 "Australians have a long tradition of ill-fated expeditions, and I felt like I had unwittingly instigated one of these." In fact, when I originally wrote these words, I had not realized that "ill-fated expedition" is a cliché phrase in Australia, especially when applied to the infamous 1860-1861 Burke and Wills expedition to the central desert of Australia; both namesakes died during this foray. However, at least some science came out of their journey: Joyce, E.B., and McCann, D.A. (editors). 2010. *Burke and Wills: The Scientific Legacy of the Victoria Exploring Expedition*. CSIRO Publishing, Collingwood, Victoria: 368 p.

p. 150 "Based on plate-tectonic reconstructions, the rocks in this part of Australia were originally formed near the South Pole, when southern Australia was connected to Antarctica. . . ." Rich, T.H., Vickers-Rich, P., 2000. *Dinosaurs of Darkness*. Indiana University Press, Bloomington, Indiana: 222 p.

p. 150 "Nearly all of the dinosaur bones and teeth in strata there were from small dinosaurs, and most of these were hypsilophodonts." Rich, T.H., and Vickers-Rich, P. 1999. The Hypsilophodontidae from southeastern Australia. *In* Tomada, Y., Rich, T.H., and Vickers-Rich, P. (editors), Proceedings of the Second Gondwana Dinosaur Symposium. *National Science Museum Monographs*, 15: 167-180.

p. 151 "In a paper they and other co-authors published in the journal *Science* in 1988, they proposed that hypsilophodonts and other dinosaurs in the region were likely endothermic. . . ." Rich, T.H., Rich, P.V., Wagstaff, B.E., McEwen-Mason, J., Douthitt, C.B., Gregory, R.T., and Felton, E.A. 1988. Evidence for low temperatures and biologic diversity in Cretaceous high latitudes of Australia. *Science*, 242, 1403-1406.

p. 151 "Not all paleontologists were so accepting of the idea that dinosaurs were less like reptiles and more like birds. . . ." Probably no book title of the time summarized those feelings better than this: Thomas, R.D.K., and Olson, E.C. 1980. *A Cold Look at the Hot-Blooded Dinosaurs*. Westview Press, Boulder, Colorado: 514 p.

p. 151 "*With the possible exception of the* Allosaurus *sp., all of the animals were small enough to have found shelter readily by burrowing.*" Quoted from Rich *et al.* (1988).

p. 155 "For one, the Cretaceous rocks of Victoria abounded with evidence for hypsilophodonts. . . ." Rich and Vickers-Rich (1999).

p. 155 "For another, such dinosaurs were too small to have migrated long distances between seasons." Bell, P., and Snively, E. 2008. Polar dinosaurs on parade: a review of dinosaur migration. *Alcheringa* 32: 271-284.

p. 155 "Thirdly, many modern polar vertebrates, from puffins to polar bears, burrow into dirt or snow to take refuge in those harsh environments." Davenport, J. 1992. *Animal Life at Low Temperatures*. Chapman & Hall, London: 246 p.

p. 157 "Gratifyingly, the editor accepted the paper the following week, which had the title. . . ." Martin (2009).

p. 157 "As of this writing, the video had the most views of any in the history of my university, and the press attention was flattering." The video includes good footage (shot by my wife, Ruth Schowalter) at Knowledge Creek, with me talking about them, at: http://www.youtube.com/watch?v=S8gahKh1l9A

p. 158 "So in an experiment straight out of the popular TV show *Mythbusters*, Woodruff joined PVC piping. . . ." Woodruff, D.C., and Varricchio, D.J. 2011. Experimental modeling of a possible *Oryctodromeus cubicularis* (Dinosauria) burrow. *Palaios*, 26: 140-151.

p. 159 ". . . the hypothesis that this *Oryctodromeus cubicularis* was a burrowing, denning dinosaur still stands." This hypothesis was further supported in a study done by Jamie Fearon, a graduate student of Varricchio's. She determined that the forelimb anatomy on *Oryctodromeus* really was different from that of other ornithopods, and was suited for digging: Fearon, J., and Varricchio, D. 2012. Comparative pectoral and forelimb morphology of Ornithopoda: does *Oryctodromeus cubicularis* exhibit specialization for digging? *Journal of Vertebrate Paleontology*, SVP Program and Abstracts, 2012: 92-93.

CHAPTER 6: BROKEN BONES, TOOTHMARKS, AND MARKS ON TEETH

p. 162 "Although this scenario was imagined, the trace fossil evidence for all such behaviors is real." (See the references related to this scene under Chapter 1 endnotes.)

p. 163 "Although this 'skin' was actually just a natural cast of the original skin, it was closely associated with the bones of *Edmontosaurus annectens*. . . ." Rothschild, B.M., and Depalma, R. 2013. Skin pathology in the Cretaceous: evidence for probable failed predation in a dinosaur. *Cretaceous Research*, 42: 44-47.

p. 164 "These unusual body fossils were normally formed first as impressions against soft sediment, and then naturally cast in sandstone, similar to how many dinosaur footprints were preserved." Martin (2006).

p. 166 "Although not all of paleopathology concerns itself with dinosaurs—much of it centers on evidence for diseases in pre-historic human remains—this science is being applied enthusiastically to dinosaurs." Rega, E. 2012. Disease in dinosaurs. *In* Brett-Surman, M.K., Holtz, T.R., Jr., and Farlow, J.O. (editors), *The Complete Dinosaur* (2nd Edition). Indiana University Press, Bloomington, Indiana: 667-711.

p. 167 "For example, one specimen of *Psittacosaurus* from Early Cretaceous rocks of China also, like the *Edmontosaurus*, has a skin impression associated with its body. . . ." Lingham-Soliar, T. 2008. A unique cross section through the skin of the dinosaur *Psittacosaursus* from China showing a complex fibre architecture. *Proceedings of the Royal Society of London, B*, 275: 775-780.

p. 168 "(This explanation for how dinosaur parts got into marine sediments is nicknamed the 'bloat-and-float' hypothesis.)" Martin (2006).

p. 168 "Amazingly, a few of these bones have toothmarks and embedded teeth from scavenging sharks. . . ." Schwimmer, D.R. 1997. Late Cretaceous dinosaurs in Eastern USA: a taphonomic and biogeographic model of occurrences. 1997 *Dinofest International Proceedings*: 203-211.

p. 168 "Meanwhile on land, lowly mammals imparted their distinctive incisor incisions on a variety of Late Cretaceous dinosaur bones from Alberta, Canada." Longrich, N.R., and Ryan, M.J. 2010. Mammalian tooth marks on the bones of dinosaurs and other Late Cretaceous vertebrates. *Palaeontology*, 53: 703-709.

p. 168 "Carrion beetles, which dine on dead bodies, or termites, which make small pits or tunnels in bones, could have made these." (1) Britt, B.B., Scheetz, R.D., and Dangerfield, A. 2008. A suite of dermestid beetle traces on dinosaur bone from the Upper Jurassic Morrison Formation,

Wyoming, USA. *Ichnos*, 15: 59-71. (2) Xing, L., Roberts, E.M., Harris, J.D., Gingras, M.K., Ran, H., Zhang, J., Xu, X., Burns, M.E., and Dong, Z. 2013. Novel insect traces on a dinosaur skeleton from the Lower Jurassic Lufeng Formation of China. *Palaeogeography, Palaeoclimatology, Palaeoecology*, 388: 58-68.

p. 169 "Among my favorite examples of these are big holes in the head shields of *Triceratops*. When paleontologists first noticed these holes, they were a bit mystified." Farke *et al.* (2009).

p. 169 "So paleontologists Andrew Farke, Ewan Wolffe, and Darren Tanke, in an attempt to make sense of these holes, took a closer look at the sizes, shapes, and placements of them. . . ." Farke *et al.* (2009).

p. 170 "As many dinosaur fans can relate . . . the genus name *Triceratops*, assigned in 1889, translates as 'three-horned face.'" Marsh, O.C. 1889. Notice of gigantic horned Dinosauria from the Cretaceous. *American Journal of Science*, 38: 173-175.

p. 170 "Paleontologists are also now sure that *Triceratops* horns changed in size and shape throughout their lives, becoming more formidable with age." Horner, J.R., and Lamm, E. 2011. Ontogeny of the parietal frill of *Triceratops*: a preliminary histological analysis. *Comptes Rendus Palevol*, 10: 439-452.

p. 171 "Paleontological artist Charles Knight (1874–1953) most famously depicted such a scenario in a 1927 mural. . . ." Simply titled *Tyrannosaurus and Triceratops*, it is probably the most iconic of Knight's gorgeous and influential artworks, and the original is in the Field Museum of Natural History in Chicago, Illinois: http://www.charlesrknight.com/Enlarge.htm?109

p. 171 "Only one tantalizing *Triceratops* skull tells of a living battle with a *Tyrannosaurus*, in which one of its brow horns and a squamosal bone were chomped and later healed." Happ, J. 2008. An analysis of predator-prey behavior in a head-to-head encounter between *Tyrannosaurus rex* and *Triceratops*. *In* Larson, P., and Carpenter, K. (editors). *Tyrannosaurus rex, the Tyrant King*. Indiana University Press, Bloomington, Indiana: 355-368.

p. 172 "In this role, their large heads would have been quite useful for recognizing the same species and perhaps gender. . . ." Padian, K., and Horner, J.R. 2011. The evolution of 'bizarre structures' in dinosaurs: biomechanics, sexual selection, social selection or species recognition? *Journal of Zoology* 283: 3-17.

p. 172 "Pachycephalosaurs, such as *Pachycephalosaurus*, *Stegoceras*, and others, are relatively rare dinosaurs and only found in Cretaceous Period rocks." (1) Makovicky, P. 2012. Marginocephalia. *In* Brett-Surman, M.K., Holtz, T.R., Jr., and Farlow, J.O. (editors), *The Complete Dinosaur* (2nd Edition). Indiana University Press, Bloomington, Indiana:

527-549. (2) Maryańska, T., Chapman, R.E., and Weishampel, D.B. 2004. Pachycephalosauria. *In* Weishampel, D.B., Dodson, P., and Osmólska, H. (editors), *The Dinosauria* (2nd Edition). University of California Press, Berkeley, California: 464-477.

p. 172 "These skullcaps . . . can be nearly 25 cm (10 in) thick. They are composed of parietals and closely associated skull bones. . . ." Makovicky (2012).

p. 173 "Nearly everybody agrees that these robust skulls must have been used for butting. . . ." Makovicky (2012).

p. 174 "In 2011, two paleontologists, Eric Snively and Jessica Theodor, had head-butting in mind. . . ." Snively, E., and Theodor, J.M. 2011. Common functional correlates of head-strike behavior in the pachycephalosaur *Stegoceras validum* (Ornithischia, Dinosauria) and combative artiodactyls. *PLoS ONE* 6(6): e21422. doi:10.1371/journal.pone.0021422.

p. 174 "They then repeated their analysis on three species of modern mammals that habitually head-smash one another. . . ." Snively and Theodor (2011).

p. 175 "In 2012 . . . two other paleontologists—Joseph Peterson and Christopher Vittore—scrutinized a *Pachycephalosaurus* skull. . . ." Peterson, J., and Vittore, C. 2012. Cranial pathologies in a specimen of *Pachycephalosaurus. PLoS ONE,* 7(4): doi: 10.1371/journal.pone.0036227.

p. 176 "In a follow-up study published in 2013, Peterson and two other paleontologists found many more examples, verifying the previous study." Peterson, J.E., Dischler, C., and Longrich, N.R. 2013. Distributions of cranial pathologies provide evidence for head-butting in dome-headed dinosaurs (Pachycephalosauridae). *PLoS ONE* 8(7): e68620; doi: 10.1371/journal.pone.0068620.

p. 176 "One is of injured but healed bony tissue in a few tail spikes of *Stegosaurus.*" Carpenter, K., Sanders, F., McWhinney, L.A., and Wood, L. 2005. Evidence for predator-prey relationships: examples for *Allosaurus* and *Stegosaurus. In* Carpenter, K. (editor), *The Carnivorous Dinosaurs.* Indiana University Press, Bloomington, Indiana: 325-350.

p. 177 "Another is of a hole in a tail vertebra of an *Allosaurus,* but one that probably healed around a piece of a *Stegosaurus* tail spike, an unwanted souvenir." Carpenter *et al.* (2005).

p. 177 "Paleontologists . . . had figured that stegosaur tail spikes were used for self-defense, but lacked further confirmation until . . . 2001 (broken spikes) and 2005 (a healed hole in the bone of a predator)." McWhinney, L.A., Rothschild, B.M., and Carpenter, K. 2001. Post-traumatic chronic osteomyelitis in *Stegosaurus* dermal spikes. *In* Carpenter, K. (editor). *The Armored Dinosaurs.* Indiana University Press, Bloomington, Indiana: 141-156.

p. 177 "Nicknamed 'Big Al' and studied by paleontologist Rebecca Hanna in the 1990s, this *Allosaurus* was unusual in two ways. . . ." Hanna began reporting on "Big Al" in the late 1990s at professional meetings, and the full results were summarized in her 2002 paper: Hanna, R.R. (2002). Multiple injury and infection in a sub-adult theropod dinosaur (*Allosaurus fragilis*) with comparisons to allosaur pathology in the Cleveland-Lloyd Dinosaur Quarry Collection. *Journal of Vertebrate Paleontology*, 22: 76-90. I also recommend checking out a still-informative and entertaining BBC documentary featuring Hanna talking about this *Allosaurus*, initially titled *The Ballad of Big Al* (2000).

p. 177 "One of the reasons for this may be an easy one: *Coelophysis* was a much smaller dinosaur than *Allosaurus*. . . ." Holtz *et al.* (2012).

p. 178 "As elaborated in a previous chapter, if a large two-legged dinosaur tripped, both its greater weight and height would conspire against it. . . ." Farlow *et al.* (1995).

p. 180 "For example, a few smaller theropods, such as the Late Cretaceous theropods *Velociraptor* and *Saurornitholestes*, had thin, curved, and beautifully serrated teeth." A good summary of theropod tooth shapes and their inferred functions is in: Fastovsky, D.E., and Weishampel, D.B. 2009. *Dinosaurs: A Concise Natural History*. Cambridge University Press, Cambridge, U.K.: 379 p.

p. 180 "Tooth forms even varied within the same dinosaur, a condition called heterodonty ('different teeth')." Martin (2006).

p. 181 "But if you looked deeper into its mouth, say, while being eaten, you would see that the teeth in the front curve more toward the rear of the mouth. . . ." Reichel, M. 2012. The variation of angles between anterior and posterior carinae of tyrannosaurid teeth. *Canadian Journal of Earth Sciences*, 49: 477-491.

p. 181 "Certain theropods had an exact number of teeth per tooth row, but that number could also be divided into two parts. . . ." Reichel (2012).

p. 182 "These marks would have been made by a pulling, lateral movement, so they would be shallow, long grooves on a bone surface. . . ." Jacobsen, A.R., and Bromley, R.G. 2009. New ichnotaxa based on tooth impressions on dinosaur and whale bones. *Geological Quarterly*, 53: 373-382.

p. 182 "Because this action would have been more perpendicular to the bone surface, these puncture toothmarks will be deeper. . . ." Jacobsen, A.R. 1998. Feeding behaviour of carnivorous dinosaurs as determined by tooth marks on dinosaur bones. *Historical Biology*, 13: 17-26.

p. 182 "Although dinosaur toothmarks were first interpreted as trace fossils near the start of the 20th century. . . ." Matthew, W.D. 1908. *Allosaurus*, a carnivorous dinosaur and its prey. *American Museum Journal*, 8: 2-5.

p. 182 "What made them think these toothmarks belonged specifically to *T. rex* stemmed from a combination of size, shape, place, and time." Erickson, G.M., Van Kirk, S.D., Su, J., Levenston, M.E., Caler, W.E., and Carter, D.R. 1996. Bite force estimation for *Tyrannosaurus rex* from tooth-marked bones. *Nature*, 382: 706-708.

p. 183 "These artificial teeth were then used to imitate a tyrannosaur bite, but one in which the force exerted by each 'bite' could be measured." Erickson *et al.* (1996).

p. 184 "Bite-force estimates came out to 6,400 to 13,400 N, which at the time were greater than those known for any living animal. . . ." Erickson *et al.* (1996).

p. 184 "In later experiments done on modern alligators and crocodiles, Erickson and other researchers found the largest American alligators (*Alligator mississippiensis*) had bite forces. . . ." Erickson, G.M., Lappin, A.K., and Vliet, K.A. 2003. The ontogeny of bite-force performance in American alligator (*Alligator mississippiensis*). *Journal of Zoology*, 260: 317-327.

p. 184 "(It was not too surprising, then, when paleontologists later found toothmarks attributable to *Deinosuchus* in dinosaur bones.)" Rivera-Sylva, H.E., Frey, E., and Guzman-Gutierrez, J.R. 2009. Evidence of predation on the vertebra of a hadrosaurid dinosaur from the Upper Cretaceous (Campanian) of Coahuila, Mexico. *Carnets de Géologie* [*Notebooks on Geology*], Letter 2009/02: 1-6.

p. 184 "This gruesome idea came about when Denver Fowler and several other paleontologists noticed, while looking at *Triceratops* bones from the Late Cretaceous of Montana. . . ." Fowler *et al.* (2012).

p. 186 "When paleontologist Ken Carpenter took a close look at this oddity, he saw signs of healing around the bone. . . ." Carpenter, K. 2000. Evidence of predatory behavior by carnivorous dinosaurs. *Gaia*, 15: 135-144.

p. 186 "Since Carpenter's study, other healed toothmarks found in *Edmontosaurus* bones, either attributed to *T. rex* or *Albertosaurus*, have both affirmed. . . ." The most recent article dealing with tyrannosaurids munching hadrosaurs, which includes references to other interpreted incidents, is: DePalma, R.A., Burnham, D.A., Martin, L.D., Rothschild, B.M., and Larson, P.L. 2013. Physical evidence of predatory behavior in *Tyrannosaurus*. *Proceedings of the National Academy of Sciences*, 110: 12560-12564.

p. 186 "So this is exactly how three paleontologists—Ray Rogers, David Krause, and Kristi Rogers—discerned that *Majungasaurus*, a large theropod from Late Cretaceous rocks of Madagascar, was a cannibal." Rogers, R.R., Krause, D.W., and Rogers, K.C. 2003. Cannibalism in the Madagascan dinosaur *Majungatholus atopus*. *Nature*, 422: 515-518.

p. 187 "Komodo dragons, crocodilians (including alligators), big cats such as lions and tigers, and bears are among the large predators that will eat their own species." Fox, L.R. 1975. Cannibalism in natural populations. *Annual Reviews of Ecology and Systematics*, 6: 87-106.

p. 187 "Apparently so, as its rocks show the region was semi-arid while dinosaurs lived there, but with pronounced wet-dry cycles." Rogers, R.R., Krause, D.W., Rogers, K.C., Rasoamiaramanana, and Rahantarisao, L. 2007. Paleoenvironment and paleoecology of *Majungasaurus crenatissimus* (Theropoda: Abelisauridae) from the Late Cretaceous of Madagascar. *Journal of Paleontology*, 27, Supplement 2 (Special Issue, Memoir 8): 21-31.

p. 188 "One of the more compelling pieces of evidence for droughts in this area during the Cretaceous comes from other trace fossils, namely lungfish burrows." Marshall, M.S., and Rogers, R.R. 2012. Lungfish burrows from the Upper Cretaceous Maevarano Formation, Mahajanga Basin, Northwestern Madagascar. *Palaios*, 27: 857-866.

p. 188 "In a 2010 study conducted by Nicholas Longrich and three other paleontologists, they examined *T. rex* bones in museum. . . ." Longrich, N.R., Horner, J.R., Erickson, G.M., and Currie, P.J. 2010. Cannibalism in *Tyrannosaurus rex*. *PLoS One*, 5(10): e13419. doi:10.1371/journal.pone.0013419.

p. 189 "This idea, first proposed by paleontologists Darren Tanke and Phil Currie in 1998, was based on healed bite marks they noted in skulls. . . ." Tanke, D.H., and Currie, P.J. 1998. Head-biting behavior in the theropod dinosaurs: paleopathological evidence. *Gaia*, 15: 167-184.

p. 190 "These scratches tell you that the hadrosaur moved its upper jaws out and to the sides while the lower jaw stayed put." Williams, V.S., Barrett, P., and Purnell, M.A. 2009. Quantitative analysis of dental microwear in hadrosaurid dinosaurs, and the implications for hypotheses of jaw mechanics and feeding. *Proceedings of the National Academy of Sciences*, 106: 11194-11199.

p. 190 "These marks, called *microwear*, were scored on dinosaur teeth when they chewed plants containing silica or plants with grit on them." Fiorillo, A.R. 1998. Dental microwear patterns of the sauropod dinosaurs *Camarasaurus* and *Diplodocus*: evidence for resource partitioning in the Late Jurassic of North America. *Historical Biology*, 13:1-16.

p. 191 "Phytoliths are common in many plants today, especially monocotyledons, which include all grasses, orchids, bamboo, palm trees, and many others." Piperno, D.R. 2006. *Phytoliths: A Comprehensive Guide for Archaeologists and Paleoecologists*. Rowman Altamira, Lanham, Maryland: 238 p.

p. 191 "Monocotyledons also got their start in the middle of the Mesozoic

Era." Daghlian, C.O. 1981. A review of the fossil record of monocotyledons. *The Botanical Review*, 47: 517-555.

p. 192 "Paleontologists who studied microwear in *Edmontosaurus* found out that this dinosaur. . . ." Williams *et al.* (2009).

p. 192 "But anatomical studies done on some sauropods now imply that some of these lengthy necks were maybe better suited for sweeping large areas—back and forth—across fields of low-lying vegetation." This research on sauropod-neck flexibility has been going on for a while, but the latest paper summarizing it (along with new findings) is: Cobley, M.J., Rayfield, E.J., and Barrett, P.M. 2013. Intervertebral flexibility of the ostrich neck: implications for estimating sauropod neck flexibility. *PLoS One*, 8: e72187. doi:10.1371/journal.pone.0072187

Chapter 7: Why Would a Dinosaur Eat a Rock?

p. 195 ". . . the Morrison Formation was rightly celebrated elsewhere in the western U.S. for yielding some of the best-loved of all dinosaurs. . . ." Foster, J. 2007. *Jurassic West: The Dinosaurs of the Morrison Formation and Their World*. Indiana University Press, Bloomington, Indiana: 416 p.

p. 196 "'Gastrolith' literally means 'stomach stone' (in Greek, *gastros* = stomach, *lithos* = stone), and they thus refer to rocks that somehow made it into the digestive tract of an animal." Wings, O. 2007. A review of gastrolith function with implications for fossil vertebrates and a revised classification. *Acta Palaeontologica Polonica*, 52: 1-16.

p. 197 "For one, they can be divided into two categories: *bio-gastroliths* and *geo-gastroliths*." Wings (2007).

p. 197 "These bio-gastroliths are deposits of calcium oxalate, which form in people's kidneys as a result of calcium imbalances. . . ." Coe, F.L., Evan, A., and Worcester, E. 2005. Kidney stone disease. *Journal of Clinical Investigation*, 115: 2598-2608.

p. 197 "Gallstones, on the other hand, are more organic than mineral and are normally composed of cholesterol, although these can sometimes have calcium mixed in as well." Attili, A.F., de Santis, A., Capri, R., Repice, A.M., and Maselli, S. 1995. The natural history of gallstones: The GREPCO experience. *Hepatology*, 21: 656-660.

p. 198 "Remarkably, some crustaceans, such as marine crabs and freshwater crayfish, secrete their own bio-gastroliths of calcium carbonate." Greenway, P. 1985. Calcium balance and moulting in the Crustacea. *Biological Reviews*, 60: 425-454.

p. 198 "In contrast, geo-gastroliths are rocks made outside of animals' bodies by normal geological processes. . . ." Wings (2007).

p. 198 "However, the width of a gastrolith seemingly never exceeds 3% the length of the animal." Wings (2007).

p. 199 "I have watched videos of crows doing the same thing, deliberately walking along and selecting pea-sized gravel to eat." One such video, titled "Crow (Corvus brachyrhynchos) Swallowing Gravel in Vancouver BC 10Apr2010," is here: http://www.youtube.com/watch?v=Gg5gGXCVNr4

p. 200 "This process, also known as *trituration*, is like having a food processor in the upper part of an animal's digestive system." Wings (2007).

p. 200 "In birds that use gastroliths for just this function, their alimentary canal, from top to bottom, goes like this: mouth, esophagus. . . ." Proctor, N.S., and Lynch, P.J. 1998. *Manual of Ornithology: Avian Structure and Function*. Yale University Press, New Haven, Connecticut: 352 p.

p. 201 "One is in predatory birds, such as hawks, which may swallow a few pea-sized rocks, keep them in their digestive tracts for a while, and then regurgitate them before hunting." Bruce, T. 2007. Observations of stone-eating in two species of neotropical falcons (*Micrastur semitorquatus* and *Herpetotheres cachinnans*). *Journal of Raptor Research*, 41: 74-76.

p. 202 "Fortunately for these birds, though, gastroliths not only help to grind these plant fibers but also separate them enough to prevent balling up and turning into impassable objects." Wings (2007).

p. 203 "Nonetheless, a few birds, mammals, and even humans sometimes ingest clay or soil as a digestive aid, a practice called geophagy." Diamond, J.M. 1999. Evolutionary biology: dirty eating for healthy living. *Nature*, 400: 120-121.

p. 203 "For example, kaolinite—a white clay common in Eocene and Cretaceous sedimentary deposits of Georgia—has been long used for medicinal purposes. . . ." Burrison, J.A. 2007. *Roots of a Region: Southern Folk Culture*. University Press of Mississippi, Jackson, Mississippi: 236 p.

p. 203 "However, ornithologists who have studied geophagy in South American birds—such as macaws and parrots. . . ." Brightsman, D.J., Taylor, J., and Phillips, T.D. 2008. The roles of soil characteristics and toxin adsorption in avian geophagy. *Biotropica*, 40: 766-774.

p. 203 "It turns out that gastroliths are indeed used in a variety of swimming vertebrates, such as pinnipeds. . . ." (1) Taylor, M.A. 1993. Stomach stones for feeding or buoyancy? the occurrence and function of gastroliths in marine tetrapods. *Philosophical Transactions: Biological Sciences*, 341: 163-175. (2) Wings (2007).

p. 204 ". . . the 'buoyancy control' *vis-á-vis* gastroliths in aquatic vertebrates was questioned in a 2003 study on alligators. . . ." Henderson, D.M. 2003. Effects of stomach stones on the buoyancy and equilibrium of a floating crocodilian: a computational analysis. *Canadian Journal of Zoology*, 81: 1346-1357.

p. 204 "Yet these trace fossils show up as concentrated masses of rocks in the skeletons of big, non-dinosaurian Mesozoic marine reptiles, such as plesiosaurs." Taylor (1993).

p. 204 "These spectacular trace fossils, evident as long, wide grooves ('gutters') preserved in Jurassic strata of Switzerland and Spain. . . ." Geister, J. 1998. Lebensspuren made by marine reptiles and their prey in the Middle Jurassic (Callovian) of Liesberg, Switzerland. *Facies*, 39: 105-124.

p. 205 "This idea got more support when two Early Cretaceous plesiosaurs described in 2005 from Queensland, Australia not only had gastroliths but also stomach contents." McHenry, C.R., Cook, A.G., and Wroe, S. 2005. Bottom-feeding plesiosaurs. *Science*, 310: 75.

p. 206 "In 2013, Laura Codorniú and several other paleontologists reported on an assortment of coarse sand to fine gravel in the body cavities of two specimens. . . ." Codorniú, L., Chiappe, L.M., and Fabricio, D.C. 2013. The first occurrence of gastroliths in pterosaurs. *Journal of Vertebrate Paleontology*, 33: 647-654.

p. 207 "A few paleontologists have suggested that if a gastrolith is found outside of the body cavity of a dinosaur . . . you should . . . instead refer to it as an *exolith*. . . ." (1) Wings (2007). (2) Wings, O., and Sander, P.M. 2007. No gastric mill in sauropod dinosaurs: new evidence from analysis of gastrolith mass and function in ostriches. *Proceedings of the Royal Society of London, B*, 274: 635-640.

p. 209 "In 2003, paleontologist Oliver Wings decided to find out under what conditions gastroliths might stay or go after the death of a dinosaur." Wings, O. 2003. Observations on the release of gastroliths from ostrich chick carcasses in terrestrial and aquatic environments. *Journal of Taphonomy*, 1: 97-103.

p. 210 "So in the early 1990s, paleontologist Kim Manley took a look at the differences of laser-generated light scattering on genuine gastroliths. . . ." (1) Manley, K. 1991. Two techniques for measuring surface polish as applied to gastroliths. *Ichnos*, 1: 313-316. (2) Manley, K. 1993. Surface polish measurements from bona fide and suspected sauropod dinosaur gastroliths, wave and stream transported clasts. *Ichnos*, 2: 167-169.

p. 210 "This technique was refined in a 1994 study by other scientists who looked at gastroliths from much more recently dead avian dinosaurs, moas from New Zealand." Johnston, R.G., Lee, W.G., and Grace, W.K. 1994. Identifying moa gastroliths using a video light scattering instrument. *Journal of Paleontology*, 68: 159-163.

p. 210 "These large, extinct flightless birds lived on both the north and south islands of New Zealand up until less than a thousand years ago. . . ." Holdaway, R.N., and Jacomb, C. 2000. Rapid extinction of the moas (Aves: Dinornithiformes): model, test, and implications. *Science*, 287: 2250-2254.

p. 211 "In 2000, scientists Christopher Whittle and Laura Onorato used a scanning electron microscope (SEM) to peer more intensely at gastroliths in general." Whittle, C.H., and Onorato, L. 2000. On the origins of gastroliths: determining the weathering environment of rounded and polished stones by scanning electron-microscope examination. *In* Lucas, S.G. and Heckert, A.B. (eds.), *Dinosaurs of New Mexico. New Mexico Museum of Natural History and Science Bulletin*, 17: 69-73.

p. 212 "First of all, despite previous attempts to diminish these important trace fossils via the almost-clever pun 'gastromyths'. . . ." Lucas, S.G. 2000. The gastromyths of "*Seismosaurus*," a Late Jurassic dinosaur from New Mexico. *In* Lucas, S.G. and Heckert, A.B. (editors), *Dinosaurs of New Mexico. New Mexico Museum of Natural History and Science Bulletin*, 17: 61-67.

p. 212 "The earliest description of possible gastroliths in a dinosaur was in 1838, when French paleontologist Jacques Amand Eudes-Deslongchamps noted about ten pebbles. . . ." Eudes-Deslongchamps, A. 1838. Mémoire sur le *Poekilopleuron bucklandi*, grande saurien fossile, intermédiare entre les crocodiles et les lézards, découvert dans les carrières de la Maladrerie, près Caen, au mois de juillet 1835. *Mémoires de la Société Linnéenne Normandie*, 6: 1-114.

p. 212 "Much later, in the late 19th century and leading up to 1900, paleontologists began spotting gastroliths in dinosaurs and marine-reptile contemporaries of dinosaurs. . . ." (1) Brown, B. 1904. Stomach stones and the food of plesiosaurs. *Science*, 20: 184-185. (2) Cannon, G.L. 1906. Sauropodan gastroliths. *Science*, 24: 116.

p. 212 "Large stones were also associated with sauropod skeletons excavated in the 1870s. . . ." Brown, B. 1907. Gastroliths. *Science*, 25: 392.

p. 213 "Sometime around 1900, Brown noticed a collection of rounded cobbles inside a hadrosaur skeleton that he identified as '*Claosaurus*.'" Brown (1907).

p. 213 "In 1907, Brown followed up this study with another paper on dinosaur gastroliths, which he very simply titled 'Gastroliths.'" Brown (1907).

p. 213 "For one, the 'gastroliths' in this particular dinosaur were probably river stones that washed into the body cavity of the hadrosaur soon after it died." Currie, P.J. 1997. Gastroliths. *In* Currie, P.J., and Padian, K. (editors), *Encyclopedia of Dinosaurs*. Academic Press, New York: 270.

p. 213 "Second, the hadrosaur was misidentified and instead was a species of *Edmontosaurus*. . . ." Horner, J.R., Weishampel, D.B., and Forster, C.A. 2004. Hadrosauridae. Weishampel, D.B., Dodson, P., and Osmólska, H. (editors), *The Dinosauria* (2nd Edition). University of California Press, Berkeley, California: 438-463.

p. 213 "Brown was also beaten to press on dinosaur gastroliths by a not-as-famous paleontologist, G.L. Cannon. . . ." Cannon (1906).

p. 213 "Other paleontologists, such as Brown's protégé Roland T. Bird, later promoted this supposed association between sauropods and gastroliths." Bird (1985).

p. 214 "Among the theropods documented thus far with gastroliths are: the Early Jurassic *Megapnosaurus* (previously known as *Syntarsus*). . . ." Most of these gastrolith-bearing theropods are listed under "Gastroliths" in the index of Weishampel *et al.*'s (2004) *The Dinosauria*.

p. 214 "These are small, effete theropods, some of which, such as *Nqwebasaurus*, had nubby teeth, and *Sinornithomimus*, which completely lacked teeth." (1) de Klerk, W.J., Forster, C.A., Sampson, S.D., Chinsamy, A., and Ross, C.F. 2000. A new coelurosaurian dinosaur from the Early Cretaceous of South Africa. *Journal of Vertebrate Paleontology*, 2: 324-332. (2) Kobayashi, Y., and Lü, J.–C. 2003. A new ornithomimid dinosaur with gregarious habits from the Late Cretaceous of China. *Acta Palaeontologica Polonica*, 48: 235-259.

p. 215 "Well, *Poekilopleuron* was not only a theropod, but also a big one. . . ." Allain, R., and Chure, D.J. 2002. *Poekilopleuron bucklandii*, the theropod dinosaur from the Middle Jurassic (Bathonian) of Normandy. *Palaeontology*, 45: 1107-1121.

p. 215 "Unfortunately, we can't study the original bones of *Poekilopleuron*, as these were lost in bombing raids during World War II." Dieter-Sues, H. 2012. European dinosaur hunters of the nineteenth and twentieth centuries. *In* Brett-Surman, M.K., Holtz, T.R., Jr., and Farlow, J.O. (editors), *The Complete Dinosaur* (2nd Edition). Indiana University Press, Bloomington, Indiana: 45-59.

p. 215 "Of the small theropods found with undoubted gastroliths in their abdominal cavities, some of these—such as *Caudipteryx*, *Sinocalliopteryx*, and *Sinosauropteryx*—also have feathers." Wings (2007).

p. 216 "The gastrolith connection . . . was bolstered by the discovery of gastroliths in the Early Cretaceous bird *Yanornis martini* of China." Zhou, Z., Clarke, J.A., Zhang, F., and Wings, O. 2004. Gastroliths in *Yanornis*: an indication of the earliest radical diet–switching and gizzard plasticity in the lineage leading to living birds? *Naturwissenschaften*, 91: 571-574.

p. 216 "In 1890, nearly a hundred years before the theropod–bird connection was firmly established, O. C. Marsh named one dinosaur *Ornithomimus velox*. . . ." Marsh, O.C. 1890. Description of new dinosaurian reptiles. *The American Journal of Science*, 39: 81-86.

p. 216 "Ornithomimids with gastroliths include the Early Cretaceous *Sinornithomimus*, the Late Cretaceous *Shenzhousaurus*. . . ." (1) Kobayashi, Y., Lü, J.–C., Dong, Z.–M., Barsbold, R., Azuma, Y., and Tomida, Y. 1999. Herbivorous diet in an ornithomimid dinosaur. *Nature*, 402: 480-481. (2) Ji, Q., Norrell, M., Makovicky, P. J., Gao, K., Ji, S., and Yuan, C. 2003. An early ostrich dinosaur and implications for ornithomimosaur phylogeny. *American Museum Novitates*, 3420: 1-19. (3) Barrett, P.M. 2005. The diet of ostrich dinosaurs (Theropoda: Ornithomimosauria). *Palaeontology*, 48: 347-358.

p. 216 "The most recent, and one of the most exciting examples of gastroliths in a ornithomimid, was . . . in a skeleton of the Late Cretaceous *Deinocheirus mirificus* of Mongolia." This dinosaur, known previously only through only a pair of disembodied and gigantic arms, was revealed by Yuong-Nam Lee and colleagues as a massive, sail-backed herbivore, and thus one of the most unusual of all theropods. Lee presented on these finds at the Society of Vertebrate Paleontology meeting in Los Angeles, California in late October, 2013: Lee, Y.-N., Barsbold, R., Currie, P., Kobayashi, Y., and Lee, H.-J. 2013. New specimens of *Deinocheirus mirificus* from the Late Cretaceous of Mongolia. *Journal of Vertebrate Paleontology*, Program and Abstracts, 2013: 161.

p. 217 "For instance, in modern ostriches and other birds that employ gastroliths to help with their food, these rocks make up about 1% of their total body mass." (1) Wings (2007). (2) Wings and Sander (2007).

p. 217 "However, for sauropods, the proportion between gastroliths and estimated body mass was about 10% that of birds." Wings and Sander (2007).

p. 217 "This huge disparity between extant and extinct gastrolith-using animals led paleontologists to conclude that these stones surely were not used for the same purposes. . . ." Wings and Sander (2007).

p. 218 "For example, one specimen of the Late Jurassic sauropod *Diplodocus hallorum* (formerly named "*Seismosaurus hallorum*") of northern New Mexico had more than 200 gastroliths. . . ." Gillette, D. 1991. *Seismosaurus halli*, gen. et sp. nov., a new sauropod dinosaur from the Morrison Formation (Upper Jurassic/Lower Cretaceous) of New Mexico, USA. *Journal of Vertebrate Paleontology*, 11: 417-433.

p. 218 "An assemblage of 115 gastroliths . . . was also packed into a small area within a skeleton of the Early Cretaceous sauropod *Cedarosaurus*. . . ." Sanders, F., Manley, K., and Carpenter, K. 2001.

Gastroliths from the Lower Cretaceous sauropod *Cedarosaurus weiskopfae*. *In* Tanke, D.H., and Carpenter, K. (editors), *Mesozoic Vertebrate Life*. Indiana University Press, Bloomington, Indiana: 166-180.

p. 218 "Other sauropods with gastroliths directly associated with their bones include: *Apatosaurus, Barosaurus*, and *Camarasaurus*. . . ." Wings and Sander (2007).

p. 218 "In one instance, 14 gastroliths were directly associated with the remains of a juvenile *Camarasaurus* scavenged by *Allosaurus*, the latter indicated by tracks, toothmarks, and dislodged teeth." Jennings and Hasiotis (2006).

p. 218 "Several specimens of a Late Triassic prosauropod from South Africa, *Massospondylus* . . . had gastroliths too." Raath, M. 1974. Fossil vertebrate studies in Rhodesia: further evidence of gastroliths in prosauropod dinosaurs. *Arnoldia*, 7: 1-7.

p. 218 "Moreover, the skeleton of another prosauropod, the Early Jurassic *Ammosaurus* of North America, had gastroliths." Whittle and Everhart (2000).

p. 218 "In short, although gastroliths are relatively rare, having been found in less than 4% of all sauropod skeletons. . . ." Wings and Sander (2007).

p. 219 "In contrast, only a few ornithischians have gastroliths. These include the . . . ankylosaur *Panoplosaurus* . . . ceratopsian *Psittacosaurus* . . . *Gasparinisaura* of Argentina. . . ." (1) Carpenter, K. 1997. Ankylosaurs. *In* Farlow, J.O., and Brett-Surman, M.K. (editors). *The Complete Dinosaur* (1st edition). Indiana University Press, Bloomington, Indiana: 307-316. (2) Hailu, Y., and Dodson, P. 2004. Basal ceratopsia. *In* Weishampel, D.B., Dodson, P., and Osmólska, H. (editors). 2004. *The Dinosauria* (2nd Edition). University of California Press, Berkeley, California: 478-493.(3) Cerda, I.A. 2008. Gastroliths in an ornithopod dinosaur. *Acta Palaeontologica Polonica*, 53: 351-355.

p. 220 "But in the past few years, behavioral scientists are gaining a greater appreciation for how some species of birds, such as crows and ravens. . . ." This is covered in much detail in Chapter 10. Or maybe you already knew that, seeing that these endnotes are, appropriately enough, at the end of the book.

p. 221 "Along those lines, in an article published by Robert Weems, Michelle Culp, and Oliver Wings in 2007. . . ." Weems, R.E., Culp, M.J., and Wings, O. 2007. Evidence for prosauropod dinosaur gastroliths in the Bull Run Formation (Late Triassic, Norian) of Virginia. *Ichnos*, 14: 271-295.

p. 221 "Despite more than 200 years of paleontological study, Mesozoic rocks of the eastern U.S. are embarrassingly shy about revealing dinosaur bones. . . ." Weishampel, D.B., and Young, L. 1998. *Dinosaurs of the East Coast*. Johns Hopkins University Press, Baltimore, Maryland: 296 p.

p. 222 "For example, only a few theropod tracks have been found thus far in Late Triassic rocks of northeastern Virginia. . . ." Weems, R.E. 1987. A Late Triassic footprint fauna from the Culpeper Basin Northern Virginia (U.S.A.). *Transactions of the American Philosophical Society,* 77: 1-79.

p. 222 "Weems and his colleagues built their case about the Bull Run gastroliths by noting a number of traits that matched those of known dinosaur gastroliths. . . ." Weems *et al.* (2007).

p. 223 "However, sauropods were rare in the Late Triassic and did not become more widespread until the Middle Jurassic. . . ." Wilson, J.A., and Curry-Rogers, K. 2012. Sauropods. *In* Brett-Surman, M.K., Holtz, T.R., Jr., and Farlow, J.O. (editors), *The Complete Dinosaur* (2nd Edition). Indiana University Press, Bloomington, Indiana: 445-481.

p. 223 "Unfortunately, the body fossil record for Late Triassic prosauropods in North America is quite scanty. . . ." Yates, A. 2012. Basal Sauropodomorpha: the "Prosauropoda." *In* Brett-Surman, M.K., Holtz, T.R., Jr., and Farlow, J.O. (editors), *The Complete Dinosaur* (2nd Edition). Indiana University Press, Bloomington, Indiana: 425-443.

p. 224 "The best match came from the Manassas Sandstone, which has conglomerates; these are sandstones that also have chunky bits, such as pebbles, gravel, and cobbles." Weems *et al.* (2007).

p. 224 "Assuming that the gastroliths are gravestones, marking the final resting places of prosauropods, then these rocks moved a minimum of about 20 km (12 mi) under dinosaur control, and probably much further." Weems *et al.* (2007).

Chapter 8: The Remains of the Day: Dinosaur Vomit, Stomach Contents, Feces, and Other Gut Feelings

p. 226 "During this walk, your dog is using its nose like you use your eyes, investigating an olfactory landscape holding thousands of invisible clues. . . ." Bradshaw, J. 2012. *Dog Sense: How the New Science of Dog Behavior Can Make You a Better Friend to Your Pet.* Basic Books, New York: 352 p.

p. 228 "For example, hippopotamuses do this by rapidly flicking their tails back and forth as they excrete, flinging poo with wild abandon." Kranz, K.R. 1982. A note on the structure of tail hairs from a pygmy hippopotamus (*Choeropsis liberiensis*). *Zoo Biology,* 1: 237-241.

p. 228 "Their digestion is also more complete than that of most mammals . . . birds excrete uric acid out of a single orifice, the cloaca." Elbroch and Marks (2001).

p. 228 "Nonetheless, raptors and other carnivorous birds are famous for their white-liquid sprays. . . ." Elbroch and Marks (2001).

p. 228 "In a paper published in 2003, two researchers, Victor Meyer-Rochow and Jozsef Gal, became intrigued with the radial patterns of feces created by two penguin species. . . ." Meyer-Rochow, V.B., and Gal, J. 2003. Pressures produced when penguins pooh—calculations on avian defaecation. *Polar Biology*, 27: 56-58.

p. 229 "In our species, vomiting mainly evicts toxic substances . . . but it can also be a reaction to severe physical or psychological trauma. . . ." Pelchat, M.L., and Rozin, P. 1982. The special role of nausea in the acquisition of food dislikes by humans. *Appetite*, 3: 341-351.

p. 229 "Probably the most well known of such deposits are cough pellets, also known as gastric pellets." Myhrvold, N.P. 2012. A call to search for fossilized gastric pellets. *Historical Biology*, 24: 505-517.

p. 230 "Tracks associated with these cough pellets normally show the coughers with their feet together (side by side) and a pellet in front of the tracks." Martin (2013).

p. 231 "Crocodilians have extremely thorough digestion. . . ." Fisher, D.C., 1981. Crocodilian scatology, microvertebrate concentrations, and enamelless teeth. *Paleobiology*, 7: 262-275.

p. 231 "So even though male ratites and some other birds, such as ducks and geese, have penises . . . these lack an enclosed urethra." Brennan, P.L.R., and Prum, R.O. 2012. The erection mechanism of the ratite penis. *Journal of Zoology*, 286: 140-144.

p. 232 "For humans, the standard descriptive method for feces is the Bristol scale, used in clinical situations." Riegler, G., and Esposito, I. 2001. Bristol scale stool form: a still valid help in medical practice and clinical research. *Techniques in Coloproctology*, 5: 163-164.

p. 232 "In between these two extremes, most of the weight of a stool (about 75%) comes from water, and of the ~25% solid stuff, almost a third of this can be from dead bacteria. . . ." Stephen, A.M., and Cummings, J.H. 1979. The microbial contribution to human faecal mass. *Journal of Medical Microbiology*, 13: 45-56.

p. 232 "This tyranny started with seed plants, which developed in the Devonian Period (about 400 million years ago), then escalated into outright slavery. . . ." Gensel, P.G., and Edwards, D. 2003. *Plants Invade the Land: Evolutionary and Environmental Persepctives*. Columbia University Press, New York: 304 p.

p. 233 "Most broadly, any former food item associated with the digestive tract of a dinosaur (or any other fossil animal, for that matter) is called a *bromalite*." These definitions have been made by so many researchers and used in so many different ways, I hereby throw up (my hands, that is) and give up attempts to untangle the literature on who defined what and when. Instead, read this: Chin, K. 2012. What did dinosaurs eat: coprolites and other direct evidence of dinosaur diets. *In* Brett-Surman, M.K., Holtz, T.R., Jr., and Farlow, J.O.

(editors), *The Complete Dinosaur* (2nd Edition). Indiana University Press, Bloomington, Indiana: 589-601.

p. 234 "*Coelophysis* was a slender, greyhound-sized theropod and is one of the most abundantly represented dinosaurs in the fossil record. . . ." Holtz (2012).

p. 234 "The 'cannibalism' hypothesis was originally interpreted and promoted by noted paleontologist Ned Colbert, who had excavated and studied *Coelophysis* since the late 1940s." Nesbitt, S.J., Turner, A.H., Erickson, G.M., and Norell, M.A. 2006. Prey choice and cannibalistic behaviour in the theropod *Coelophysis*. *Biology Letters*, 2: 611-614.

p. 234 "In 2002, paleontologist Robert Gay took a closer look at these *Coelophysis* specimens and found that the 'juvenile *Coelophysis*' bones were misidentified." Gay, Robert J. 2002. The myth of cannibalism in *Coelophysis bauri*. *Journal of Vertebrate Paleontology* 22(3): 57A.

p. 234 "Another study done by Sterling Nesbitt and others in 2006 confirmed that it was *Hesperosuchus* in the belly of the beast, not another *Coelophysis*." Nesbitt *et al.* (2006).

p. 234 "In yet another study in 2010, Gay pointed out that supposed 'stomach contents' in one *Coelophysis* made up a larger volume than its original stomach, so this was not an enterolite, either." Gay, R. 2010. Evidence related to the cannibalism hypothesis in *Coelophysis bauri* from Ghost Ranch, New Mexico. In *Notes on Early Mesozoic Theropods*. Lulu Press: 9-24.

p. 234 "Interestingly, this deposit, because of its position and jumble of bones, was interpreted as a regurgitalite." Rinehart, L.F., Lucas, S.G., Heckert, A.B., Spielmann, J.A., and Celesky, M.D. 2009. The paleobiology of *Coelophysis bauri* (Cope) from the Upper Triassic (Apachean) Whitaker quarry, New Mexico, with detailed analysis of a single quarry block. *New Mexico Museum of Natural History & Science*, Bulletin 45: 260 p.

p. 235 "For instance, in 2011, Jingmai O'Connor and two other paleontologists reported a specimen of *Microraptor* . . . with bird bones between its ribs." O'Connor, J., Zhou, Z., and Xu, X. 2011. Additional specimen of *Microraptor* provides unique evidence of dinosaurs preying on birds. *Proceedings of the National Academy of Sciences*, 108: 19662-19665.

p. 235 "A couple of Early Cretaceous theropods from China, *Sinosauropteryx* and *Sinocalliopteryx*, also have intriguing enterolites." (1) Dong, Z., and Chen, P. 2000. A tiny fossil lizard in the stomach content of the feathered dinosaur *Sinosauropteryx* from northeastern China. *Vert PalAsiatica*, 38 (supplement): 10. (2) Xing, L., Bell, P.R., Persons, W.S., Ji, S., Miyashita, T., Burns, M.E., Ji, Q., and Currie, P.J. 2012. Abdominal contents from two large Early Cretaceous compsognathids (Dinosauria: Theropoda) demonstrate feeding on Confuciusornithids and Dromaeosaurids. *PLoS One*, 7: e44012. doi:10.1371/journal.pone.0044012.

p. 235 "Another *Sinosauropteryx* contained jaws from three different small mammals. . . ." Hurum, J.H., Luo, Z.-X., and Kielan-Jaworowska, Z. 2006. Were mammals originally venomous? *Acta Palaeontologica Polonica*, 51: 1-11.

p. 236 "Fortunately for Mesozoic mammals, at least one struck a blow against its dinosaurian oppressors. . . ." Hu, Y.M., Meng, J., and Wang, Y.Q. 2005. Large Mesozoic mammals fed on young dinosaurs. *Nature*, 433: 149-152.

p. 236 "Based on other stomach contents, another feathered theropod, the Early Cretaceous *Sinocalliopteryx* of China, also ate birds and non-avian dinosaurs." Xing *et al.* (2012).

p. 236 "Ably representing North American theropods and their meals, David Varricchio—introduced in a previous chapter—discovered a Late Cretaceous tyrannosaurid (*Daspletosaurus*) in Montana with a few anomalous bones in its abdominal region." Varricchio, D.J. 2001. Gut contents from a Cretaceous tyrannosaurid: implications for theropod dinosaur digestive tracts. *Journal of Paleontology*, 75: 401-406.

p. 237 "The first specimen of this dinosaur was discovered in a clay pit mine in 1983; paleontologists Alan Charig and Angela Milner named it several years later, in 1986." Charig, A.J., and Milner, A.C. 1986. *Baryonyx*, a remarkable new theropod dinosaur. *Nature*, 324: 359-361.

p. 237 "In a more detailed study of this specimen published in 1997, Charig and Milner concluded that all of these traits made *Baryonyx* well suited for . . . eating fish." Charig, A.J., and Milner, A.C. 1997. *Baryonyx walkeri*, a fish-eating dinosaur from the Wealden of Surrey. *Bulletin of the Natural History Museum of London, Geology Series* 53: 11-70.

p. 238 "(As learned previously, paleontologists were further encouraged to adopt this formerly strange idea of fish-eating theropods when they found thousands of theropod swim tracks in Early Jurassic rocks of Utah.)" Milner *et al.* (2006).

p. 238 "This same specimen of *Baryonyx* included lots of other ichnological extras, such as the remains of a juvenile ornithopod (identified as *Iguanodon*), gastroliths, and its own broken bones." Charig and Milner (1997).

p. 238 "Oddly enough, they mention 'gastroliths' in their 1986 paper, but only 'an apparent gastrolith' in the 1997 one." Charig and Milner (1986, 1997).

p. 238 "Only a few such trace fossils are known in dinosaurs, but one of the best is also in the most complete dinosaur from an entire continent, *Minmi paravertebrata*." Molnar, R.E., and Clifford, H.T. 2001. An ankylosaurian cololite from the Lower Cretaceous of Queensland,

Australia. *In* Carpenter, K. (editor), *The Armored Dinosaurs*. Indiana University Press, Bloomington, Indiana: 399-412.

p. 239 "It turns out this specimen of *Minmi* was deposited in a Cretaceous sea, and no river was anywhere near its body when it was laid to rest in a shallow marine grave." Molnar and Clifford (2001).

p. 240 "One regurgitalite inferred from Late Triassic rocks contains pterosaur remains, but is thought to have come from a large fish rather than a land-dwelling animal." Dalla Vecchia, F.M., Muscio, G., and Wild, R. 1989. Pterosaur remains in a gastric pellet from Upper Triassic (Norian) of Rio Seazza Valley (Udine, Italy). *Gortania-Atti Museo Friul. St. Nat.*, 10: 121-132.

p. 240 "I also mentioned one example interpreted in 2009, which was apparently fossilized beside the mouth of a *Coelophysis*." Rinehart *et al.* (2009).

p. 241 "Only one instance of fossilized dinosaur vomit, containing a mix of dinosaur and turtle bones in Early Cretaceous rocks of Mongolia, is inferred to have come from an actual dinosaur." I mentioned this interpreted regurgitalite in my textbook (Martin, 2006), but unfortunately have not been able to hunt down the original reference, which I recall was in a Russian journal.

p. 241 "Another possible dinosaurian up-chuck is embodied in a mass of four birds from the Early Cretaceous of Spain." Sanz, J.L., Chiappe, L.M., Fernandez-Jalvo, Y., Ortega, F., Sanches-Chillon, B., Poyato-Ariza, F.J., and Perez-Moreno, B.P. 2001. An Early Cretaceous pellet. *Nature*, 409: 998–999.

p. 242 "These deposits, found in Late Jurassic rocks of the U.K., consist of concentrated collections of belemnite shells." Walker, S.E., and Brett, C.E. 2002. Post-Paleozoic patterns in marine predation: was there a Mesozoic and Cenozoic marine predatory revolution? *In* Kowalewski, M., and Kelley, P.H. (editors), *The Fossil Record of Predation*. Paleontological Society Special Publication, 8: 119-193.

p. 243 "This urolite was mentioned in a poster presentation at a paleontology meeting in 2002 and understandably garnered much media attention. . . ." McCarville, K., and Bishop, G.A. 2002. To pee or not to pee: evidence for liquid urination in sauropod dinosaurs. *Journal of Vertebrate Paleontology*, 22 (Supplement to 3): 85.

p. 244 "Known as the Purgatoire site (after the Purgatoire River, which runs through the area), it is one of the most spectacular dinosaur tracksites in the western U.S. . ." Lockley (1991).

p. 245 "Given the controversy over the Morrison urolite, I was much relieved to later find out that paleontologists Marcelo Fernandes, Luciana Fernandes, and Paulo Souto had also interpreted dinosaur urolites. . . ." Fernandes, M.A., Fernandes, L.B., and Souto, P.B. 2004. Occurrence of urolites related to dinosaurs in the Lower Cretaceous of the Botucatu

Formation, Paraná Basin, São Paulo State, Brazil. *Revista Brasileira de Paleontologia*, 7: 263-268.

p. 246 "Instead, they took two liters of water and poured it from 80 cm (2.6 ft) above and onto a loose sand surface with a slope having the same angle (about 30°) as the fossil ones." Fernandes *et al.* (2004).

p. 248 "Cincinnati's a great city with a lot going for it, but it has the wrong age rocks (Ordovician Period, 450 million years old) and wrong rocks. . . ." Davis, R.A., and Meyer, D.L. 2009. *A Sea Without Fish: Life in the Ordovician Sea of the Cincinnati Region*. Indiana University Press, Bloomington, Indiana: 368 p.

p. 249 "This means carnivorous dinosaur scat had a better chance of preserving than that coming from insectivores or herbivores. . . ." Chin, K. 2002. Analyses of coprolites produced by carnivorous vertebrates. *Paleontological Society Papers*, 8: 43-50.

p. 249 "Second, anaerobic bacteria in the feces could have assisted in preserving it, in which their metabolic processes caused chemical reactions that made more minerals precipitate, and do so rapidly (geologically speaking)." Hollocher, T.C., Chin, K., Hollocher, K.T., and Kruge, M.A. 2001. Bacterial residues in coprolite of herbivorous dinosaurs: role of bacteria in mineralization of feces. *Palaios*, 16: 547-565.

p. 249 "And third, rapid burial, such as from a nearby river flood, would have prevented fresh droppings from getting eaten, poked, prodded, sniffed, trampled, washed away, or otherwise damaged." Thulborn, R.A., 1991. Morphology, preservation and palaeobiological significance of dinosaur coprolites. *Palaeogeography, Palaeoclimatology, Palaeoecology*, 83: 341-366.

p. 250 "Of dinosaur coprolites identified thus far, the best understood ones are attributed to the Late Cretaceous hadrosaur *Maiasaura* of Montana." Chin, K. 2007. The paleobiological implications of herbivorous dinosaur coprolites from the Upper Cretaceous Two Medicine Formation of Montana: why eat wood? *Palaios*, 22: 554-566.

p. 250 "Other large coprolites are credited to Late Cretaceous tyrannosaurids, such as *Tyrannosaurus rex*; one of these is more than twice the length of a 12-inch sub sandwich." (1) Chin, K., Tokaryk, T.T., Erickson, G.M., and Calk, L.C. 1998. A king-sized theropod coprolite. *Nature*, 393: 680-682. (2) Chin, K., Eberth, D.A., Schweitzer, M.H., Rando, T.A., Sloboda, W.J., and Horner, J.R. 2003. Remarkable preservation of undigested muscle tissue within a Late Cretaceous *Tyrannosaurus* coprolite from Alberta, Canada. *Palaios*, 18: 286-294.

p. 251 "For example, Early Cretaceous coprolites from Belgium have bone fragments in them." Poinar, G., Jr., and Boucot, A.J. 2006. Evidence of intestinal parasites of dinosaurs. *Parasitology*, 133: 245-250.

p. 251 "This seemingly un-dinosaur-like behavior, which is extremely common in modern birds, was proposed for the bizarre theropod *Mononykus*. . . ." Senter (2005).

p. 251 "Some beetles roll balls of this nutritious stuff as take-out, which they push into burrows, lay eggs on them, and seal off the burrow." Bertone, M., Green, J., Washburn, S., Poore, M., Sorenson, C., and Watson, D.W. 2005. Seasonal activity and species composition of dung beetles (Coleoptera: Scarabaeidae and Geotrupidae) inhabiting cattle pastures in North Carolina (USA). *Annals of the Entomological Society of America*, 98: 309-321.

p. 252 "As is often the case in ichnology, body parts had little to do with this discovery, which Chin and Gill documented in 1996." Chin, K., and Gill, B.D. 1996. Dinosaurs, dung beetles, and conifers: participants in a Cretaceous food web. *Palaios*, 11: 280-285.

p. 252 "Some of the burrows were open, but in others the insects had actively filled them, having packed a mixture of sediment and dung behind them and leaving distinctly visual 'plugs.'" Chin and Gill (1996).

p. 253 "Dung flies are relatively small and normally just lay their eggs on feces. . . ." Wiegmann, B., and Yeates, D.K. 2013. *The Evolutionary Biology of Flies*. Columbia University Press, New York: 512 p.

p. 253 "Dung beetles today employ three different strategies in handling feces: tunneling, dwelling, or rolling." Martin (2013).

p. 254 "In a paper published in 2001, geochemist Thomas Hollocher, Karen Chin, and two other colleagues detected both abundant body fossils and chemical signatures of anaerobic bacteria in the coprolites." Hollocher, T.C., Chin, K., Hollocher, K.T., and Kruge, M.A. 2001. Bacterial residues in coprolite of herbivorous dinosaurs: role of bacteria in mineralization of feces. *Palaios*, 16: 547-565.

p. 254 "Once these researchers examined thin sections of the coprolites under microscopes, they realized that the calcite in the coprolites was probably precipitated in two stages. . . ." Hollocher *et al.* (2001).

p. 254 "The fossilized wood lacked lignin, a connective tissue that holds wood fibers together." Chin (2007).

p. 257 "We did not know it at the time, but Karen Chin had already finished studying these snails and surmised how they got into feces in the first place." Chin, K., Hartman, J.H., and Roth, B. 2009. Opportunistic exploitation of dinosaur dung: fossil snails in coprolites from the Upper Cretaceous Two Medicine Formation of Montana. *Lethaia*, 42: 185-198.

p. 258 "Speaking of ecosystems, individual dinosaurs harbored their own diverse microflora and microfauna . . . called a *microbiome*." Because people tend to be anthropocentric (go figure), most literature about microbiomes focuses on them in humans. But here's one in which the authors show how these microorganisms can influence non-human animal behavior: Ezenwa, V.O., Gerardo, N.M., Inouye, D.W., Medina, M., and Xavier, J.B. 2012. Animal behavior and the microbiome. *Science*, 338: 198-199.

p. 258 "Not all of this flora and fauna are necessarily harmful to the host, though; for example, some gut bacteria help to produce vitamins K and B12. . . ." Hill, M.J. 1997. Intestinal flora and endogenous vitamin synthesis. *European Journal of Cancer Prevention*, 6: S43-45.

p. 258 "Parasites fall into two broad categories: *endoparasites* (living inside), such as tapeworms, and *ectoparasites* (living outside), such as ticks." Gunn, A., and Pitt, S.J. 2012. *Parasitology: An Integrated Approach*. John Wiley & Sons, New York: 442 p.

p. 259 "In 2006, two paleontologists, George Poinar and Art Boucot, documented three types of endoparasites in a dinosaur coprolite. . . ." Poinar and Boucot (2006).

p. 259 "The coprolites, first identified in 1903, ranged from 2 to 5 cm (1–2 in) wide and 11 to 13 cm (4.3–5 in) long, or human-sized." Poinar and Boucot (2006).

p. 260 "The first known coprolite attributed to a tyrannosaurid came out of fluvial (river) deposits in the Frenchman Formation of Saskatchewan." Chin *et al.* (1998).

p. 261 "Furthermore, toothmarks in *Triceratops* and other dinosaur bones show that *T. rex* punctured bone." Erickson *et al.* (1996).

p. 261 "Despite the impressive size of this *T. rex* coprolite, it was surpassed by one from the Late Cretaceous Dinosaur Park Formation of Alberta, reported in 2003." Chin *et al.* (2003).

p. 263 "Reported in 2005 by paleontologist Kurt Hollocher and three colleagues, these coprolites, like the tyrannosaurid ones, had bone fragments and were cemented with apatite." Hollocher, K.T., Alcober, O.A., Colombi, C.E., and Hollocher, T.C. 2005. Carnivore coprolites from the Upper Triassic Ischigualasto Formation, Argentina: chemistry, mineralogy, and evidence for rapid initial mineralization. *Palaios*, 20: 51-63.

p. 264 "The conventional wisdom about grasses is that they first evolved from non-grassy flowering plants during the Cenozoic Era. . . ." Kellogg, E.A. 2001. The evolutionary history of the grasses. *Plant Physiology*, 125: 1198-1205.

p. 264 "The first detailed report on these coprolites was by Prosenjit Ghosh and five of his colleagues in 2003." Ghosh, P., Bhattacharya, S.K., Sahni, A., Kar, R.K., Mohabey, D.M., and Ambwani, K. 2003. Dinosaur coprolites from the Late Cretaceous (Maastrichtian) Lameta Formation of India: isotopic and other markers suggesting a C_3 plant diet. *Cretaceous Research*, 24: 743-750.

p. 265 "Plants, through different means of photosynthesis, take in carbon and nitrogen isotopes in distinct ways, which is then reflected by their stable-isotope ratios." Todd, E., Dawson, T.E., Mambelli, S., Plamboeck, A.H., Templer, P.H., and Tu, K.P. 2002. Stable isotopes in plant ecology. *Annual Review of Ecology and Systematics*, 33: 507-559.

p. 265 "Additionally, C_3 plants make up nearly 90% of all modern vegetation, and include grasses." Heldt, H.W. 2005. *Plant Biochemistry*. Academic Press, New York: 630 p.

p. 266 "So in 2005, Vandana Prasad and four of his colleagues hit the jackpot, finding grass phytoliths in the same Late Cretaceous coprolites from India studied by Ghosh and others." Prasad, V., Strömberg, C.A.E., Alimohammadian, H., and Sahni, A. 2005. Dinosaur coprolites and the early evolution of grasses and grazers. *Science*, 310: 1177-1180.

CHAPTER 9: THE GREAT CRETACEOUS WALK

p. 267 *"Because of the sparse and uneven record of dinosaurs in Australia, their fossil footprints are more valuable here than anywhere else on Earth."* Rich, T.H., and Vickers-Rich, P., 2003. *A Century of Australian Dinosaurs*. Queen Victoria Museum and Art Gallery, Launceston, Australia: 124 p.

p. 268 "It was a time when Australia was close to the South Pole, and dinosaurs presumably walked across broad floodplains of rivers that coursed through its circumpolar valleys." Rich and Vickers-Rich (2000).

p. 270 "In North America, the U.S., Canada, and Mexico have tens of thousands of dinosaur tracks and plenty of other trace fossils. . . ." (1) Lockley (1991). (2) Lockley and Hunt (1995). (3) Lockley and Meyer (2000).

p. 270 "Dinosaurs certainly lived there, as evidenced by theropod and ornithopod bones and teeth recovered from Early Cretaceous strata (120–105 *mya*) in Victoria, and theropod, ornithopod, and sauropod bones in central Queensland." (1) Rich *et al.* (1988); (2) Rich, T.H., Vickers-Rich, P., and Gangloff, R.A. 2002. Polar dinosaurs. *Science*, 295: 979-980.

p. 271 "In the 1980s, Greg and his father, David Denney, assisted Tom, Pat Vickers-Rich, and a crew of volunteers. . . ." Rich and Vickers-Rich (2000).

p. 274 "For instance, some aquatic insect larvae dig burrows in sediments under very shallow water or on the surfaces of emergent sand bars. . . ." (1) Ratcliffe, B.C., and Fagerstrom, J.A. 1980. Invertebrate lebensspuren of Holocene floodplains: their morphology, origin and paleontological significance. *Journal of Paleontology*, 54: 614-630. (2)

Charbonneau, P., and Hare, L. 1998. Burrowing behavior and biogenic structures of mud-dwelling insects. *Journal of the North American Benthological Society*, 17: 239-249.

p. 274 "Insects and other invertebrates in polar environments cannot burrow into frozen sediment." Danks, H.V. 2007. Aquatic insect adaptations to winter cold and ice. In *Aquatic Insects: Challenges to Populations*, Proceedings of the Royal Entomological Society's 24th Symposium: 1-19.

p. 274 "Along those lines, I had published a paper in 2009 about physical sedimentary structures . . . and traces . . . on the North Slope of Alaska." Martin, A.J. 2009. Neoichnology of an Arctic fluvial point bar, Colville River, Alaska (USA). *Geological Quarterly*, 53: 383-396.

p. 275 "In 1980, a little more than thirty years before Tom, Greg, and I stepped foot on Milanesia Beach together, Tom and Pat Vickers-Rich found this track." Rich and Vickers-Rich (2003).

p. 275 "Still, I was lucky enough to have discovered possible dinosaur burrows and invertebrate trace fossils there, the latter nearly identical to the ones we were seeing that morning at Milanesia Beach." Martin (2009a).

p. 276 "This track later became the template for thousands of reproductions used for science education in Victoria, and photographs of it appeared in many popular articles and books." Rich and Vickers-Rich (2003).

p. 277 "The two tracks from 105-million-year-old rocks west of Melbourne—Knowledge Creek and Skenes Creek—were attributable to small ornithopod dinosaurs, like *Leaellynasaura*." Rich and Vickers-Rich (1999).

p. 277 "Large-sized theropods, about the size of a smaller version of *Allosaurus*, likely made the two 115-million-year-old tracks in rocks east of Melbourne." (1) Molnar, R.E., Flannery, T., and Rich, T.H. 1981. An allosaurid theropod dinosaur from the Early Cretaceous of Victoria, Australia. *Alcheringa*, 5: 141-146. (2) Molnar, R.E., Flannery, T., and Rich, T.H. 1985. Aussie *Allosaurus* after all. *Journal of Paleontology*, 59, 1511-1513.

p. 278 ". . . I was traveling in uncharted territory for dinosaur trace fossils, an ichnological analog to 'The Ghastly Blank,' an endearing term applied to the desert interior of Australia." Actually, it was not such an endearing phrase in the 19th century, when Burke and Wills applied it to describe the region they explored and died in: Joyce and McCann (2010).

p. 283 "Nearly a year later, in early June 2011, Tom . . . succeeded in carrying out his quixotic goal." In a video produced by Museum Victoria, titled *First Victorian Dinosaur Trackway*, Rich describes the recovery of these two sandstone blocks: http://www.youtube.com/watch?v=TpMLuPWHTDA

p. 286 "They had a merry adventure doing this, climbing over hill and dale, while I, doing my best Marlin Perkins *Wild Kingdom* impression, stayed safely on the beach. . . ." One of the recurring tropes of the TV show *Mutual of Omaha's Wild Kingdom* (1963–1985), which I watched religiously as a child, was that the show's host, Marlin Perkins (a zoologist), would stay in his vehicle (usually a jeep) while his assistant, Jim Fowler, would imperil himself while facing various deadly animals.

p. 287 "Sure enough, exactly one year later, on June 14, 2011, my coauthors and I . . . received the good news that our scientific article had been accepted. . . ." Martin, A.J., Rich, T.H., Hall, M., Vickers-Rich, P., and Vasquez-Prokopec, G. 2012. A polar dinosaur-track assemblage from the Eumeralla Formation (Albian), Victoria, Australia. *Alcheringa: An Australasian Journal of Palaeontology*, 36: 171-188.

Chapter 10: Tracking the Dinosaurs Among Us

p. 290 "They can be 2 m (6.6 ft) tall and weigh as much as 80 kg (175 lbs), making them among the heaviest birds alive." Corlett, R.T., and Primack, R.B. 2011. *Tropical Rain Forests: An Ecological and Biogeographical Comparison*. John Wiley & Sons, New York: 336 p.

p. 290 "Given their weights, and that they are capable of running 45 km/hr (28 mph) and leaping more than a meter into the air. . . ." Although cassowaries are formidable animals, nearly all incidences of their attacking humans were indirectly caused by people, either through habitual feeding or trying to harm them: Kofron, C.P. 1999. Attacks to humans and domestic animals by the southern cassowary (*Casuarius casuarius johnsonii*) in Queensland, Australia. *Journal of Zoology*, 249: 375-381.

p. 291 "Incidentally, if you are chased by a cassowary and try to escape it by jumping in a nearby water body, be aware that they are also excellent swimmers. . . ." Although many popular accounts attest to cassowary swimming abilities, I had a tough time finding it documented in peer-reviewed literature. Nevertheless, it is mentioned in: Taylor, S. 2012. *John Gould's Extinct and Endangered Birds of Australia*. National Library Australia, Sydney, Australia: 247 p.

p. 291 "In fact, they are so comfortable in water that, much like human hot-tub or pool parties, their mating rituals sometimes happen while immersed." Again, I only found this factoid cited in secondary sources, and—sadly enough—the only photos of mating cassowaries I could find on the Internet showed them on land. Clearly we need more aquatically inclined nature photographers to document this, albeit very carefully.

p. 291 "Cassowaries are omnivorous, eating fungi, forest invertebrates (insects and snails), and small mammals, but they are primarily known as terrific fruit eaters." Bradford, M.G., Dennis, A.J., and Westcott, D. 2008. Diet and dietary preferences of the Southern cassowary (*Casuarius casuarius*) in North Queensland Australia. *Biotropica*, 40: 338-343.

p. 292 "Like other ratites, cassowaries are ground nesters, in which the males scratch out a meter-wide shallow depression in the forest floor that they rim with vegetation." Taylor (2012).

p. 292 "Once the eggs hatch, a male cassowary's parental responsibilities do not end. Instead, they spend at least seven months tending to the cassowary chicks. . . ." Moore, L.A. 2007. Population ecology of the southern cassowary *Casuarius casuarius johnsonii*, Mission Beach north Queensland. *Journal of Ornithology*, 148: 357-366.

p. 293 "Moreover, about fifteen families of flowering plants there are directly descended from plants that were alive when Australia was still attached. . . ." Metcalfe, D.A., and Ford, A.J. 2009. A re-evaluation of Queensland's Wet Tropics based on "primitive" plants. *Pacific Conservation Biology*, 15: 80-86.

p. 293 "For example, *Austrobaileya scandens*, a perennial flowering vine that composes part of the shady undergrowth in the rainforest. . . ." Feild, T.S., Franks, P.J., and Sage, T.L. 2003. Ecophysiological shade adaptation in the basal angiosperm, *Austrobaileya scandens* (Austrobaileyaceae). *International Journal of Plant Sciences*, 164: 313-324.

p. 294 "A cyclone that hit this area just a little more than a year before had ripped open parts of the canopy, illuminating the way." Cyclone Larry, a Category 4 tropical storm, made landfall on the Queensland coast on the morning of March 20, 2006, with its eye just south of Mission Beach.

p. 294 "Had I been eavesdropping on these avian conversations, they first would have alerted me to two people walking on the trail about a hundred meters behind us." I realize such a specific interpretation based on songbird sounds and behaviors might sound a little too "woo-woo" for some of my more skeptical readers. But in my experience, it works well, and is increasingly backed up by peer-reviewed scientific research, such as: Evans, C.S., Evans, L., and Marler, P. 1993. On the meaning of alarm calls: functional reference in an avian vocal system. *Animal Behaviour*, 46: 23-28. For a less technical introduction to this topic, look at: Young, J. 2013. *What the Robin Knows: How Birds Reveal the Secrets of the Natural World*. Mariner Books, New York: 272 p.

p. 294 "The low-frequency boom was subtle enough that at first I dismissed it." Cassowaries have the lowest frequency call known for any bird, almost below the range of human hearing, as documented by: Mack, A.L., and Jones, J. 2003. Low-frequency vocalizations by cassowaries (*Casuarius* spp.). *The Auk*, 120: 1062-1068.

p. 295 "Even in fiction, such as during the unforgettable scene of the first

Tyrannosaurus attack in the movie *Jurassic Park*, this theropod's petrifying roar. . . ." Hill, K. 2013. The animals hiding in a T. rex's roar. *Scientific American Blogs*, April 10, 2013: http://blogs.scientificamerican.com/guest-blog/2013/04/10/the-animals-hiding-in-a-t-rexs-roar/

p. 297 "Bird traces include their tracks, cough pellets, beak marks . . . as well as talon marks, nests, burrows, gastroliths, droppings, food caches, dust baths, and broken seashells. . . ." Elbroch and Marks (2001).

p. 298 "For example, a Cherokee story tells how the Appalachian Mountains were made from the beating of a vulture's wings against the earth. . . ." Krech, S., III, 2009. *Spirits of the Air: Birds and American Indians in the South*. University of Georgia Press, Athens, Georgia: 245 p.

p. 298 "The Ngarrindjeri people of Australia have a Dreamtime story about a rivalry between the emu (*Dromaius novaehollandiae*) and brush-turkey (*Alectura lathami*). . . ." Ziembicki, M. 2010. *Australian Bustard*. CSIRO Publishing, Collingwood, Australia: 102 p.

p. 298 "Ancient Egyptians revered a heron-like god, Bennu. . . ." Birds and other animals are frequently used symbols in world religions. For a good introduction to this topic, take a look at: Kemmerer, *Animals and World Religions*. Oxford University Press, Oxford, U.K.: 346 p.

p. 298 "Birdwatching is among the most popular of all outdoor activities in the U.S., enjoyed by about 60 million people." This might be an exaggeration, as surveys by the U.S. Fish and Wildlife Service place it at about 46 million, whereas the Audubon Society says 51 million. Regardless of the number, though, watching birds (and by default, dinosaurs) is very likely the most popular nature-related outdoor activity for Americans: USFWS Report 2006-4. 2009. *Birding in the United States: A Demographic and Economic Analysis*. Addendum to the 2006 National Survey of Fishing, Hunting, and Wildlife-Associated Recreation: 20 p.

p. 298 "In terms of an unconscious homage to birds, nearly every rock concert in the southeastern U.S. since 1973 has witnessed someone at some point yelling the words 'Free Bird!'" For those of you not into this cultural joke, it refers to the title of a popular song by a 1970s southern U.S. rock band, Lynyrd Skynyrd, a name in which both parts also rhyme with "bird." According to rock musicians who play in the southern U.S., this is by far the most-requested song by their audiences.

p. 299 "First discovered in 1861, *Archaeopteryx lithographica*, from Late Jurassic (150 *mya*) rocks of Germany, was long regarded as the oldest known bird. However, some paleontologists now regard *Archaeopteryx* as more of a non-avian dinosaur. . . ." (1) Xu, X., Zhao, Q., Norell, M., Sullivan, C., Hone, D., Erickson, G., Wang, X., Han, F., and Guo, Y.

2009. A new feathered maniraptoran dinosaur fossil that fills a morphological gap in avian origin. *Chinese Bulletin of Science*, 54: 430-435. (2) Godefriot, P., Cau, A., Dong-Yu, H., Escuillié, F., Wenhao, W., and Dyke, G. 2013. A Jurassic avialan dinosaur from China resolves the early phylogenetic history of birds. *Nature*, doi:10.1038/nature12168.

p. 299 "As far as bird trace fossils are concerned, the oldest suspected bird tracks come from Late Jurassic–Early Cretaceous rocks in various places. . . ." Lockley, M.G., Yang, S.Y., Matsukawa, M., Fleming, F., and Lim, S.K. 1992. The track record of Mesozoic birds: evidence and implications. *Philosophical Transactions: Biological Sciences*, 336: 113-134. This record, of course, has improved considerably since 1992.

p. 299 "For example, bird tracks are preserved in Cretaceous rocks formed in previously polar environments of Alaska as well as a few I recently co-discovered with colleagues in Victoria, Australia." (1) Fiorillo, A.R., Hasiotis, S.T., Kobayashi, Y., Breithaupt, B.H., and McCarthy, P.J. 2011. Bird tracks from the Upper Cretaceous Cantwell Formation of Denali National Park, Alaska, USA: a new perspective on ancient northern polar vertebrate biodiversity. *Journal of Systematic Paleontology*, 9: 33-49. (2) Martin, A.J., Vickers-Rich, P., Rich, T.H., and Hall, M. 2013. Oldest known avian footprints from Australia: Eumeralla Formation (Albian), Dinosaur Cove, Victoria. *Palaeontology*, online version at: DOI: 10.1111/pala.12082.

p. 299 "Explaining the evolutionary history of birds is a massive undertaking, and other people have written excellent, lengthy books on this topic, such as Thor Hanson's *Feathers: The Evolution of a Natural Miracle* (2011) and *Living Dinosaurs: The Evolutionary History of Modern Birds* (2011). . . ." Another very good book I should have mentioned here that directly connects bird evolution to theropod dinosaurs is: Chiappe, L.M. 2007. *Glorified Dinosaurs: The Origin and Early Evolution of Birds*. Wiley-Liss, Wilmington, Delaware: 192 p.

p. 300 "For example, if you took a trip back in time to the Eocene of North America, you might meet up with *Gastornis* (formerly known as *Diatryma*). . . ." Mathew, W.D., and Granger, W. 1917. The skeleton of *Diatryma*, a gigantic bird from the Lower Eocene of Wyoming. *Bulletin of the American Museum of Natural History*, 37: 307-336.

p. 300 "Paleontologists now suggest its huge beak was better suited for large fruits and nuts or for scavenging." One analysis showed that *Gastornis* was capable of crushing bones with its beak, but later work, including its probable habitats, has caused paleontologists to lean more toward herbivory in this giant bird: Naish, D. 2012. Birds. *In* Brett-Surman, M.K., Holtz, T.R., Jr., and Farlow, J.O. (editors), *The Complete Dinosaur* (2nd Edition). Indiana University Press, Bloomington, Indiana: 379-423.

p. 300 "For an example of the latter, the largest flying bird known from the fossil record is the giant teratorn (*Argentavis magnificens*) from the Miocene Epoch (about 15 million years ago) of South America." Naish (2012).

p. 300 "On land, terror birds—known by paleontologists as phorusrhacids—lived in South America and the southern part of North America from about 60 to only 2 million years ago." Naish (2012).

p. 301 "Among these terror birds was the Miocene *Kelenken guillermoi* of Argentina, which possessed the largest skull of any known bird: 71 cm (28 in) long, just smaller than that of an *Allosaurus*." Bertellia, S., Chiappe, L.M., and Tambussi, C. 2007. A new phorusrhacid (Aves: Cariamae) from the middle Miocene of Patagonia, Argentina. *Journal of Vertebrate Paleontology*, 27: 409-419.

p. 301 "Like the South American *Kelenken* that also lived during the Miocene, *Dromornis* stood about 3 m (10 ft) tall. . . ." Murray, P.F., and Vickers-Rich, P. 2004. *Magnificent Mihirungs: The Colossal Flightless Birds of the Australian Dreamtime*. Indiana University Press, Bloomington, Indiana: 416 p.

p. 301 "For example, the elephant birds of Madagascar, which consisted of species of *Aepyornis* and *Mullerornis*. . . ." Hume, J.P., and Walters, M. 2012. *Extinct Birds*. T. & A.D. Poyser, London: 544 p.

p. 301 "Sadly, all elephant birds were extinct by the end of the 18th century, likely victims of a disastrous blend. . . ." Burney, D.A., and Flannery, T.F. 2005. Fifty millennia of catastrophic extinctions after human contact. *Trends in Ecology & Evolution*, 20: 395-401.

p. 302 "New Caledonia, a Pacific island with a main landmass that was connected to Australia and New Zealand during the Mesozoic, had both *Sylviornis neocaledoniae* . . . and *Megapodius molistructor*. . . ." Steadman, D.W. 1999. The biogeography and extinction of megapodes in Oceania. *Zoologische Verhandelingen*, 327: 7-21.

p. 302 "Archaeologists at first thought huge earthen mounds on New Caledonia . . . were human burial grounds. . . ." Hansell, M.H. 2004. *Bird Nests and Construction Behavior*. Cambridge University Press, Cambridge, U.K.: 280 p.

p. 302 "Other sizeable birds on islands included: the giant swans of Malta. . . ." (1) Hume and Walters (2012). (2) Naish (2012).

p. 302 "All of these birds had something else in common, which was their rapid extinction soon after humans came in contact with them, found them delicious, introduced egg predators, and changed their habitats." Burney and Flannery (2005).

p. 302 "For instance, moa-nola diets are known through their coprolites, which showed these were important grazers in Hawaiian ecosystems that lacked mammals as herbivores." James, H.F., and Burney, D.A., 1997. The diet and ecology of Hawaii's extinct flightless waterfowl: evidence from coprolites. *Biological Journal of the Linnean Society*, 62: 279-297.

p. 303 "Bones from about 12,000 to 100,000 years ago there include remains of both the largest known stork (*Leptoptilos robustus*) and a diminutive species of hominin (human relative), *Homo floresiensis*, nicknamed the 'hobbit.'" (1) Meijer, H.J.M., and Due, R.A. 2010. A new species of giant marabou stork (Aves: Ciconiiformes) from the Pleistocene of Liang Bua, Flores (Indonesia). *Zoological Journal of the Linnean Society*, 160: 707-724. (2) Brown, P., Sutikna, T., Morwood, M.J., Soejono, R.P., Jatmiko, S.E.W., and Due, R.A. 2004. A new small-bodied hominin from the Late Pleistocene of Flores, Indonesia. *Nature*, 431: 1055-1061.

p. 303 "More likely menu items for these storks would have been rodents of unusual size, such as the Flores giant rat (*Papagomys armandvillei*). . . ." Zijlstra, J.S., van den Hoek Ostende, L.W., and Due, R.A. 2008. Verhoeven's giant rat of Flores (*Papagomys theodorverhoeveni*, Muridae) extinct after all? *Contributions to Zoology*, 77: 25-31.

p. 303 "These consist of holes in the skull of the 'Taung child,' which belonged to a juvenile *Australopithecus africanus*." Berger, L.W., and McGraw, W.S. 2007. Further evidence for eagle predation of, and feeding damage on, the Taung child. *South African Journal of Science*, 103: 496-498.

p. 303 "Some paleoanthropologists even speculate that frequent eagle attacks would have selected for more cooperative behavior and larger body size in hominins, thus deterring these predators and contributing to greater stature." Berger and McGraw (2007).

p. 304 "Both can be summarized as the 'island rule,' although this can be split into 'island gigantism' and 'island dwarfism,' respectively." Meiri, S., Cooper, N., and Purvis, A. 2008. The island rule: made to be broken? *Proceedings of the Royal Society of London, B*, 275: 141-148.

p. 304 "Back in the Cretaceous, plate-tectonic shifting caused the main landmasses of New Zealand to split from eastern Australia, taking its dinosaurs with it." Molnar, R.E., and Wiffen, J. 1994. A Late Cretaceous polar dinosaur fauna from New Zealand. *Cretaceous Research*, 15: 689-706.

p. 304 "For example, before humans showed up about eight hundred to a thousand years ago, Haast's eagles (*Harpagornis moorei*) were the top carnivores in New Zealand. . . ." Naish (2012).

p. 305 "Fortuitously enough, paleontologist Sir Richard Owen—who coined the word 'dinosaur'—first studied these fabulous flightless birds in the early 19th century. . . ." Dawson, G. "On Richard Owen's Discovery, in 1839, of the Extinct New Zealand Moa from Just a Single Bone." *In* Felluga, D.F. (editor), *BRANCH: Britain, Representation and*

Nineteenth-Century History. Extension of *Romanticism and Victorianism on the Net*: http://www.branchcollective.org/?ps_articles=gowan-dawson-on-richard-owens-discovery-in-1839-of-the-extinct-new-zealand-moa-from-just-a-single-bone

p. 305 "Thanks to their coprolites, we even know what plants they ate and in which ecosystems, described in a 2013 study done by Jamie Wood and other scientists." Wood, J.R., Wilmshurst, J.M., Richardson, S.J., Rawlence, N.J., Wagstaff, S.J., Worthy, T.H., and Cooper, A. 2013. Resolving lost herbivore community structure using coprolites of four sympatric moa species (Aves: Dinornithiformes). 2013. *Proceedings of the National Academy of Sciences*, published ahead of print September 30, 2013: doi:10.1073/pnas.1307700110.

p. 305 "Like the elephant birds in Madagascar, though, they fell victim to the ways of humans, and were extinct by the end of the 15th century." Burney and Glannery (2005).

p. 305 "Among these birds are the flightless kiwis, consisting of five species of *Apteryx*." Colbourne, R. 2006. Kiwi (*Apteryx* spp.) on offshore New Zealand islands: populations, translocations and identification of potential release sites. *DOC Research and Development Series*, 208: 24 p.

p. 305 "Female kiwis also stand out from other avians by having two ovaries, sharing nesting burrows with males. . . ." Robertson, H., and Heather, B. 2001. *Hand Guide to the Birds of New Zealand*. Oxford University Press, Oxford, U.K.: 168 p.

p. 306 "Although we normally think of parrots as subtropical–tropical birds flitting about in rainforests, keas live in alpine environments. . . ." Robertson and Heather (2001).

p. 306 "Male kakapos dig a series of half-meter (20 in) wide bowl-like depressions that they use like megaphones to project their mating calls." Powlesland, R.G., Lloyd, B.D., Best, H.A., and Merton, D.V. 2008. Breeding biology of the Kakapo *Strigops habroptilus* on Stewart Island, New Zealand. *Ibis*, 134: 361-373.

p. 307 "Bird tracks can be divided into four main categories based on overall form, all of which are derived variations of theropod dinosaur feet: anisodactyl, palmate, totipalmate, and zygodactyl." Bird tracks actually can be further subdivided on the basis of differences in foot anatomy, but for the sake of simplicity, this four-fold classification for their tracks works well, summarized by Elbroch and Marks (2001).

p. 308 "Bird trackway patterns fall into five behavioral groupings: diagonal walking (or running), hopping, skipping, standing, and flying." (1) Elbroch and Marks (2001). (2) Martin (2013).

p. 310 "Thus, despite having anatomies that differ from Mesozoic dinosaurs, it is no wonder that dinosaur ichnologists still turn to birds . . . as their default models for how bipedal dinosaurs moved and made tracks." Farlow *et al.* (2012).

p. 310 "In a 2013 study, paleontologists using CT scans of non-avian and avian theropod skulls (including that of *Archaeopteryx*) suggested. . . ." Balanoff, A.M., Bever, G.S., Rowe, T.B., and Norell, M.A. 2013. Evolutionary origins of the avian brain. *Nature*, 2013; published online before print version, DOI: 10.1038/nature12424.

p. 310 "Based on other anatomical traits, like long fingers with claws, other small feathered theropods, such as the Late Jurassic *Scansoriopteryx* (*Epidendrosaurus*) and *Epidexipteryx* of China, look like they were better adapted for climbing trees. . . ." (1) Some people consider *Scansoriopteryx* and *Epidendrosaurus* as synonymous, in which different names are used for the same genus of dinosaur. The original article naming it was by Czerkas and Yuan (2002), mentioned in the endnotes for Chapter 4. (2) Zhang *et al.* (2008), also cited in Chapter 4.

p. 310 "However, a few feathered theropods were far too big to have either flown or climbed trees, such as the Early Cretaceous *Yutyrannus huali*. . . ." Xu, X., Wang, K., Zhang, K., Ma, Q., Xing, L., Sullivan, C., Hu, D., Cheng, S., and Wang, S. 2012. A gigantic feathered dinosaur from the Lower Cretaceous of China. *Nature*, 484: 92-95.

p. 312 "Nothing quite says 'flight' like a right–left pair of bird tracks with no other tracks in front of or behind them." Martin (2013).

p. 312 "At first only two hypotheses were offered, with pithy summary titles: 'ground up' and 'trees down.'" Martin (2006).

p. 313 "Called the 'wing-assisted incline running' (WAIR) hypothesis, the researchers who developed this model—mostly Kenneth Dial and his colleagues. . . ." (1) Dial, K.P. 2003. Wing-assisted incline running and the evolution of flight. *Science*, 299: 402-405. (2) Dial, K.P., Randall, R.J., and Dial, T.R. 2006. What use is half a wing in the ecology and evolution of birds? *BioScience*, 56: 437-445.

p. 313 "However, some Early Cretaceous (about 120-million-year-old) small-shorebird tracks from Korea, reported by Amanda Falk and others in 2009. . . ." (1) Falk, A.R. 2009. *Interpreting Behavior from Early Cretaceous Bird Tracks and the Morphology of Bird Feet and Trackways*. Unpublished M.S. thesis, University of Kansas, Lawrence, Kansas: 140 p. (2) Falk, A.R., Hasiotis, S.T., and Martin, L.D. 2010. Feeding traces associated with bird tracks from the Lower Cretaceous Haman Formation, Republic of Korea. *Palaios* 25: 730-741.

p. 313 "In 2013, Pat Vickers-Rich, Tom Rich, and I interpreted two of three closely associated anisodactyl tracks from Early Cretaceous rocks of Victoria, Australia as bird tracks, with the third coming from a non-avian theropod." Martin *et al.* (2013).

p. 314 "I first noticed this phenomenon during a hot summer in 2004 while walking along the edge of a pond on San Salvador Island, Bahamas." Martin, A.J. 2005. Avian tracks as initiators of mudcracks: models for similar effects of nonavian theropods? *Journal of Vertebrate Paleontology*, 25 [Supplement to No. 3]: 89A.

p. 314 "In the summer of 2007, while doing field work at a dinosaur dig site on the North Slope of Alaska, I noticed bird tracks . . . in the muddy patches of a river floodplain." Martin (2009b).

p. 315 "The same apparent causal relationship between theropod tracks and mudcracks shows up in Early Jurassic rocks of southwestern Utah, in the same strata holding thousands of dinosaur swim-tracks. . . ." Milner *et al.* (2006).

p. 315 "Examples of scrape nests are those of penguins or some shorebirds, such as oystercatchers (species of *Haematopus*) or plovers (mostly species of *Pluvialis* and *Charadrius*), which use their feet to scratch out slight depressions into which they deposit a few eggs." Martin (2013).

p. 315 "The largest bald-eagle nest yet measured was 3 m (10 ft) wide, 6 m (20 ft) deep, and weighed more than two tons." Largest Bird's Nest, *Guinness Book of World Records*: http://www.guinnessworldrecords.com/records-10000/largest-birds-nest/

p. 315 "Unlike the . . . nests people may normally see in urban settings, these nests are made collectively, shared, and reused by hundreds of weaver birds." Crook, J.H. 1963. A comparative analysis of nest structure in the weaver birds (Ploceinae). *Ibis*, 105: 238-262.

p. 316 "Not to be outdone, the mallee fowl (*Leipoa ocellata*) of Australia constructs gigantic incubation mounds for its eggs." Weathere, W.W., Seymour, R.S., and Baudinette, R.V. 1993. Energetics of mound-tending behaviour in the malleefowl, *Leipoa ocellata* (Megapodiidae). *Animal Behaviour*, 45: 333-341.

p. 316 "Although bird nests are almost as diverse as birds themselves, they can be placed into eight basic kinds, based on where they are located or their overall shape: scrape, platform, crevice, cup (or saucer), spherical, pendant, cavity, burrow, and mound." (1) Elbroch and Marks (2001). (2) Martin (2013).

p. 316 "The earliest known dinosaur nests, made by the Early Jurassic sauropodomorph *Massospondylus* of South Africa, were definitely on ground surfaces." Reisz *et al.* (2012).

p. 316 "At the other end of the Mesozoic Era, Late Cretaceous *Maiasaura*, *Troodon*, and titanosaur nests were bowl-like depressions surrounded by raised rims." Varricchio (2011).

p. 316 "A burrow entombing the small Cretaceous ornithopod *Oryctodromeus* and two of its offspring may or may not have been a nest site, although it was likely used as a den for raising its young." Varricchio *et al.* (2007).

p. 316 "Moreover, based on dinosaur egg porosities, nearly all were likely partially buried. . . ." This statement is based on a personal correspondence with David Varricchio in December 2013, but also is the main point of this article: Deeming, D.C. 2006. Ultrastructural and functional morphology of eggshells supports the idea that dinosaur eggs were incubated buried in a substrate. *Palaeontology*, 49: 171-185.

p. 317 "In one instance, egg predation from feral hogs on a Georgia barrier island was so pervasive that wild turkeys began nesting in the trees there." Fletcher, W.O., and Parker, W.A. 1994. Tree nesting by wild turkeys on Ossabaw Island, Georgia. *The Wilson Bulletin*, 106: 562.

p. 317 "At least one vase-shaped structure described from a petrified log from central Europe, dating from the Miocene Epoch (about 23 to 5 *mya*), is remarkably similar to woodpecker nests." Mikuláš, R., and Zasadil, B. 2004. A probable fossil bird nest, ?*Eocavum* isp., from the Miocene wood of the Czech Republic. 4th International Bioerosion Workshop, Prague, Czech Republic, Abstracts Book: 49-51.

p. 318 "In North America, belted kingfishers (*Megaceryle alcyon*), bank swallows (*Riparia riparia*), and rough-winged swallows (*Stelgidopteryx serripennis*) burrow into soft, sandy bluffs adjacent to rivers or other bodies of water for their nesting and raising of young." (1) Elbroch and Marks (2001). (2) Martin (2013).

p. 318 "In Europe, Africa, and elsewhere in the world, all species of bee-eaters make bank burrows, sometimes by the thousands at nesting sites." Casas-Crivillé, A., and Valera, F. 2005. The European bee-eater (*Merops apiaster*) as an ecosystem engineer in arid environments. *Journal of Arid Environments*, 60: 227-238.

p. 318 "The Atlantic puffin (*Fratercula arctica*) of North America digs lengthy burrows. . . ." Hornung, M. 1982. Burrows and burrowing of the puffin (*Fratercula arctica*). *Bangor Occasional Paper* 10. Institute of Terrestrial Ecology, Bangor, Maine: 1-30.

p. 318 ". . . the little penguin (*Eudyptula minor*) of Australia and New Zealand, as well as the Magellanic penguin (*Spheniscus magellanicus*) of southern South America, also excavate burrows." (1) Dann, P. 1991. Distribution, population trends and factors influencing the population size of Little Penguins *Eudyptula minor* on Phillip Island, Victoria. *Emu*, 91: 263-272. (2) Stokes, D.L., and Boersma, P.D. 1991. Effects of substrate on the distribution of magellanic penguin (*Sphenicus magellanicus*) burrows. *The Auk*, 108: 923-933.

p. 318 "Imagine carrying out a mouthful of sediment at a time or scratching with your feet to make a tunnel many times longer than your body: like *The Shawshank Redemption*, but without even the benefit of a little rock hammer." Spoiler alert if you have never seen the movie *The Shawshank Redemption* (1994), starring Tim Robbins and Morgan Freeman: one of the main characters, Andy Dufrense (Robbins)—wrongfully imprisoned for a crime he did not commit—breaks out of prison by chipping away at his cell wall with a small geologic hammer for years, making a long tunnel that allowed his escape.

p. 319 "For example, burrowing owls may take over appropriately sized mammal or gopher-tortoise burrows." Martin, D.J. 1973. Selected aspects of burrowing owl ecology and behavior. *The Condor*, 75: 446-456.

p. 319 "Rough-winged swallows might usurp kingfisher burrows, and vice-versa." Martin (2013).

p. 319 "A male–female pair of Atlantic puffins, which normally mate for life, might return to the same burrow they used the previous year for nesting. . . ." Creelman, E., and Storey, A.E. 1991. Sex differences in reproductive behavior of Atlantic puffins. *The Condor*, 93: 390-398.

p. 319 "Complicating these underground neighborhoods is another bird, the Manx shearwater (*Puffinus puffinus*), which also nests in burrows. . . ." James, P.C. 1986. How do Manx shearwaters *Puffinus puffinus* find their burrows? *Ethology*, 71: 287-294.

p. 320 "Extreme examples are hummingbirds, some of which can produce thousands of wing beats per minute, or Arctic terns (*Sterna paradisaea*). . . ." (1) Kircher, J.C. 1999. *Neotropical Companion: An Introduction to the Animals, Plants, and Ecosystems of the New World Tropics*. Princeton University Press, Princeton, New Jersey: 451 p. (2) Egevang, C., Stenhouse, I.J., Phillips, R.A., Petersen, A., Fox, J.W., and Silk, J.R.D. 2010. Tracking of Arctic terns *Sterna paradisaea* reveals longest animal migration. *Proceedings of the National Academy of Sciences*, 107: 2078-2081.

p. 320 "For instance, sandpipers do double taps—bang-bang—which leave distinctive two-holed patterns. In contrast, plovers keep their beaks on a beach surface while running along at high speed, making a more continuous and connected series of holes." (1) Elbroch and Marks (2001). (2) Martin (2013).

p. 321 "With clams, oysters, and other bivalves, more subtle traces of bird predation might be on the shells themselves." Martin (2013).

p. 322 "In urban environments of Japan, carrion crows (*Corvus corone*) . . . place these on the crosswalks of busy streets for cars to run over and break." Nihei, Y. 1995. Variations of behavior of carrion crows *Corvus corone* using automobiles as nutcrackers. *Japan Journal of Ornithology*, 44: 21-35.

p. 322 "Insect-eating birds, such as grackles or starlings, systematically insert their beaks into soil or ground vegetation to find their food. Northern flickers in particular are both persistent and insidious in their ground probing." Elbroch and Marks (2001).

p. 322 "Seeing that ant eating has been proposed as a lifestyle for some theropods, such as the Late Cretaceous *Xixianykus zhangi* of China. . . ." Xu, X., Wang, D.-Y., Sullivan, C., Hone, D.W.E., Han, F.-L., Yan, R.-H., and Du, F.-M. 2010. A basal parvicursorine (Theropoda: Alvarezsauridae) from the Upper Cretaceous of China. *Zootaxa,* 2413: 1-19.

p. 324 "Peacocks, in order to show off the full extent of their tail feathers, perform a slow pirouette, often defining a tight circle." For one of the most exquisite descriptions of a peacock strutting its stuff, I recommend reading *Living with a Peacock,* written by famed Southern gothic writer Flannery O'Connor. Her essay was originally published in *Holiday* magazine, September 1961, but also was in one of her anthologies (*Mystery and Manners,* 1969) under the title *The King of Birds.*

p. 324 "Male plovers also make distinctive pre-mating trackways by high stepping (also called 'marking time') and placing one foot directly in front of the other." Elbroch and Marks (2001).

p. 324 "Sometimes nicknamed the 'amorous architects,' male bowerbirds build a variety of structures. . . ." Frith, C.B., and Frith, D.W. 2004. *The Bowerbirds: Ptilonorhynchidae.* Oxford University Press, Oxford, U.K.: 508 p.

p. 325 "Some bowerbirds even use perspective in their handiwork . . . the only examples of such artistic renderings known outside of humans." Endler, J.A. 2012. Bowerbirds, art and aesthetics: are bowerbirds artists and do they have an aesthetic sense? *Communicative and Integrative Biology,* 5: 281-283.

p. 325 "Fortunately, a lack of evidence for Mesozoic bowers did not stop the imaginations of screenwriters and computer-graphics artists for the Discovery Channel, in the TV series *Dinosaur Revolution* (2011). . . ." This fun and ichnologically rich scene was in the first episode of this four-part series.

p. 325 "This scene starred a computer-generated image of the large feathered theropod *Gigantoraptor,* an oviraptorosaur from the Late Cretaceous (70 *mya*) of Mongolia." Xu, X., Tan, Q., Wang, J., Zhao, X., and Tan, L. 2007. A gigantic bird-like dinosaur from the Late Cretaceous of China. *Nature,* 447: 844-847.

p. 327 "The trace that results from a dust bath is a shallow semicircular depression slightly wider than the wingspan of the bird; sparrows make ones that are only about 10 to 15 cm (4–6 in) across, whereas turkeys' dust baths are more than 50 cm (20 in) wide." (1) Elbroch and Marks (2001). (2) Martin (2013).

p. 327 "On the opposite end of the size spectrum from sparrows, ostriches in Africa make dust baths by sitting down. . . ." Several videos of ostriches dust-bathing are online, but probably my favorite (at the time of this writing) for its insights on the traces that might be left by a similar-sized feathered dinosaur is the following, titled "Ostrich having a dust bath": http://www.youtube.com/watch?v=e868pFsd2PI

p. 328 "However, paleontologist Darren Naish, in a 2013 blog post, made such a link by reviewing sunbathing in modern birds. . . ." Naish, D. 2013. It's hot and sunny, so birds lie down and sunbathe. *Tetrapod Zoology* (*Scientific American* blog), http://blogs.scientificamerican.com/tetrapod-zoology/2013/07/22/sunbathing-birds-2013/: posted July 22, 2013. Other references listed in this post include: (1) Hauser, D. 1957. Some observations on sunbathing in birds. *The Wilson Bulletin*, 69: 78-90. (2) Johnston, R.F. 1965. Sunbathing by birds. *The Emu*, 64: 325-326. (3) Horsfall, J. 1984. Sunbathing: is it for the birds? *New Scientist*, 103: 28-31.

p. 330 "This was verified with chickens (*Gallus gallus*), which are often impugned as the dumbest of all birds and thus deserving of roasting pans." Karakashian, S.J., Gyger, M., and Marler, P. 1988. Audience effects on alarm calling in chickens (*Gallus gallus*). *Journal of Comparative Psychology*, 102: 129-135.

p. 330 "This sort of learning, recall, and teaching ability has been documented in American crows." (1) Marzluff, J.M., Walls, J., Cornell, H.N., Withey, J.C., and Craig, D.P. 2010. Lasting recognition of threatening people by wild American crows. *Animal Behaviour*, 79: 699-707. (2) Cornell, H.N., Marzluff, J.M., and Pecoraro, S. 2012. Social learning spreads knowledge about dangerous humans among American crows. *Proceedings of the Royal Society of London, B*, 279: 499-508. (3) These works were preceded by a book that provides an overview of the complexity of American crow behavior: Marzluff, J.M., and Angell, T. 2005. *In the Company of Crows*. Yale University Press, New Haven, CT: 408 p. I also recommend watching the *Nature* documentary, *A Murder of Crows* (2010), aired by PBS and available free online.

p. 330 "Other birds that can learn individual human faces include pigeons, magpies (*Pica pica*), and northern mockingbirds (*Mimus polyglottus*). . . ." (1) Stephan, C., Wilkinson, A., and Huber, L. 2012. Have we met before? pigeons recognise familiar human faces. *Avian Biology Research*, 5: 75-80. (2) Lee, W.Y., Lee, S., Choe, J.C., and Jablonski, P.G. 2011. Wild birds recognize individual humans: experiments on magpies, *Pica pica*. *Animal Cognition*, 14: 817-825. (3) Levey, D.J., *et al.* 2009. Urban mockingbirds quickly learn to identify individual humans. *Proceedings of the National Academy of Sciences*, 106: 8959-8962.

p. 330 "How about superb fairy-wrens (*Malurus cyaneus*) giving their chicks a 'password' (a single note) as a cue for feeding, but while the chicks are still snugly inside their eggs?" Colombelli-Négrel,

D., Hauber, M.E., Robertson, J., Sulloway, F.J., Hoi, H., Griggio, M., and Kleindorfer, S. 2012. Embryonic learning of vocal passwords in superb fairy-wrens reveals intruder cuckoo nestlings. *Current Biology*, 22: 2155-2160.

p. 330 "Nonetheless, when this was first documented starting in the 1960s, it was an eye-opener for behavioral biologists." (1) Van Lawick-Goodall, J., and Van Lawick-Goodall, H. 1966. Use of tools by the Egyptian vulture, *Neophron percnopterus*. *Nature*, 212: 1468-1469. (2) Alcock, J. 1972. The evolution of the use of tools by feeding animals. *Evolution*, 26: 464-472. (3) Bentley-Condit, V.K., and Smith, E.O. 2010. Animal tool use: current definitions and an updated comprehensive catalog. *Behaviour*, 147: 185-32A.

p. 331 "Of these birds, New Caledonian crows are the most impressive of tool users and problem solvers." Chappell, J., and Kacelnik, A. 2002. Tool selectivity in a non-primate, New Caledonian crow (*Corvus moneduloides*). *Animal Cognition*, 5: 71-78.

p. 331 "Woodpecker finches latch on to cactus spines or twigs with their beaks and then manipulate these like fine surgical tools to extract insect larvae from tight spots in trees." Tebbich, S., Taborsky, M., Fessl, B., and Blomqvist, D. 2001. Do woodpecker finches acquire tool-use by social learning? *Proceedings of the Royal Society of London, B*, 268: 2189-2193.

p. 331 "Egyptian vultures and bristle-thighed curlews share the practice of grabbing a rock with their beaks and throwing these at eggs to break them open." (1) Thouless, C.R., Fanshawe, J.H., and Bertram, C.R. 1989. Egyptian vultures *Neophron percnopterus* and ostrich *Struthio camelus* eggs: the origins of stone-throwing behavior. *Ibis*, 131: 9-15. (2) Marks, J.S., and Hall, C.S. 1992. Tool use by bristle-thighed curlews feeding on albatross eggs. *The Condor*, 94: 1032-1034.

p. 331 "Male palm cockatoos grasp sticks in their beaks, which they then drum against trees to alert females that they are in the area. . . ." Wood, G.A. 1988. Further field observations of the palm cockatoo *Probosciger aterrimus* in the Cape York Peninsula, Queensland. *Corella*, 12: 48-52.

p. 331 "Brown-headed nuthatches acquire short pieces of bark or sticks, which they lever against the bark on tree trunks to expose insects." Morse, D.H. 1968. The use of tools by brown-headed nuthatches. *The Wilson Bulletin*, 80: 220-224.

p. 331 "These actually use mammal feces as a tool, by picking up pieces of bison or cattle dung, placing these in front of their burrows. . . ." Levey, D.J., Duncan, R.S., and Levins, C.F. 2004. Animal behaviour: use of dung as a tool by burrowing owls. *Nature*, 431: 39.

p. 331 "Lastly, a few species of herons, egrets, and seagulls go fishing by employing feathers, berries, and even bread as lures." (1) Robinson,

S.K. 1994. Use of bait and lures by green-backed herons in Amazonian Peru. *The Wilson Bulletin*, 106: 567-569. (2) Post, R.J., Post, C.P.K., and Walsh, J.K. 2009. Little egret *(Egretta garzetta)* and grey heron *(Ardea cinerea)* using bait for fishing in Kenya. *Waterbirds*, 32: 450-452. (3) Ruxton, G.D., and Hansell, M.H. 2011. Fishing with a bait or lure: a brief review of the cognitive issues. *Ethology*, 117: 1-9.

p. 331 "So do birds have memes—culturally acquired behaviors that modify over time—as well as genes?" Blackmore, S. 2000. *The Meme Machine*. Oxford University Press, Oxford, U.K.: 264 p.

Chapter 11: Dinosaurian Landscapes and Evolutionary Traces

p. 334 "Sinclair Oil Corporation encouraged this illusion by sponsoring dinosaur exhibits at the Chicago and New York World's Fairs. . . ." To this day, Sinclair is still making this connection, and they are quite proud of how this marketing-by-dinosaurs was ahead of its time: http://www.sinclairoil.com/history/symbol_01.html

p. 334 "It turns out that nearly all petroleum is from algae, most of which were deposited and buried in marine environments. . . ." Stoneley, R. 1995. *Introduction to Petroleum Exploration for Non-Geologists*. Oxford University Press, Oxford, U.K.: 119 p.

p. 334 "Indeed, some of the most prolific petroleum reservoirs in the world are filled with oil that post-dates the end-Cretaceous extinction of dinosaurs." Naim, A.E.M., and Alsharhan, A.S. 1997. *Sedimentary Basins and Petroleum Geology of the Middle East*. Elsevier, Amsterdam: 878 p.

p. 337 "At this meeting, Tony Thulborn . . . gave a talk simply titled 'Giant Tracks in the Broome Sandstone (Lower Cretaceous) of Western Australia.'" Thulborn, T., 2002. Giant tracks in the Broome Sandstone (Lower Cretaceous) of Western Australia. *In* Brock, G.A., and Talent, J.A. (editors), *First International Palaeontological Congress*, Abstracts No. 63, Geological Society of Australia, Sydney, 154-155.

p. 337 "In 2012, Thulborn elaborated on that idea in an article titled 'Impact of Sauropod Dinosaurs on Lagoonal Substrates in the Broome Sandstone (Lower Cretaceous), Western Australia.'" Thulborn, R.A. 2012. Impact of sauropod dinosaurs on lagoonal substrates in the Broome Sandstone (Lower Cretaceous), Western Australia. *PLoS One*, 7(5): e36208. doi:10.1371/journal.pone.0036208.

p. 338 "Such grand disturbances of pliable mud or sand . . . are sometimes called 'dinoturbation.'" Lockley (1991).

p. 339 "Elephants consist of three species: the African bush elephant (*Loxodonta africana*), African savannah elephant (*L. cyclotis*), and Indian elephant (*Elephas maximus*)." Roca, A.L., Georgiadis, N., Pecon-Slattery, J., and O'Brien, S.J. 2001. Genetic evidence for two species of elephant in Africa. *Science*, 293: 1473-1477.

p. 339 "Thanks to fossil trackways recently discovered in the United Arab Emirates which show a series of parallel and overlapping tracks (group behavior) crossed by one trackway (a lone male). . . ." Bibi, F., Kraatz, B., Craig, N., Beech, M., Schuster, M., and Hill, A. 2012. Early evidence for complex social structure in Proboscidea from a late Miocene trackway site in the United Arab Emirates. *Biology Letters*, 23: 670-673.

p. 339 "Elephant trails can also form depressions deep enough for water to flow along them, creating canals that connect previously isolated rivers or ponds." Mosepele, K., Moyle, P.B., Merron, G.S., Purkey, D.R., and Mosepele, B. 2009. Fish, floods, and ecosystem engineers: aquatic conservation in the Okavango Delta, Botswana. *BioScience*, 59: 53-64.

p. 340 "In a study by geologist Daniel DeoCampo published in 2002, he documented how hippos in Tanzania. . . ." DeoCampo, D.M. 2002. Sedimentary structures generated by *Hippopotamus amphibius* in a lake-margin wetland, Ngorongoro Crater, Tanzania. *Palaios*, 17: 212-217.

p. 340 "This channelization via hippo traces, in which their trails eventually turn into new river channels, is also well documented in the Okavango Delta of Botswana." McCarthy, T.S., Ellery, W.N., and Bloem, A. 1998. Some observations on the geomorphological impact of hippopotamus (*Hippopotamus amphibious* L.) in the Okavango Delta, Botswana: *African Journal of Ecology*, 36: 44-56.

p. 340 "In this respect, in 2006, two geologists—Lawrence Jones and Edmund Gustason—did indeed propose that avulsion features in the Morrison Formation of east-central Utah were likely caused by sauropod trails that created 'channels' for the flow of floodwaters." Jones, L.J., and Gustason, E.R. 2006. Dinosaurs as possible avulsion enablers in the Upper Jurassic Morrison Formation, east-central Utah. *Ichnos*, 13: 31-41.

p. 341 "These rivers include the Nile in eastern Africa, the Amazon of South America, the Macleay and Murrumbidgee of Australia, and the Colorado River of the western U.S., among others." Here are a few references on the antiquity of river valleys, and how at least parts of their original drainage systems might have coincided with a dinosaurian presence: (1) Gani, N.D, Abdelsalam, M.G., Gera, S., and Gani, M.R. 2009. Stratigraphic and structural evolution of the Blue Nile Basin, northwestern Ethiopian Plateau. *Geological Journal*, 44: 30-56. (2) Potter, P.E. 1990. The Mesozoic and Cenozoic paleodrainage of South America: a natural history. *Journal of South American Earth Sciences*, 10: 331-344. (3) Oilier, C.D. 1982. The Great Escarpment of eastern Australia: tectonic and geomorphic significance. *Journal of the Geological*

Society of Australia, 29: 13-23. (4) Flowers, R.M., and Farley, K.A. 2012. Apatite ^{4}He/^{3}He and (U-Th)/He evidence for an ancient Grand Canyon. *Science*, 338: 1616-1619.

p. 341 *"If sauropods were as wary as elephants in negotiating sloping terrain, they would naturally have tended to walk on the lower and safer ground. . . ."* Quoted verbatim from Thulborn (2012).

p. 342 "But geologists actually apply these two words to different mass movements of earth material. . . ." Clague, J.J., and Stead, D. 2012. *Landslides: Types, Mechanisms and Modeling*. Cambridge University Press, Cambridge, U.K.: 429 p.

p. 342 "In an article by geologist David Loope published in 2006, he showed how Early Jurassic dinosaurs caused sand-dune surfaces to collapse as they walked across them." Loope, D.B. 2006. Dry-season tracks in dinosaur-generated grainflows. *Palaios*, 21: 132-142.

p. 343 "Perhaps the most surprising conclusion of his investigation was that many of the dinosaur tracks had been made on dry sand, which was always regarded as a poor medium for preserving tracks." Loope (2006).

p. 343 "Other geologists had reckoned that the climate for that area during the Early Jurassic was monsoonal, so Loope figured that the tracks were made and preserved during the dry season, which was likely in the winter." Loope (2006).

p. 343 "This sort of movement is called *grainflow*, in which sand grains flow as if they are in water, but in this instance are cushioned by air between the grains." Lowe, D.R. 1976. Grain flow and grain flow deposits. *Journal of Sedimentary Petrology*, 46: 188-199.

p. 344 "In cross-sections of the sandstone, Loope found theropod and prosauropod tracks directly associated with grainflows. . . ." Loope (2006).

p. 345 "High concentrations of dinosaur eggs elsewhere in the world, such as in South Africa, Spain, and India. . . ." Carpenter, K. 2000. *Eggs, Nests, and Baby Dinosaurs: A Look at Dinosaur Reproduction*. Indiana University Press, Bloomington, Indiana: 352 p.

p. 345 "In some instances, birds in breeding colonies, such as Australasian gannets (*Morus serrator*). . . ." Wingham, E.J. 1984. Breeding biology of the Australasian Gannet *Morus serrator* (Gray) at Motu Karamarama, Hauraki Gulf, New Zealand. I. The egg. *Emu*, 84: 129-136.

p. 346 "In an article published in 2012, Jenni Scott, Robin Renaut, and Bernhart Owen described the terrain-altering effects of flamingo nesting grounds. . . ." Scott, J.E., Renaut, R.W., and Owen, R.B. 2012. Impacts of flamingos on saline lake margin and shallow lacustrine sediments in the Kenya Rift Valley. *Sedimentary Geology*, 277: 32-51.

p. 346 "Flamingos mine the lakeshore mud, which they do by dragging it with their beaks or scooping up mud in their bills and spitting it onto the nest." Scott *et al.* (2012).

p. 347 "Jack Horner originally proposed that the Late Cretaceous hadrosaur *Maiasaura* of Montana may have had nesting colonies. . . ." (1) Horner and Makela (1979). (2) Horner (1982).

p. 347 "Near the *Maiasaura* nests are *Troodon* nests, which are rimmed, bowl-like depressions that outwardly resemble flamingo nests." Varricchio *et al.* (1999).

p. 347 "How about the recently extinct mound-nesting birds of New Caledonia, *Sylviornis neocaledoniae* and *Megapodius molistructor*, which made mound nests so big that archaeologists at first mistook these for human burial mounds?" Hansell (2004).

p. 348 "In at least one recent documentary (*Dinosaur Revolution*, 2011), *Tyrannosaurus rex* was depicted as a mound nester with an accordingly huge mound for its eggs." The tyrannosaur mound-nesting scenes were in Episode 4 ("End Game") of the *Dinosaur Revolution* series.

p. 349 "For example, in humans, a flatus produced by swallowed air that made its way through the digestive tract. . . ." Suarez, F., Furne, J., Springfield, J., and Levitt, M. 1997. Insights into human colonic physiology obtained from the study of flatus composition. *American Journal of Physiology: Gastrointestinal and Liver Physiology*, 272: G1028-G1032.

p. 349 "Most of the unpleasant smells we associate with flatulence come from sulfurous gases, such as hydrogen sulfide (H_2S)." Suarez, F.L., Springfield, J., and Levitt, M.D. 1998. Identification of gases responsible for the odour of human flatus and evaluation of a device purported to reduce this odour. *Gut*, 43: 100-104.

p. 350 "As everyone now knows (with the exception of a few billionaires), the overproduction of certain gases . . . has insulating qualities in our atmosphere. . . ." Shortly after I wrote this sentence, the latest IPCC (Intergovernmental Panel on Climate Change) report (2013) on global-climate change, titled *Climate Change 2013: The Physical Science Basis*, confirmed a 95% certainty that this change was human-caused, and this is agreed upon by about 97% of climatologists. For those who say we need to "teach the controversy" and "let's hear both sides of the issue," there is no controversy, and there is no other side.

p. 350 "Although carbon dioxide is often vilified for its role in climate change, methane in smaller concentrations has a greater impact. . . ." Howarth, R.W., Santoro, R., and Ingraffea, A. 2011. Methane and the greenhouse-gas footprint of natural gas from shale formations. *Climatic Change*, 106: 679-690.

p. 350 "In a 2012 paper, three scientists—David Wilkinson, Euan Nisbet, and Graeme Ruxton—asked themselves this: Did dinosaur-produced methane contribute to global warming?" Wilkinson, D.M., Nisbet, E.G., and Ruxton, G.D. 2012. Could methane produced by sauropod dinosaurs have helped drive Mesozoic climate warmth? *Current Biology*, 22: R292-R293.

p. 351 "Instead, it was warm throughout almost its entire 185 million years, from the start of the Triassic Period to the end of the Cretaceous Period." Price, G.D., Twitchett, R.J., Wheeley, J.R., and Giuseppi, B. 2013. Isotopic evidence for long term warmth in the Mesozoic. *Nature Scientific Reports* 3, 1438: doi:10.1038/srep01438

p. 351 "For instance, the jury is still out on whether or not some herbivorous dinosaurs . . . used fermentation in their hindguts to digest their food." (1) Farlow, J.O. 1987. Speculations about the diet and digestive physiology of herbivorous dinosaurs. *Paleobiology*, 13: 60-72. (2) Hummel, J., and Clauss, M. 2011. Sauropod feeding and digestive physiology. *In* Klein, N., Remes, K., and Gee, C. 2011. *Biology of the Sauropod Dinosaurs: Understanding the Life of Giants*. Indiana University Press, Bloomington, Indiana: 11-33.

p. 351 "Many modern animals, from termites to cattle, have this symbiotic relationship. . . ." (1) Brune, A., and Ohkuma, M. 2011. Role of the termite gut microflora in symbiotic digestion. *In* Bignell, D.E., Roisin, Y., and Lo, N. (editors), *Biology of Termites: A Modern Synthesis*. Springer, Berlin: 439-475.

p. 351 "In this instance, the rarity of gastroliths in most herbivorous dinosaurs is now regarded as a good reason to think that sauropods and other hefty herbivores used hindgut fermentation." Wings and Sander (2007).

p. 353 "Robert Bakker was the first dinosaur paleontologist to suggest that dinosaur feeding might have 'invented' flowering plants (*angiosperms*). . . ." Bakker, R.T. 1978. Dinosaur feeding behaviour and the origin of flowering plants. *Nature*, 274: 661-663.

p. 353 "Among the first paleobotanists to start examining whether flowering plant and land-dwelling vertebrate evolution went hand in hand (or rather, leaf in mouth) were Scott Wing and Bruce Tiffney." (1) Wing, S.L., and Tiffney, B.H. 1987a. Interactions of angiosperms and herbivorous tetrapods through time. *In* Friis, E.M., Chaloner, W.G., and Crane, P.R. (editors), *The Origin of Angiosperms and their Biological Consequences*. Cambridge University Press, Cambridge, U.K.: 203-224. (2) Wing, S.L., and Tiffney, B.H. 1987b. The reciprocal relationship of angiosperm evolution and tetrapod herbivory. *Review of Palaeobotany and Palynology*, 50: 179-210.

p. 354 "In a 2001 study done by dinosaur paleontologist Paul Barrett and paleoecologist Katherine Willis. . . ." Barrett, P.B., and Willis, K.J. 2001. Did dinosaurs invent flowers?: dinosaur-angiosperm coevolution revisited. *Biological Reviews*, 76: 411-447.

p. 354 "Similarly, another review done in 2008 by Graeme Lloyd and eight other paleontologists examined how what they called the 'Cretaceous Terrestrial Revolution'. . . . " Lloyd, G.T., Davis, K.E., Pisani, D., Tarver, J.E., Ruta, M., Sakamoto, M., Hone, D.W.E., Jennings, R., and Benton, M.J. 2008. Dinosaurs and the Cretaceous terrestrial revolution. *Proceedings of the Royal Society of London, B*, 275: 2483–2490

p. 356 "Some of these traits might include different root systems. . . ." Lambers, H., Chapin, F.S., III, and Pons, T.L. 1998. *Plant Physiological Ecology*. Springer, Berlin: 540 p.

p. 356 "As a hint of how dinosaur feces might have affected soils and plants during the Mesozoic, paleoecologists, in a 2013 study done on extinctions of large herbivorous mammals in the Amazon River Basin. . . ." Doughty, C.E., Wolf, A., and Malhi, Y. 2013. The legacy of the Pleistocene megafauna extinctions on nutrient availability in Amazon soils. *Nature Geoscience*, doi:10.1038/ngeo1895d

p. 357 "Likewise, whenever great poopers die out, one should also mourn for dung beetles, which die with them. . . ." Johnson, C.N. 2009. Ecological consequences of Late Quaternary extinctions of megafauna. *Proceedings of the Royal Society, B*: 276: 2509-2519.

p. 358 "One of the most surprising results of these reintroductions was how riparian ecosystems in Yellowstone improved, with greater and more vigorous plant growth that approached their recent, pre-colonial state, when wolves naturally inhabited this area." Ripple, W.J., Beschta, R.L., Fortin, J.K., and Robbins, C.T. 2013. Trophic cascades from wolves to grizzly bears in Yellowstone. *Journal of Animal Ecology*, DOI: 10.1111/1365-2656.12123.

p. 358 "First, with regard to wolves, their favorite item on the menu—one that they will pick nearly every time if given a choice—is elk (*Cervus canadensis*)." Creel, S., Christianson, D., Liley, S., and Winnie, J.A., Jr. 2007. Predation risk affects reproductive physiology and demography of elk. *Science*, 315: 960.

p. 358 "But once wolves were back in the neighborhood, salads that were once taken leisurely plummeted, riparian plant communities bounced back. . . ." (1) Ripple, W.J., and Beschata, R.L. 2003. Wolf reintroduction, predation risk, and cottonwood recovery in Yellowstone National Park. *Forest Ecology and Management*, 184: 299-313. (2) Ripple *et al.* (2013).

p. 358 "This fear factor even caused elk to have smaller families, as the added stress of possible predation triggered hormones that decreased female-elk fertility." Creel *et al.* (2007).

p. 358 "This effect even helped wolves' super-friends, grizzly bears, which then had lots more berries to eat during lean times." Ripple *et al.* (2013).

p. 360 "Coincidentally (or not), early birds and flowering plants expanded and diversified at about the same time, which was in the middle of the Cretaceous Period (about 100–125 *mya*)." Zhou, Z., Barrett, P.M., and Hilton, J. 2003. An exceptionally preserved Lower Cretaceous ecosystem. *Nature*, 421: 807-814.

p. 360 "In some of the earliest studies done of this phenomenon, ecologists estimated that about half to 90% of all fruited trees of modern forests are adapted for birds and mammals to eat them." Howe, H.F. 1986. Dispersal by fruit-eating birds and mammals. *In* Murray, D.M. (editor), *Seed Dispersal*. Academic Press, New York: 123-144.

p. 362 "This even happens in the guts of alligators and crocodiles, some of which eat a surprising amount of fruit." Platt, S.G., Elsey, R.M., Liu, H., Rainwater, T.R., Nifong, J.C., Rosenblatt, A.E., Heithaus, M.R., and Mazzotti, F.J. 2013. Frugivory and seed dispersal by crocodilians: an overlooked form of saurochory? *Journal of Zoology*, 291: 87-99.

p. 362 "But do not chew and swallow, say, a cup of apple seeds: every seed contains a small amount of cyanide. . . ." Silverton, J. 2009. *An Orchard Invisible: A Natural History of Seeds*. University of Chicago Press, Chicago, Illinois: 224 p.

p. 363 "Charles Darwin even thought of this, as he wondered how the isolated Galapagos Islands off the coast of South America had managed to acquire such thriving plant communities." Grant, K.T., and Estes, G.B. 2009. *Darwin in Galápagos: Footsteps to a New World*. Princeton University Press, Princeton, New Jersey: 362 p.

p. 363 "Animal immigrants could include members of these birds' microbiomes . . . but other accidental tourists—such as snails, larval insects, or larval crustaceans—also can attach to bird feet." Green, A.J., and Figuerola, J. 2005. Recent advances in the study of long-distance dispersal of aquatic invertebrates via birds. *Diversity and Distributions*, 11: 149-156.

p. 364 "Based on fossil tracks from Korea, we know that palmate bird feet had evolved by about 120 *mya* (Early Cretaceous). Moreover, webbed bird tracks became more common and bigger throughout the rest of the Cretaceous." Kim, J.Y., Lockley, M.G., Seo, S.J., Kim, K.S., Kim, S.H., and Bark, K.S. 2012. A paradise of Mesozoic birds: the world's richest and most diverse Cretaceous bird track assemblage from the Early Cretaceous Haman Formation of the Gajin Tracksite, Jinju, Korea. *Ichnos*, 19: 28-42.

p. 365 "Yes, that's right, Charles Darwin thought of flighted birds taking both seeds and small invertebrates to new homes, and repeating these actions over many generations." Grant and Estes (2009).

p. 366 "*In my opinion these footsteps (with which subject your name is certain to go down to long future posterity) make one of the most curious discoveries of the present century. . . .*" Letter from Charles Darwin to Edward Hitchcock, November 6, 1845. Entire contents of correspondence transcribed at: https://www.darwinproject.ac.uk/letter/entry-925

p. 366 "As of this writing, despite continued discoveries of theropod and ornithopod tracks, almost no dinosaur bones are known from the Connecticut River Valley." Weishampel and Young (1998).

p. 367 "One of the most excitedly received dinosaur books in recent years was *All Yesterdays* (2012). . . ." Conway, J., Koseman, C.M., Naish, D., and Hartman, S. 2012. *All Yesterdays: Unique and Speculative Views of Dinosaurs and Other Prehistoric Animals.* Irregular Books: 100 p.

p. 367 "That book's popularity then led to a sequel published in late 2013, titled *All Your Yesterdays. . . .*" Koseman, C.M. 2013. *All Your Yesterdays: Extraordinary Visions of Extinct Life by a New Generation of Palaeoartists.* Irregular Books: 180 p.

Acknowledgments

Ideally, books are traces of an author's thoughts. But in reality they are composite traces built, shaped, and birthed by a huge number of tracemakers, with many influencing an author well before words are imagined, spoken, or written and coalesce with coherence and meaning. I am extremely grateful to all of these tracemakers for their help in making this book real, but also must note—with great frustration—that I cannot possibly name all who left their imprints on my consciousness or otherwise aided in its physical manifestation before the book started and while it was in process.

Of course, that shouldn't stop me from trying. So I'll start with the most important people, who are the bookends of my life, my parents and my wife. Chronologically, the first are those who made me exist in the first place, my parents, Veronica and Richard Martin. Neither went to college, and our family only had enough money to pay the bills each month. Yet they still understood the intrinsic value of my enthusiasm for natural history, and they encouraged my learning from it from an early age. All of those hours spent in our back yard, trips to the countryside for hunting or fishing, and jaunts to public libraries, are starting to pay off. Thank you, Mom and Dad. (Side note: Thanks also to public libraries and librarians everywhere. We love you, need you, and want to read all your books.)

At the other end of my life—which I plan to continue as long as possible, and with her—is the one person who was most responsible for nurturing and encouraging my creative growth, Ruth Schowalter. I am in a state of still-stunned disbelief that she agreed to my writing another book so quickly on the heels of one that occupied about 35% of our lives together. Fortunately, that daunting percentage diminishes with each passing year, and this more recent spate of writing and editing was made shorter and far easier with the growing realization of the depth of her wisdom, love, and support that has always been there. Thank you,

Ruth, for all you have done, whether together at home or in the field, or apart from each other. I could not have started, done, or completed this book without you.

The idea for a book on dinosaur ichnology had been germinating for quite a while in my mind, but I needed to gain a little more knowledge and experience before taking it on. For instance, in 1996, back when HTML was still carved out by hand with hammers and chisels, I wrote and posted a crude Web site simply titled *Dinosaur Trace Fossils*, which summarized what I then knew about this topic. (Let's just say it was a short site.) In 2006, many of the same themes appeared in a chapter on dinosaur trace fossils in my college-level dinosaur textbook (*Introduction to the Study of Dinosaurs*). Soon after that, another book, but one more overtly ichnological, took over my life: *Life Traces of the Georgia Coast*, which took about four years (2008 to 2012) from book-proposal acceptance to holding the book in my hands. That tome, like a sauropod trackway reshaping a landscape, cleared the way for me to write something more focused and fun: namely, the one you're holding now. My appreciation is thus extended to editors Nancy Whilton (Wiley-Blackwell) and Robert (Bob) Sloan and James (Jim) Farlow (Indiana University Press) for their assistance with the aforementioned books, which made this one more probable.

That's where Sydney (Sid) Perkowitz enters the story. A friend and colleague of mine at Emory University (the place where I occasionally work), he graciously met with me several times to share his experiences with writing on scientific topics for broader audiences. During one of those meetings, he suggested that I contact his literary agent, Laura Wood (FinePrint Literary Management) for possible representation. Laura agreed to give me a try, took a look at my initial (and now cringe-worthy) book proposal, and shaped it into something that worked splendidly, attracting the attention of multiple publishers. (Obviously, Pegasus Books was the best fit.) She then guided this naïve academic through the world of trade publishing, and she was there with me all of the way from proposal to final book. Thank you, Laura and Sid, for these big breaks in what I hope will be a long career of sharing my scientific passions with those who care about them.

Meanwhile, at Pegasus, my editor Jessica Case was a superb partner in the development of the book, from offering a contract to final page-turner. Her feedback, reassurances, and expert guidance were invaluable, helping me to find my voice as a writer. I've often heard that one of the best compliments an author can give to an editor is that her or his changes to writing later become nearly invisible, and Jessica certainly fulfilled this ideal. Sure, most of my darlings were killed along the way, bearing non-healed toothmarks and showing up in coprolites. Nonetheless, most deserved to die, and she spared the lives of a few that pleaded their case. Other good folks at Pegasus, who were indispensable for the honing and crafting of the manuscript, as well as ensuring its good looks, were Deb Anderson (copyediting), Maria Fernandez (interior design), and Phil Gaskill (proofreading). Thank you, Jessica, Deb, Maria, and Phil.

I am extremely grateful to the two outside reviewers of the book manuscript, James (Jim) Kirkland and David (Dave) Varricchio, whose input greatly improved it, taking parts of it from "truthiness" to (dare I say it?) almost scholarly. Both I have known for quite a while, especially Dave, who went to graduate school with me at the University of Georgia in the late 1980s. While there, we both benefited from the tutelage of one of the best ichnologists of all time, Robert (Bob) Frey, and we do our best to carry on what we learned from him. Jim has also always been a straightforward, honest, and generous friend, whose encyclopedic knowledge of all things dinosaurian—bones, traces, environments, or otherwise—continues to awe me. Thank you, Jim and Dave; and by extension let us acknowledge, remember, and honor our mentors, who helped to make us what we are today.

One point of jealousy, though: both Jim and Dave live in western U.S. states that abound with dinosaur trace fossils, whereas here in Georgia I must content myself with too-infrequent flights out there or flights of fancy here. Helping with the latter are my paleo-friends here in the southeastern U.S., Andrew (Andy) Rindsberg, Stephen (Steve) Henderson, Sally Walker, Melanie DeVore, Patricia (Trish) Kelley, and Gale Bishop. There are many more who should be included on this list, but the preceding have been the ones with me longest in my continued learning in a region sadly lacking in dinosaur trace fossils, but otherwise rich in natural history. Steve Henderson in particular was a huge help in our Emory University effort to educate students about dinosaurs, whether through field courses, on-campus classes, or guest lectures. Thanks to my paleo-friends.

As loyal readers noticed, much of this book is inspired by and takes place in Australia, a place most people in the world have not visited, nor might they have known for its dinosaurs and dinosaur trace fossils. My connection with Australia and its trace fossils began serendipitously with a sabbatical at Monash University (Melbourne) in 2006. My host and mentor there, Patricia (Pat) Vickers-Rich, still doesn't quite know what she started by inviting me there, but it's been a fun ride since. Her husband and fellow paleontologist, Thomas (Tom) Rich, was a huge help in my Down Under explorations, and I am very thankful to him for acquiring funds to bring me over in 2010 for a month of field time there, nicknamed "The Great Cretaceous Walk." Other Australians I must thank for all they have done during my visits there include David (Dave) Pickering, Lesley and Gerry Kool, Michael and Naomi Hall, Greg and Deb Denney, Mike Cleeland, Doris Seegets-Villiers, David and Judy Elliott, and Trish Sloan. I am also extremely lucky to have the gorgeous art of Australian artist Peter Trusler gracing the cover. Peter is one of best paleo-artists alive, and may he continue to produce fine art in that state of being for a long time. As one might reckon, this book is also a big love letter to all of Australia, a place with an incredible four-billion-year history recorded in its rocks, still beckoning us to plumb its mysteries: my thanks to an entire continent. Good on ya, mates!

The ichnologists who have taught me throughout the many years of tracing ancient lives are far too many to name, so I will honor four senior practitioners from outside of the U.S. who have been the most influential and encouraging of my ichnological obsessions. These ichnologists are Richard Bromley, Murray Gregory, Dolf Seilacher, and Alfred Uchman: *Tak*, thanks, *danke*, and *dziękuję*!

For training received and applied since to tracking modern animals, my enduring gratitude is extended to the instructors of the Tracker School (New Jersey), Wilderness Awareness School (Washington), and A Naturalist's World (Montana). The ancient science of tracking is still alive because of people like them, and it has been even more thrilling to apply tracking principles to animals gone from the earth long before our species evolved.

Because I am not really a dinosaur paleontologist (but could easily play one on TV), I've depended on the collective knowledge of many paleontologists who work directly and often with dinosaur fossils, whether these are body or trace. A sample of these people includes (in alphabetical order): Lisa Buckley, Karen Chin, Peter Falkingham, Andrew (Andy) Farke, Jim Farlow, Patrick Getty, Lee and Ashley Hall, Thomas (Tom) Holtz, John (Jack) Horner, Frankie Jackson, Yoshi Katsura, Glen Kuban, Martin Lockley, Heinrich Mallinson, Andrew Milner, Jason Moore, Emma Rainforth, Jenni Scott, Glen Storrs, Hans-Dieter Sues, and Tony Thulborn. I also would be remiss in not thanking Brian Switek for his exquisitely crafted books, correspondences, and conversations we've had about dinosaurs, which all taught me a few things about how to write about dinosaurs for a popular audience. Many thanks to all of my dino-phile friends.

Finally, I always remember the dinosaur tracemakers—both those long dead and those still living among us as birds—for donating so much tangible evidence of their active, breathing existence. We live on a planet composed of many worlds separated by time, but these threads bring together a great tapestry of life, unfinished yet continually interwoven by its traces.

Index

A

Adélie penguin (*Pygoscelis adeliae*), 228
Aepyornis (elephant bird), 301
African crowned eagles (*Stephanoaetus coronatus*), 303
Albertosaurus, 186, 189, 260
Alcheringa (Australian paleontology journal), 287
Alectura lathami (Australian brush turkey), 105, 298
Alexander, R. McNeil (physicist), 30–31, 34–35, 64
Alexander's formula, 30, 35
Alligator mississippiensis (American alligator), 124, 184
Alligator sinensis (Chinese alligator), 124
Allosaurus, 50, 63, 73, 150–51, 176–78, 195, 214–15, 218, 244, 277, 301, 348, 357
American alligator (*Alligator mississippiensis*), 124, 184
American Museum of Natural History, 42, 49, 62, 96
Anchiornis huxleyi, 299
angiosperms (flowering plants), 353–57, 361. *See also* flowering plants
animal communities, evolution in, 9
anisodactyl, 307, 309, 314
ankylosaurs, 18–19, 23, 27, 38, 47, 88, 179, 192, 213, 225, 239–40, 251, 348, 351–52, 355, 369
Antarctica, 18, 23, 150, 270, 293
Archaeopteryx lithographica, 299
Argentavis magnificens (giant teratorn), 299–300
Argentina, 12, 18, 25, 43, 94, 96, 101, 109, 112, 206, 219, 263, 301, 345
arthropods, 259, 327
Aurornis xui, 299
Australia, 12, 43, 60–61, 73, 150, 153–59, 184, 239, 267–68, 270–73, 276–84, 286–88, 293, 298–99, 301–2, 304, 316, 318, 324, 330, 336–37, 341, 347, 361. *See also* Queensland, Australia; Victoria, Australia
Australian brush turkey (*Alectura lathami*), 105, 298
Austrobaileya scandens *(perennial flowering* vine), 293

B

baseline gait, 28
bee hummingbird (*Mellisuga helenae*), 307
beetles
  about, 8, 108, 360
  carrion, 109, 168
  Cretaceous dung, 253, 256, 258
  dung, 6–7, 14, 251–53, 256, 263, 331, 357
  wood-boring, 254
bipedal. *See also* quadrupedal; three-toed; two-legged
  dinosaurs, 27, 29, 38, 63, 70, 86, 88, 175, 178, 310
  facultatively, 27
  trackways, 32, 70
Bird, Roland T. (paleontologist), 42, 46, 48, 213
birds
  carnivorous, 228, 300
  co-evolution of flowering plants and, 363
  in Cretaceous Period, 360
  early, 216, 360
  evolution of, 215, 297–307, 313, 316–17, 360–62, 366
  flightless birds are modern theropods, 36
  in Mesozoic Era, 57, 299–300, 306, 361
  non-avian dinosaurs, evolution from, 15, 361–62
bladed bumps (denticles), 180, 182, 261
bowerbirds, 293, 324–25
brooding behavior, 106
burrows, fossil, 12, 128

C

Camarasaurus, 50, 117, 218
Canada, 10, 23, 47, 189, 219, 260, 270, 333
  Alberta, 9, 163, 168, 260
  British Columbia, 23
cannibalism, 163, 187, 189, 234
carnivorous
  birds, 228, 300
  dinosaurs, 13, 130, 166, 182, 249–50, 261, 358
  theropods, 163, 214, 251–52, 263
carrion beetles, 109, 168
cassowaries, 289–300, 307, 309, 325–26, 346, 361
Caudipteryx, 214–15, 217
Cenozoic Era (65 *mya*)
  birds, evolution of, 361
  flowering plants in, 264, 359–61
ceratopsian conflict, 6, 172

ceratopsians, 5–6, 8, 10, 18, 20–21, 23, 27, 35, 38, 45, 47, 52, 55, 70, 87–88, 92, 96, 128, 162, 169–73, 178–79, 185, 192, 213, 219, 251, 305, 348, 351
CGI. *See* computer generated imagery (CGI)
China, 16, 38, 40, 43, 47, 51, 94, 96, 98, 124, 147–48, 167, 189, 216, 219, 235–36, 310, 322
Chinese alligator (*Alligator sinensis*), 124
chinstrap penguin (*Pygoscelis antarctica*), 228
Choteau (Montana), 252
Citipati, 38, 93, 96–97, 106
clades/cladogram of dinosaurs, 18–19, 27, 55, 92, 215, 266
clams, 5, 52, 205, 239, 320–21
Claosaurus, 213
clawmarks, 71–72, 292, 312, 315
climate change, 301, 335, 349–50, 352. *See also* global warming
cocoons
  about, 8, 99, 107–8, 188
  fossil insect, 108–9, 256
  Late Cretaceous fossil, 122
co-evolution. *See also* Darwin, Charles; evolution
  of flowering plants and birds, 363
  of grasses and mammals, 264, 354
cololites (fossil intestinal contents), 226, 233, 238, 266
Colorado (USA), 23, 43, 46–47, 114, 243
combat (dinosaurs in), 52, 162, 172–74
computer generated imagery, 37, 165, 367
Confuciusornis, 236
coprolites (fossil feces), 13–14, 42, 221, 226, 233, 242, 246–66, 269–71, 278, 302–3, 305, 335, 351, 355, 357, 359, 369
crawling on hands and knees, 27
Cretaceous Period (70 *mya*). *See also* Early Cretaceous Period; Late Cretaceous Period; Middle Cretaceous Period
  about, 3, 41, 134, 351
  angiosperm seeds, 354, 361
  ankylosaurs, 239
  dinosaurs, 12, 54, 60, 168, 188, 238, 261, 270, 305, 355
  dung beetles, 253, 256, 258
  early birds, 360
  ecosystems, 187, 253, 258, 364
  egg-shaped mass of sandstone, 134
  flowering plants in, 293, 353–57, 360–63
  Frenchman Formation, 260
  herbivorous dinosaurs, 355
  meteorite (65 *mya*, near Yucatan peninsula, Mexico), 131
  in Montana and Alberta, Canada, 9
  98 million years ago, 56
  95 million years ago, 133
  non-avian dinosaurs, 264, 293
  105 million years ago, 268
  theropod dinosaurs, 290
  titanosaurs, 118
Cretaceous–Paleogene boundary, 55
crocodiles (*Crocodylus porosus*), 184
"crouching" traces, 38
Cuba, 302, 307
Cygnus falconeri, 302

D

Darwin, Charles, 363–66. *See also* evolution
Daspletosaurus, 180, 189, 236–38, 262
Davenport Ranch (Texas), 46
deinonychosaurs, 16
Deinosuchus, 184
denticles (bladed bumps), 180, 182, 261
Devonian Period (400 *mya*), 232
diagonal-walking pattern/trackway, 38, 310, 336
dinosaur(s). *See also* non-avian dinosaurs
  ancestors of, 14
  behavior of, 12–13, 18, 27, 33, 142, 162, 177, 193, 196, 220, 233, 269, 353, 355, 366–68, 370
  bipedal, 27, 29, 38, 63, 70, 86, 88, 175, 178, 310
  butt, 39
  cared for their young, 85
  clade of, 27, 55
  in Cretaceous Period, 12, 54, 60, 168, 188, 238, 261, 270, 305, 355
  ecosystems, 226
  eggs, 12, 85, 90–95, 97, 102, 252, 345, 348
  evolution of, 14–15, 18–19, 85, 92, 214, 242–43, 362
  family, 46, 95, 287
  feces ("manna from heaven"), 14, 252–53, 266, 356–57
  group-oriented, 47
  herbivorous, 13, 126, 190–92, 214, 295, 351–55
  ichnology, 21, 69, 74, 335
  live birth of young (viviparity), 91
  lives, tracing, 9–16
  macroevolutionary studies of, 26
  meat-eating, 214
  in Mesozoic Era, 15, 37, 45, 87, 95, 149, 173, 234, 266, 306, 310, 333, 346
  originated (235 *mya*), 25
  proto-dinosaurs, 26
  quadrupedal, 19, 27, 35, 38, 88
  sauropod, 7–8, 337
  sex lives of, 4, 16, 115
  sitting traces, 41, 44
  stampede of, 10, 65–67, 73–74, 79, 270
  swim tracks of, 11, 44, 79, 238, 315
  three-toed, 22, 64–65, 67, 69–70, 72–74, 79, 284
  toothmarks of, 13–14, 50, 159, 162, 165–66, 168, 170, 178–80, 182–89, 218, 221, 224, 226, 233–35, 240, 249, 260–61, 269–71, 278, 303, 333, 335, 359, 369, 371

trace fossils of, 9–10, 14–16, 38, 63, 80, 102, 108, 120, 158, 161, 176, 178, 197, 207, 212, 216, 224
trackways of, 28, 30–31, 34–36, 38, 52, 65, 244, 277, 280, 287, 310
two-legged, 11, 27, 86, 178
Dinosaur Cove (Australia), 154, 271, 276
"Dinosaur Death Pits from the Jurassic of China," 51
Dinosaur Revolution (TV series), 325, 348
Dinosaur Stampede (ABC film), 75
"Dinosaur Stampede in the Cretaceous of Queensland" (Thulborn and Wade), 65
"Dinosaur Trackways in the Winton Formation (Mid-Cretaceous) of Queensland" (Thulborn and Wade), 65
dinosauromorph tracks, 26
Diplodocus hallorum, 218
Dromaeosaurids, 21, 48
Dromaeosaurus (predatory theropods), 3–4, 8, 16
Dromaius novaehollandiae, 36, 298
Dromornis, 301
"duckbill" dinosaurs (hadrosaurs), 6–8, 14, 41–45, 55, 162, 189–92, 213, 225, 236–37, 254–55, 352, 355
dung beetles, 6–7, 14, 250–54, 256, 263, 331, 357

E

Early Cretaceous Period (120-135 *mya*), 219, 236, 293, 353, 355. *See also* Cretaceous Period
Early Jurassic Period (190–180 *mya*)
Ammosaurus, 218
dinosaur eggs, oldest known, 92
dinosaur swim tracks, 79
dinosaur tracks, 365
dinosaurs gave birth to live young, 91
feathered theropods, 40
Megapnosaurus, 214
period, 22
prosauropod *Massospondylus*, 96, 118, 219, 223, 316
rocks, 11, 16, 38, 41, 44, 92, 118, 148, 238, 315, 349
sandstones, 315, 342
theropod, 40
theropod swim tracks, 237
trace fossil, 39
trackway, 53
Early Triassic Period (245 *mya*), 26
ecosystems
about, 9, 14, 45, 53–54, 107, 131, 187, 231, 253, 258, 274, 292, 300–302, 305, 317, 332, 335, 355, 359–60
ancient, 14, 274
Cretaceous, 187, 364
dinosaur, 226
forest, 317, 357
freshwater, 339
Hawaiian, 302
land, 350
of Madagascar, 187
marine, 335
Mesozoic, 45, 85, 193, 266, 359, 362
ocean, 352
polar, 274
riparian, 358
savannah, 253
terrestrial, 335, 339, 353–54
top-niche predators in, 300
Edmontosaurus, 6–7, 13–14, 162–63, 166–67, 176, 185–86, 192, 213
egg-laying, 12, 91, 105, 116
Elephas maximus (Indian elephant), 43–44
emus of Australia (*Dromaius novaehollandiae*), 36, 298
enterolites (fossil stomach contents), 226, 233, 235, 238, 256, 266, 269
Eoraptor, 25–26, 263
evolution. *See also* Darwin, Charles
of amniotes, 91
of angiosperms, 354–55, 357
in the animal communities, 9
of birds, 215, 219, 297–308, 313, 316–17, 360–62, 366
of cannibalism, 163, 187
of carnivory, 263
of Chinese and American alligators, 124–25
co-evolution of flowering plants and birds, 363
co-evolution of grasses and mammals, 264, 354
"Cretaceous Terrestrial Revolution" and, 354
dinosaur behavior and, 196
of dinosaur feet, 19–21
of dinosaur parenting, 100, 120
of dinosaurs, 14–15, 18–19, 85, 92, 214, 242–43, 362
dinosaurs originated (235 *mya*), 25
of endoparasites, 259
of feces and seed plants, 232
of flight in dinosaurs, 312
of flowering plants, 232, 353, 355, 357, 360
of gastroliths, 219
of grasses in Mesozoic ecosystems, 266
of herbivorous dinosaur species, 354
of land plants, 353, 357, 360
of land-dwelling vertebrates, 353
live birth and, 91
macroevolutionary studies of dinosaurs, 26
of mammals, 362
of modern dinosaurs (chickens), 92
of modern ecosystems, 332
of nesting in dinosaurs, 118
of non-avian dinosaurs, 362
of pachycephalosaurs, 173
of plant communities in northern Queensland, 293

of plant-eating dinosaurs, 191
of pollinators, 360
of *Pterodaustro*, 206
sexual selection and, 323
of terrestrial ecosystems, 335
of theropods, 317, 429
of tree-dwelling mammals, 360
of *Triceratops* horns, 170–71
trituration and, 200
of *Troodon* egg care, 93
of vertebrates burrow, 128, 144
of whales and sea turtles, 204
evolutionary theory, 67, 365

F
face-biting, 189
Feathers: The Evolution of a Natural Miracle (Hanson), 299
floodplains, 1–8, 43, 53, 90, 107, 153, 190, 252, 260, 262, 268, 274–78, 280, 284, 286–87, 299, 311, 314, 339, 341, 357
Flores giant rat (*Papagomys armandvillei*), 303
flowering plants (angiosperms). *See also* monocotyledons; non-flowering seed plants
angiosperm, 353–57, 361
in Cenozoic Era (65 *mya*), 264, 359–61
co-evolution of birds and, 363
in Cretaceous Period (70 *mya*), 293, 353–57, 360–63
evolution of, 232
in Jurassic Period, 232, 393
in Mesozoic Era, 232, 353, 356
modern, 353
forest ecosystems, 317, 357
fossil burrows, 12, 128
four-legged. *See also* two-legged
animals, 30, 68
gait, 27
males, 86
toothy tracemakers, 370
freshwater ecosystems, 339

G
Gasparinisaura, 219
Gastornis (flightless bird), 300
gastroliths ("stomach stones"), 4, 13
gender differences, 46, 88, 114
Genyornis (bird), 301
geologic record, 34, 39, 42, 63, 88, 99, 138, 157, 198, 246, 319, 323, 325, 338, 344
ghost lineage, 26
giant teratorn (*Argentavis magnificens*), 299–300
Gigantoraptor (feathered theropod), 325–26, 329
Gigantosaurus, 37, 357, 368
Glen Rose, Texas, 42, 48
global warming, 348, 350–52, 354. *See also* climate change
Gorgosaurus, 189, 262
Great Ocean Walk (Australia), 268
grinding stones, 352
ground nests, 3, 11–12, 97, 109, 112, 228, 345, 347, 369
Guanlong wucaii, 51
gymnosperms (non-flowering seed plants), 353, 355–56. *See also* angiosperms

H
hadrosaurs ("duckbill" dinosaurs), 6–8, 14, 41–45, 55, 162, 189–92, 213, 225, 236–37, 254–55, 352, 355
Hawaiian ecosystems, 302
head-butting, 173–74
Hell Creek Formation of Montana, 189
herbivorous dinosaurs, 13, 126, 190–92, 214, 295, 351–55
Herrerasaurus, 25, 263
Hitchcock, Edward (paleontologist), 40, 366
Homo floresiensis, 303
Homo sapiens, 303
hypsilophodonts, 126, 150–51, 155

I
ichnology (study of trace fossils), 9, 21, 57, 60, 63, 66, 71, 88, 101, 138, 154, 160
ichthyosaurs, 91–92, 239, 242
Indian elephants (*Elephas maximus*), 43–44
iridium (rare element), 54
ischial callosity, 39

J
Jurassic Park (movie), 36, 48, 66, 74, 258, 295, 303
Jurassic Period, 41, 239
flowering plants in, 232
Late (145-150 *mya*), 232
Middle (165 *mya*), 47

K
Kelenken guillermoi, 301
Kuban, Glen (paleontologist), 49

L
land ecosystems, 350
Lark Quarry (dinosaur tracksite, Queensland, Australia), 56–57, 60, 62, 66–70, 73–83, 270, 277, 337
Late Cretaceous Period (65 *mya*), 18, 23, 54, 118, 219. *See also* Cretaceous Period
Late Jurassic Period (145-150 *mya*)
Allosaurus, 150, 176, 214, 244
flowering plants, evolution of, 232
hypsilophodont dinosaurs, small, 130
Mamenchisaurus, 51
Morrison Formation, 195, 243
Nqwebasaurus, 214
rocks, 114, 242

INDEX

sauropod *Brachiosaurus*, 241
sauropod Diplodocus hallorum, 218
sauropod tracksite, 46
Sinraptor, 189
small dinosaurs, 41
theropod tracks and sauropod bones, 49–50
theropod trackway, 52
Late Triassic Period (230 *mya*)
about, 22–23, 25
Allosaurus, 177
Coelophysis, 177, 240
dinosaur coprolite, 247, 263
dinosaur tracks, 50
dinosaurs, arrival of true, 25
dinosaurs eggs, 92
Eodromeus ("dawn runner"), 25
Eoraptor, 263
gastroliths Bull Run Formation, 221, 224
Herrerasaurus, 263
Maiasaura coprolites, 263
Massospondylus, 118, 218–19
ornithopod tracks, 365
Prosauropods, 22
regurgitalite, 240
saurischians, 219
sauropods, 23, 43, 223
theropod, fossil, 365
theropod *Coelophysis bauri*, 234
theropod coprolites, 263
theropod tracks, 222
tyrannosaurid coprolites, 263
warm climate of, 351
Leaellynasaura, 277, 283, 425
lesser rhea (*Rhea pennata*), 366
Limusaurus, 51
Living Dinosaurs: The Evolutionary History of Modern Birds (Hanson), 299
lungfish, 128–29, 188

M

Maiasaura peeblesorum, 95, 100–101, 107, 109, 117–18, 142, 250–58, 263, 316, 345, 347
Majungasaurus (large theropod), 186–87, 234
mallee fowl (*Leipoa ocellata*), 316, 347
Mamenchisaurus, 51
mammals
about, 12, 28, 48, 55, 85–86, 95, 129, 132
burrowing, 142, 144
Cretaceous, 62
egg-laying, 91
head-butting, 174
hoofed, 175
incisors of, 168
nipples and, 115
small, 3, 8, 128, 164, 169
mammoths, 44
marginocephalians, 18–19
marine ecosystems, 335
marine reptiles, 91, 204–5, 208, 217
meat eaters, 180
meat-eating dinosaurs, 214
Megapnosaurus, 214
Megapodius molistructor, 302, 347
Mesozoic Era (65–70 *mya*)
arthropods, 259
birds, 57, 299–300, 306, 361
bones of vertebrates/invertebrates, 168
bowers, 325
burrows, 320
cannibalism, 234
dinosaur behavior, 353
dinosaur eggs and, 95, 332
dinosaur nests, 317
dinosaurs, 15, 37, 45, 87, 95, 149, 173, 234, 266, 306, 310, 333, 346
ecosystems, 45, 85, 193, 266, 350, 359
flowering plants, 232, 353, 356
fossil burrows, 128
global warming, 351–52
grasses, 264
mammals, 140, 236, 362
marine reptiles, 204
monocotyledons and, 191
non-avian dinosaurs, 361, 370
"nuclear winter," 54
plants, 293
regurgitalites, 242
river deposits, 340
rocks, 17, 99, 162, 195–96, 212, 221, 269, 271, 299, 348
sauropods, 258
snow, 344
theropods, 36, 45, 214, 291, 332, 345
titanosaur nests, 316
trace fossils, 271, 359
meteorite crater (near Yucatan peninsula, Mexico), 54, 131
"meteorite winter," end-Cretaceous, 132
Mexico, 43, 54, 270
Microraptor, 52, 88, 235, 310
Middle Cretaceous Period (70 *mya*), 360. *See also* Cretaceous Period
Middle Jurassic Period (170 *mya*), 23, 44, 47, 51, 212, 223
Middle Triassic Period (235 *mya*), 25–26, 68
mixing stones, 352
monocotyledons, 191. *See also* flowering plants
Mononykus (feathered theropod dinosaurs), 2, 4
Montana, USA, 12, 68, 95–96, 100–101, 107, 109, 122, 124, 126, 133–35, 137–41, 147–48, 152, 154–55, 157, 184, 188, 236, 250, 345, 347
Mullerornis (elephant bird), 301
Museum of the Rockies (Montana), 188
Muttaburrasaurus langdoni, 57, 73

INDEX

N

natural selection, 6, 215, 304, 312, 354
neoichnology, 140
nest structures, 12, 109–10, 118–19, 345
New Mexico, 43, 55, 218, 234
New Zealand, 210, 302, 304–6, 309, 318, 345
nodosaurs, 18–19, 251, 351
non-avian dinosaurs, 54, 98. *See also* dinosaur(s)
  avian dinosaurs and, 305
  brain CT scans, 310
  Cenozoic Era, 264
  courtship traces of, 323
  Cretaceous Period, 264, 293
  earliest birds evolved from, 362
  Early Cretaceous Period, 236
  end-Cretaceous Period, 98, 131–32, 260, 264, 293, 304
  extinction of, 299
  Mesozoic Period, 361
  petroleum deposits from, 334
  sunbathing and, 328
  theropods, 55
non-flowering seed plants (gymnosperms), 353, 355–56. *See also* flowering plants
North Pole, 14, 53
North Slope of Alaska, 18, 274, 314
Nqwebasaurus, 214
nuclear winter, 54

O

ocean ecosystems, 352
Olympic racewalkers, 31
ornithomimids ("ostrich mimics"), 13, 216–17, 251, 305
ornithopods, 2
  bipedal, 88
  burrowing, 8, 130
  burrows/burrowing, 4, 8, 130, 367, 369
  cladogram for dinosaurs and, 19
  dinosaur toothmarks, 179
  ecological role of, 305
  eggs or egg clutches of, 92
  gymnosperms and, 353
  herbivorous dinosaurs, 351
  herding of, 47
  hypsilophodonts, 155
  large, 8
  nesting grounds, 345
  polar environment adaptation, 150–51
  quadrupedal, 88
  sauropods and, 351
  size of, 57
  social lives of, 47
  theropods and, 2, 11, 18–19, 29, 35–36, 57, 59, 63–64, 72, 79, 165, 246, 306
  tracks and nests of, 306
  tracks of, 20–21, 47, 70–72, 88
  urolites and coprolites, 246
Oryctodromeus cubicularis (burrowing dinosaur), 11–12, 98, 145, 148, 152, 154–55, 158–59, 269, 316, 319
ostriches (*Struthio camelus*), 36, 209
ostrich-like theropods (*Struthiomimus*), 4, 13
Oviraptor, 93–94, 96–97, 106, 325

P

Pachycephalosaurus ("bone-headed dinosaurs"), 18, 21, 172–75, 178, 251
Paleogene Period (65 *mya*), 54
paleontology, 9, 56, 61, 64, 67–68, 79–80, 123, 129, 134, 160, 243, 254, 264, 311, 352, 369
Panoplosaurus, 219
Papagomys armandvillei (Flores giant rat), 303
plant communities, 8
  in the Amazon River Basin, 293, 356
  on the Galapagos Islands, 363
  in northern Queensland, 293
  during the Pleistocene, 356
  riparian, 358
  river–floodplain, 357
plant eaters, 214–15, 369
plesiosaurs, 204–5, 211–12, 239
Poekilopleuron, 212, 215
Poland, 14, 26, 38, 68
polar ecosystems, 274
polar environments, 14, 131, 151, 155, 157, 274, 299
post-Cretaceous monkey, 132
Pratt Museum of Natural History (Amherst College), 40
predatory theropods, 3, 12, 15, 304–5, 338, 357, 359
prosauropods, 18–20, 22, 25, 27, 96, 118, 218–19, 223–24, 333, 343–44
proto-dinosaurs, 26
Psittacosaurus, 96, 128, 167, 219, 236
Pterodaustro guinazui, 206
pterosaurs, 2–3, 26, 206, 300
Pygoscelis adeliae (Adélie penguin), 228
Pygoscelis antarctica (chinstrap penguin), 228

Q

quadrupedal. *See also* bipedal; three-toed; two-legged
  animals, 28
  ankylosaurids, 20, 70, 88
  ceratopsians, 20, 35, 70, 88
  diagonal walking pattern, 89
  dinosaurs, 19, 27, 35, 38, 88
  facultatively, 27
  nodosaurids, 20
  ornithopods, 88
  prosauropods, 20
  sauropods, 20, 35, 70, 86, 88
  stegosaurs, 20, 70
  trackways, 32

INDEX

Queensland, Australia, 11, 35, 44, 56, 60–62, 66–67, 73, 80, 82, 205, 238–39, 270, 289, 293. *See also* Australia
Queensland Museum (Australia), 61–62

R
regurgitalites (fossil puke), 233, 235, 240, 242, 260, 266, 268
riparian ecosystems, 358

S
salamanders, 5, 230
sanderlings (*Calidris alba*), 309
sauropod breeding, 13
sauropod dinosaurs, 7–8, 337
Saurornitholestes, 180
savannah ecosystems, 253
sea-turtle nest, 326
sediment-rimmed nests, 12
sexual dimorphism, 88
Shenzhousaurus, 214, 216
Sinocalliopteryx, 214–15, 235–36
Sinosauropteryx, 214–15, 235
Sinraptor, 189
snails, 5–7, 52, 200, 205, 251, 256–58, 266, 291, 320, 322, 325, 364–65
Society of Vertebrate Paleontology, 277
South Pole, 14, 53
southern cassowary (*Casuarius casuarius*), 289
southwestern Utah, 11, 39, 44, 315
Spain, 11, 23, 44, 55, 71, 77, 96, 114, 204, 241, 345
Spinosaurus, 37, 88, 368
Stegoceras, 172, 174
stegosaurs, 13, 18–20, 23, 38, 70, 179, 213, 251, 348, 351–52, 369
Stephanoaetus coronatus (African crowned eagles), 303
"stomach stones" (gastroliths), 4, 13
stork (*Leptoptilos robustus*), 303
Struthio camelus (ostriches), 36, 209
Struthiomimus (ostrich-like theropods), 4, 13
swim tracks, 11, 44, 79, 238, 315
Sylviornis neocaledoniae, 302, 347
synergism, 8
Syntarsus, 214

T
T. rex. *See Tyrannosaurus rex (T. rex)*
terrestrial ecosystems, 335, 339, 353–54
Therizinosaurs, 20–21, 96, 251
theropods. *See also Allosaurus; Oviraptor;* pachycephalosaurs; *Struthiomimus*
  about, 2, 4, 38
  aggressive behaviors of, 178
  aquatic food source, 45
  asphyxiation and fossilization of, 51
  bipedal, 29, 35
  birds evolved from, 15
  brooding behavior, 106
  carnivorous, 162–63, 165, 180, 214
  Citipati, 38, 93, 96–97, 106
  cladogram for dinosaurs and, 19
  claw marks, 12
  Cretaceous, 73
  Daspletosaurus, 180, 189, 236–38, 262
  deinonychosaurs, 16
  Dromaeosaurids, 21, 48
  Dromaeosaurus, 3–4, 8, 16
  in Early Cretaceous rocks, 11
  eggs laid in the nests, 3, 92
  evolution of, 317
  face-biting by, 189
  feathered, 40, 162
  feathers, 106
  flightless birds are modern, 36
  fossils of, 51, 106
  gastroliths and, 214
  Guanlong wucaii, 51
  heterodonty in, 180–81
  Late Cretaceous, 180
  Limusaurus, 51
  mating traces, 88
  meat eaters, 180
  Microraptor, 52, 88, 235, 310
  Mononykus-like, 2
  nest structure, 96
  non-avian, 55
  ornithopods and, 2, 11, 18–19, 29, 35–36, 57, 59, 63–64, 72, 79, 165, 246, 306
  pack hunting by, 15–16, 45
  "palms" turned inward, 39
  predation and scavenging to feed, 186
  predatory, 3, 12, 15, 304–5, 338, 357, 359
  Saurornitholestes, 180
  Spinosaurus, 88
  "stomach stones," 13
  teeth, 180
  Therizinosaurs, 20–21
  three-toed track, 88
  tracks, 20, 44, 47, 49–50
  Triceratops defense against, 172
  Troodon, 3, 12, 106
  two-legged, 57, 70
  tyrannosaur-sized, 73
  Tyrannosaurus, 180
  urolites and coprolites, 246
  vegetation eating, 179
  Velociraptor, 16, 180
  weighed 8 tons, 86
Thescelosaurus, 2–4, 7
three-toed. *See also* bipedal
  bird tracks, 22
  Camarasaurus tracks, 50
  cassowary tracks, 291–93
  and clawed footprints, 7

dinosaur tracks, 22, 64–65, 67, 69–70, 72–74, 79, 284
emu or brush-turkey tracks, 61
feet, 309
ornithopod tracks, 22, 75, 79
theropod tracks, 42, 53, 79, 88, 336
tracks of pranksters, 309
thyreophorans, 18–19
titanosaurs, 12, 101, 109–11, 115–18, 128, 264–65
toothmarks on bones, 13
totipalmate feet, 307–9, 364
track tectonics, 33
trackways. *See also* Alexander's formula
about, 7, 16, 27–29
of animals moving at high speeds, 241
ankylosaur, 47
bipedal dinosaur, 32, 35, 38, 70
bird, 309, 312, 324
cassowary, 292
Ceratopsian, 368
dinosaur, 30, 34, 49, 52, 244, 277
dromaeosaur, 48
elephant, 339
herding-sauropod, 244
of insects, 241
narrow gauge, 29, 336
ornithopod, 47, 310
pachycephalosaur, 173–74, 368
pattern, 28
quadrupedal dinosaur, 32, 35
by running dinosaurs, 35
running-dinosaur, 35
sauropod, 31, 43, 46–47, 244, 368
"stalking theropod," 48
theropod, 15, 47, 244, 368
toad, 230
Triassic dinosauromorph tracks, 26
Triassic Period. *See* Early Triassic Period; Late Triassic Period; Middle Triassic Period
Triceratops (three-horned), 1–6, 10, 13–14, 31, 54, 66, 131, 141, 162, 169–72, 182–85, 261, 348
Troodon, 3, 8, 12, 93–94, 96, 101–11, 117–18, 122, 126, 128, 130, 188, 235, 256, 316, 345, 347, 352
trophic cascade, 54, 131
trot (trotting), 5, 10, 24, 28–33, 35
Two Medicine Formation of Montana, 101, 252, 256, 258
two-legged. *See also* bipedal; four-legged
dinosaurs, 11, 27, 86, 178
gait, 27
theropods and ornithopods, 57
tyrannosaurid, 236, 250, 260, 262–63
Tyrannosaurus rex (T. rex), 5–7, 14, 36–37, 41, 54–55, 66, 73–74, 131, 148, 170–72, 177, 180, 182–89, 213, 215, 234, 250, 257, 260–62, 295, 311, 334, 348, 368

U
undertracks, 33, 53, 71, 280
United Kingdom, 43, 365
urolites (fossilized structures formed by urination), 226, 242–43, 245–46, 368
Utah, 38–40, 43–44, 47–48, 52, 55, 79, 218, 303, 315, 340, 342–43
Utahraptor, 303

V
Velociraptor, 16, 48, 66, 87, 180
Victoria, Australia, 12, 68, 149, 157, 159, 267–69, 276–77, 299, 314. *See also* Australia
viviparity (live birth), 91

W
wasps, 8, 108–9, 274, 354, 360
webbed feet, 308, 364
webbing, 42, 307–8
wide gauge trackways, 29
wood-boring insects, 254–55, 322
Wyoming Dinosaur Center, 50

Y
Yellowstone National Park (Wyoming), 357–58

Z
Zimbabwe, 43, 47
zygodactyl, 307–9